II International Conference "Gas Discharge Plasma and Synthesis of Nanostructures" (GDP-NANO 2021)

AA007407

Journal of Physics: Conference Series Volume 2270

Online
1 - 5 December 2021

ISBN: 978-1-7138-5523-1
ISSN: 1742-6588

Printed from e-media with permission by:

Curran Associates, Inc.
57 Morehouse Lane
Red Hook, NY 12571

Some format issues inherent in the e-media version may also appear in this print version.

This work is licensed under a Creative Commons Attribution 3.0 International Licence.
Licence details: http://creativecommons.org/licenses/by/3.0/.

No changes have been made to the content of these proceedings. There may be changes to pagination and minor adjustments for aesthetics.

Printed with permission by Curran Associates, Inc. (2023)

For permission requests, please contact the Institute of Physics at the address below.

Institute of Physics
Dirac House, Temple Back
Bristol BS1 6BE UK

Phone: 44 1 17 929 7481
Fax: 44 1 17 920 0979

techtracking@iop.org

Additional copies of this publication are available from:

Curran Associates, Inc.
57 Morehouse Lane
Red Hook, NY 12571 USA
Phone: 845-758-0400
Fax: 845-758-2633
Email: curran@proceedings.com
Web: www.proceedings.com

TABLE OF CONTENTS

2ND INTERNATIONAL CONFERENCE "GAS DISCHARGE PLASMA AND SYNTHESIS OF NANOSTRUCTURES 2021
B A Timerkaev, I I Fairushin

PEER REVIEW STATEMENT

INFLUENCE OF EMITTING NANOPARTICLES CONCENTRATION TO THE DENSITY OF FREE ELECTRON IN THE THERMAL DUSTY PLASMA 1
I. I. Fairushin

JUSTIFICATION OF THE CHOICE OF THE OPTIMAL DESIGN OF EQUIPMENT FOR ELECTROLYTE-PLASMA POLISHING 4
B K Tazmeev, O N Sokolenko, V M Pogosyan, R N Tazmeeva

EFFECT OF SILICON FULLERENES ON CHANGES IN THE HEAT CAPACITY OF NACL ELECTROLYTES 9
S S Rajabova, T R Tilloeva, M M Safarov, M A Zaripova

PROPERTIES OF ELECTRIC DISCHARGE OF A JET ANODE AND AN ELECTROLYTIC CATHODE 13
M F Akhatov, R K Galimova, R R Mardanov, A A Nizameev, N A Loginov

THE DEPENDENCE OF THE SURFACE ROUGHNESS OF ALUMINUM NITRIDE FILMS ON THE PROCESSING METHODS WHEN MAGNETRON SPUTTERING IS USED 17
L Baranova, V Strunin, B Baysova

SURFACE TENSION OF COLLOID SYSTEMS OF CASTOR OIL AND BENZENE AT P=0.101 MPA 22
A R Rajabov, D S Hakimov, M M Safarov

SPECTRAL STUDIES OF INDUCTIVELY COUPLED PLASMA CHARACTERISTICS OF LOW PRESSURE DISCHARGES FOR TWO CONFIGURATIONS OF VACUUM CHAMBERS 26
Yu A Shemakhin, E S Samsonova, E F Voznesenskiy, Y O Zhelonkin, E Yu Shemakhin, S A Fadeev, T N Terentev, R A Maksimov, V A Tyurin

NANODIAMONDS FROM FUEL 32
B A Timerkaev, R B Shakirov, A A Zalyalieva, R K Gevorgyan, R R Shaikhattarov

STUDY OF THE EFFECT OF PLASMA TREATMENT ON THE MECHANICAL PROPERTIES OF MINERAL FIBER THREADS 37
M V Antonova, A R Ibatullina, A S Parsanov, I V Krasina

THE EFFECT OF JET ELECTRIC DISCHARGE ON THE STRENGTH CHARACTERISTICS OF THE SURFACE 42
M F Akhatov, R R Kayumov, R R Mardanov, I M Loginova

HYSTERESIS PHENOMENA AND THE EFFECT OF REORIENTATION IN A POLYMER-LIQUID CRYSTAL SYSTEM UNDER THE INFLUENCE OF LASER RADIATION AND UNIAXIAL DEFORMATION 45
B I Makhsudov, I K Fayzulloev, M K Egamov

NON-THERMAL PLASMA INSTALLATION AS A PRE-TREATMENT METHOD OF BARLEY SEEDS (HORDEUM VULGARE L.) 53
D Petrukhina, S Gorbatov, O Tkhorik, I Medzhidov, V Tikhonov, V Shishko, I Ivanov, A Tikhonov, N Gluchshenko, V Kharlamov

APPLICATION OF THE HARRINGTON FUNCTION IN SUBSTANTIATING THE CHOICE OF THE BEST EQUIPMENT FOR PLASMA PROCESSING OF MATERIALS 57
B K Tazmeev, V V Tsybulevsky, R N Tazmeeva, G K Tazmeev

EXPERIMENTAL INSTALLATION FOR STUDY THE LOW-POWER ICRF PLASMA AT LOW PRESSURES WITH EQUIPMENT FOR EXPERIMENT DATA SYNCHRONIZATION 63
T N Terentev, V S Zheltukhin, A Yu Shemakhin, I K Nekrasov, E S Samsonova, E Yu Shemakhin, I E Tsokalo, V A Tyurin

INVESTIGATION OF THERMAL CONDUCTIVITY OF NANOSTRUCTURES USING A MATHEMATICAL MODEL 67
H. S. Juraev, G.H. Juraeva

ONE-DIMENSIONAL CONTINUUM FINITE-DIFFERENCE MODEL OF THE DYNAMICS OF A DUSTY MEDIUM IN AERODYNAMIC, ELECTRIC AND GRAVITATIONAL FIELDS 73
D A Tukmakov

STUDY OF THE EFFECT OF LOW-TEMPERATURE PLASMA BETWEEN ELECTROLYTIC ELECTRODES ON POLYMER SURFACES 79
R R Kayumov, A A Nizameev, R R Mardanov, M F Akhatov

PLASMOCHEMICAL METHOD OF POLYANILINE FILM OBTAINED 82
M P Danilaev, V A Kuklin, I N Sidorov

CREATION OF PLASMA FLOW FROM A LIQUID ELECTROLYTE CATHODE WITH AN ARBITRARY ORIENTATION IN SPACE 87
G K Tazmeev, K K Tazmeev, B K Tazmeev

INVESTIGATION OF VOLT-FARAD CHARACTERISTICS OF THIN MULTIFERRROIC FILMS 91
Q B Saidzoda, A A Semenov, M M Safarov

FEATURES OF A BURNING ELECTRIC ARC SUPERIMPOSED ON GAS DISCHARGE WITH LIQUID ELECTROLYTE CATHODE 97
K K Tazmeev, A K Tazmeev

STUDIES OF A MONATOMIC LENNARD-JONES SYSTEM AT SLOW COOLING BY MOLECULAR DYNAMICS SIMULATIONS AND REGRESSION ANALYSIS 101
A R Farkhutdinov

A STABLE NUMERICAL ALGORITHM FOR INVESTIGATING THE PROPERTIES OF TWO-DIMENSIONAL LASER WAVEGUIDES IN MULTILAYER QUANTUM-DIMENSIONAL SEMICONDUCTOR HETEROSTRUCTURES 106
H. Sh. Juraev, N.O. Mamatkulova, B.I. Makhsudov

SYNTHESIS OF NICKEL COATINGS WITH HIGH SURFACE AREA ON THE CERAMIC POWDERS 115
M V Morozov, K Y Nagulin, R N Mansurov, A K Gilmutdinov

EFFECT OF TEMPERATURE, PRESSURE, AND FRACTION OF NANOPARTICLES ON CHANGES IN THE THERMAL CONDUCTIVITY OF COLLOID NANOFLUIDS 121
T R Tilloeva, M M Safarov, M A Zaripova

DENSITY OF TOLUENE-BASED NANOFLUIDS 125
Q Mirzoeva, J Matlabi, M M Safarov

ELECTROPHYSICAL PARAMETERS OF AC PLASMA SYSTEM 129
I I Oshenko, S A Smirnov

LOW-CURRENT ARC IN PLASMA FLOW FROM DISCHARGE WITH LIQUID ELECTROLYTE CATHODE 133
K K Tazmeev, R N Tazmeeva, B K Tazmeev

INVESTIGATION OF ELECTRIC DISCHARGE WITH LIQUID ELECTRODES UNDER INFLUENCE ON CARBON FIBER 137
K I Khazeev, R R Kayumov, A A Nizameev, M F Akhatov

SYNTHESIS OF MICRODIAMONDS AND GERMANIUM NANOTUBES IN THE ARGON-GERMANIUM ARC 140
B A Timerkaev, A A Abdullin, R M Suleimanov, R R Shaikhattarov, R K Gevorgyan

UNIVERSAL STRUCTURAL AND DYNAMIC FEATURES IN METALS NEAR THEIR MELTING POINTS 145
R R Khairullina, R M Khusnutdinov

A MODEL OF THE POSITIVE COLUMN OF LONGITUDINAL GLOW DISCHARGE IN PLANAR CHANNEL WITH ADDITIONAL GAS INJECTION THROUGH THE SIDE WALLS 149
I G Galeev, T Y Asadullin, D R Valeeva

ON THE QUESTION OF THE APPLICABILITY OF THE PRINCIPLE OF THERMODYNAMIC SIMILARITY IN LIQUID ALKALI METALS 153
A A Tsygankov

STRUCTURAL TRANSFORMATIONS IN BINARY NI62NB38 ALLOY AT ULTRAHIGH TEMPERATURES 158
Dinar T. Yarullin, Maria A. Doronina

ON THE ACCURACY OF ANALYTICAL SOLUTION OF THE POISSON-BOLTZMANN EQUATION IN THE LINEAR APPROXIMATION 162
I. I. Fairushin

CREATION OF A FLEXIBLE THIN-FILM THERMOELECTRIC CONVERTER FOR THE RANGE -50...150 DEGREES CELSIUS 166
O P Plotnikova, D V Kirillov, N A Tumanov, D S Manegin, V A Riazanov

LOCAL DENSITY DYNAMICS IN A SUPERCRITICAL LENNARD-JONES FLUID 169
R A Khabibullin

THE DEVELOPMENT OF STATISTICAL ANALYSIS METHODS FOR THE STUDY OF CORRELATIONS AND STATISTICAL MEMORY EFFECTS IN THE RECORDED DATA OF PHYSICAL EXPERIMENTS 173
S A Demin, V A Yunusov

SPATIAL DISTRIBUTIONS OF X-RAY RADIATION DURING FERRITE SURFACE BREAKDOWN 178
S N Andreev, A V Bernatskiy, G A Dushkin, S N Tskhai, I N Tilikin, S A Pikuz, S Y Savinov

INFLUENCE OF ION-PLASMA TREATMENT ON THE SURFACE MORPHOLOGY OF EPITAXIAL LEAD-TIN TELLURIDE FILMS 183
Y D Belov, S P Zimin, I I Amirov, V V Naumov

CLASSICAL AND QUANTUM WARM DENSE ELECTRON GAS DYNAMIC CHARACTERISTICS: ANALYTIC PREDICTIONS 188
J. Ara, A.V. Filinov, I.M. Tkachenko

THE STUDY OF STATISTICAL FEATURES OF THE EVOLUTION OF COMPLEX PHYSICAL SYSTEMS USING ADAPTIVE MACHINE LEARNING METHODS 196
V A Yunusov, S A Demin, A A Elenev

PORE FORMATION IN SIMPLE SYSTEMS SIMULATED BY MOLECULAR DYNAMICS 201
R A Khabibullin

INVESTIGATION OF THE STRUCTURE OF AN ALUMINUM NITRIDE FILM FOR A BRAGG REFLECTOR OBTAINED BY MAGNETRON SPUTTERING 204
L Baranova, B Baysova, V Strunin

USE OF NON-THERMAL PLASMA FOR DECONTAMINATION OF TITANIUM IMPLANTS 208
E. Sysolyatina, A. Petryakov, M. Abdulkadieva, E. Vasilieva, M. Chernukha, L. Avetisyan, T. Alekseeva, V. Barsuk, S. Ermolaeva, N. Titov, Yu. Akishev

CONTROL OF THE DISTRIBUTION OF THE INTERNAL CHARACTERISTICS OF THE DISCHARGE USING SUPERSONIC GAS PUMPING 218
B A Timerkaev, O A Petrova, R S Shamsutdinov, A I Sayfutdinov

SPATIAL STRUCTURE OF GAS DYNAMIC CHARACTERISTICS IN A GLOW DISCHARGE WITH A SUPERSONIC AXISYMMETRIC GAS FLOW 224
B A Timerkaev, O A Petrova, R S Shamsutdinov, A I Sayfutdinov

ELECTRIC ARC SYNTHESIS OF SILICON NANOSTRUCTURES 229
B A Timerkaev, V Y Turutin, V S Feltsinger, A A Zalyalieva, R R Vakhitov, R S Shamsutdinov

PRODUCTION OF HYDROGEN FROM HEAVY HYDROCARBONS 234
B A Timerkaev, G R Ganieva, B R Shakirov, A A Zalyalieva, R S Shamsutdinov

HYDROPHILIZATION OF THE SURFACE OF A NATURAL HYDROPHOBIC MATERIAL BY NON-EQUILIBRIUM LOW-TEMPERATURE PLAZMA 238
G R Rakhmatullina, V P Tikhonova, D K Nizamova, R F Akhverdiyev, K F Bigeyeva

APPLICATION OF PLASMA METHODS FOR THE SYNTHESIS OF NANOSTRUCTURES 241
A A Masyanov

FIRST-PRINCIPLE MOLECULAR DYNAMICS STUDY OF METHANE HYDRATE 245
M B Yunusov, R M Khusnutdinov

EXPERIMENTAL AND COMPUTATIONAL STUDY OF ARC DISCHARGES IN POWERFUL ALTERNATE ARC HEATER 250
V V Mironov, A S Kozaev, A S Negodyaev

A NEW APPROACH TO ESTIMATING KINETIC RATE FACTORS OF PHASE TRANSFORMATIONS 256
Dinar T. Yarullin

INVESTIGATION OF A HIGH-FREQUENCY MAGNETRON SPUTTERING SYSTEM OPERATION MODES DURING COPPER THIN FILMS DEPOSITION 260
N A Tumanov, D V Kirillov, E V Vorob'ev

MODIFICATION OF THE MATERIALS BASED ON PROTEIN FIBERS IN A LOW-PRESSURE HIGH-FREQUENCY PLASMA FLOW 264
G R Rakhmatullina, E A Pankova, O V Fukina, M Khayytov, L V Chapaeva

CONFIDENCE OF NANO-MEASUREMENTS OF HF-TI-N MULTILAYER PVD CONDENSATES FROM THE PLASMA PHASE 267
M M Mironov, M M Grebenshchikova

THREE-COLOR PYROMETRY STUDY OF RADIATION AND GAS DYNAMIC PROCESSES IN A HETEROGENEOUS PLASMA-VORTEX RING 271
L Y Volodin

ELECTRIC DISCHARGE GENERATOR OF LONG-LIVED PLASMA FORMATIONS IN ATMOSPHERIC PRESSURE AIR 277
L Y Volodin, A S Kamrukov

OPTIMIZATION OF THE CONDITIONS FOR THE SYNTHESIS OF ZINC OXIDE NANOPARTICLES BY PLASMA CHEMICAL METHOD 282
S M Shavaleeva, Z A Uktamov, E A Erlingayte, K S Nuriddinov

ABOUT GRAPHITE EVAPORATION DYNAMICS CAUSED BY RADIATION IN NEAR ZONE OF MAGNETOPLASMA COMPRESSOR DISCHARGE 287
A V Pavlov, T S Shchepanyuk, E O Chebykin, A S Skriabin, V D Telekh

INTERACTION OF POWERFUL PLASMA JETS WITH MO/SI MULTILAYERS 292
A S Skriabin, V D Telekh, A V Pavlov, A G Turyanskiy, V M Senkov, P A Tsygankov, F F Parada-Becera

FINITE-DIFFERENCE SCHEME AND ALGORITHM FOR SOLVING A TWO-DIMENSIONAL SYSTEM OF MAXWELL'S EQUATIONS FOR INDUCTIVELY COUPLED PLASMA 295
A V Gerasimov, A P Kirpichnikov, R N Gaynullin, Y D Lapteva, T M Pham

COMPARISON OF THE SMALLEST SQUARES AND SMALLEST MODULES METHODS IN MODELING PROCESSING OF MATERIALS BY PLASMA OF A GAS-VAPOR DISCHARGE WITH LIQUID ELECTRODES 300
R R Takseitov, R K Galimova, Z Y Yakupov

Author Index

2nd INTERNATIONAL CONFERENCE "GAS DISCHARGE PLASMA AND SYNTHESIS OF NANOSTRUCTURES 2021

B A Timerkaev[1], I I Fairushin[2,3]

[1] Kazan national research technical university named after A. N. Tupolev-KAI, Kazan, Russia
[2] Kazan Federal University, Kazan, Russia
[3] Joint Institute for High Temperatures of RAS, Moscow, Russia

btimerkaev@gmail.com, fairushin_ilnaz@mail.ru

In the city of Kazan from December 1 to December 4, 2021 on the basis of the Kazan National Research Technical University named after V.I. A.N. Tupolev-KAI hosted the II International Conference "Gas Discharge Plasma and Synthesis of Nanostructures" was held. The main organizers of this conference are KNRTU-KAI, KFU, Academy of Science Republic of Tatarstan. This event has become a notable event for the community of scientists and researchers involved in the physics of gas discharges and the synthesis of nanostructured materials. The conference discussed issues related to fundamental, problematic issues in the physics of gas discharges, such as the theory of arc plasma, high-frequency discharges, microwave discharges, glow discharge. Nevertheless, the main attention was paid to the problems of using gas discharge plasma for the synthesis of nanostructures. It is gratifying that there are results in this direction that are ready for industrial application. This year, the topic of the conference additionally includes a section on the physics of non-ideal and dusty plasma. On this topic, several plenary reports were made at once, which were presented by leading world scientists. The conference became a platform where completely new fundamental results that are at the forefront of science were presented. The vast majority of the presented works were presented for the first time.

Content from this work may be used under the terms of the Creative Commons Attribution 3.0 licence. Any further distribution of this work must maintain attribution to the author(s) and the title of the work, journal citation and DOI.
Published under licence by IOP Publishing Ltd

Journal of Physics: Conference Series **2270** (2022) 011001 doi:10.1088/1742-6596/2270/1/011001

Organizing Committee

- Boris Akhunovich Timerkaev (Chairman)
- Fairushin Ilnaz Izailovich (Deputy Chairman)
- Sofronitsky A.O. (Scientific secretary of the conference organizing committee)
- Dautov G.Yu.
- Fayzrakhmanov I.A.
- Galeev I.G.
- Gerasimov A.V.
- Israfilov I.Kh.
- Khrapak S.A.
- Kudryavtsev A.A.
- Kustova E.V.
- Lebedev Yu.A.
- Mokshin A.V.
- Petrov O.F.
- Ramazanov K.N.
- Saifutdinov A.I.
- Salakhov M.Kh.
- Schneider M.N.
- Shaehov M.F.
- Shaikhitdinov R.Z.
- Smolanov N.A.
- Strunin V.I.
- Shemakhin A.Yu.
- Tagirov L.R.
- Tagirov M.S.
- Tazmeev Kh.K.
- Tkachenko I.M.
- Vasiliev M.M.

Scientific Committee

- Boris Akhunovich Timerkaev (Co-chairman)
- Gilmutdinov Albert Harisovich (Co-chairman)
- Fairushin Ilnaz Izailovich (Deputy Chairman:)
- Galeev I.G.
- Israfilov I.Kh.
- Kudryavtsev A.A.
- Kustova E.V.
- Lebedev Yu.A.
- Mokshin A.V.
- Ramazanov K.N.
- Saifutdinov A.I.
- Shemakhin A.Yu.
- Schneider M.N.
- Tagirov M.S.
- Tazmeev Kh.K.

Peer Review Statement

All papers published in this volume have been reviewed through processes administered by the Editors. Reviews were conducted by expert referees to the professional and scientific standards expected of a proceedings journal published by IOP Publishing.

- **Type of peer review:** Single Anonymous
- **Conference submission management system:** Morressier
- **Number of submissions received:** 64
- **Number of submissions sent for review:** 64
- **Number of submissions accepted:** 64
- **Acceptance Rate (Submissions Accepted / Submissions Received × 100):** 100
- **Average number of reviews per paper:** 1
- **Total number of reviewers involved:** 6
- **Contact person for queries:**
  Name: Fairushin Ilnaz
  Email: fairushin_ilnaz@mail.ru
  Affiliation: Kazan Federal University

Content from this work may be used under the terms of the Creative Commons Attribution 3.0 licence. Any further distribution of this work must maintain attribution to the author(s) and the title of the work, journal citation and DOI.
Published under licence by IOP Publishing Ltd

Influence of emitting nanoparticles concentration to the density of free electron in the thermal dusty plasma

I. I. Fairushin[1, 2]

[1]Kazan Federal University, Kazan, Russia
[2]Joint Institute for High Temperatures, Russian Academy of Sciences, Moscow, Russia

fairushin_ilnaz@mail.ru

Abstract. On the basis of a statistical model describing the emission of electrons from the surface of dust particles in thermal plasma, a nonmonotonic character of the dependence of the concentration of emitted electrons on the concentration of grains has been established. The nature of the influence of the equilibrium temperature of the system and the size of dust particles on this effect is studied.

1. Introduction

Thermal plasma containing particles of condensed matter in the form of nanoparticles is widely distributed in nature, is used in many technological processes and is one of the varieties of complex plasma [1–2]. The presence of particles in thermal plasma leads to their charging. One of the key mechanisms of charge acquisition by nano and microparticles in a high temperature gas is thermionic emission [3–12]. The acquisition of an electric charge by nanoparticles due to the emission of electrons strongly affects the electrophysical properties of the system as a whole and the electrical conductivity in particular.

The most common approach to describing the process of charging dust particles as a result of thermionic emission is the Richardson model [3–5]. This model is based on the Poisson-Boltzmann equation and the Richardson formula for the electron concentration near the surface of a solid [3]. An alternative approach is the description method based on the "jelly" model for condensed particles. For micron particles in plasma, this model was first applied in [6–9]. The advantage of this model is that it can take into account the features of the internal structure of particles. Thus, hollow particles were considered in [9].

2. Theoretical background

The buffer gas surrounding the particles is assumed to be neutral at all considered temperatures of the system. As a dust subsystem, we consider nanoparticles, which are balls of an n-type semiconductor [6, 7], with particle radius R. Let us take the value of the charge inside a sphere with radius R as the charge of a dust particle. Then, under the condition that the buffer gas is not ionized for concentration of free electrons in the space between particles n_{ed} we get:

$$n_{ed} = \frac{3\bar{V}_d}{a^3}(n_i - n_{e0})(acha - sha). \quad (1)$$

Content from this work may be used under the terms of the Creative Commons Attribution 3.0 licence. Any further distribution of this work must maintain attribution to the author(s) and the title of the work, journal citation and DOI.
Published under licence by IOP Publishing Ltd

Here $\overline{V}_d$ is the volume fraction of dust particles in the plasma, n_i is the concentration of ions of the positive background of dust particles, n_{e0} is the concentration of free electrons in the geometric center of the dust particle, the dimensionless parameter a is given as follows

$$a = \sqrt{\frac{q^2R^2n_{e0}}{\varepsilon_0 kT}}, \tag{2}$$

where R – is the dust particle radius, T – is the absolute temperature of the system, ε_0 – electric constant, k – is the Boltzmann constant. The value n_{e0} is calculated using the following relationship:

$$n_{e0} = n_i\left[1 - \frac{a(\lambda-1)\mathrm{ch}(a(\lambda-1)) - (1-a^2\lambda)\mathrm{sh}(a(\lambda-1))}{a\lambda\mathrm{ch}(a\lambda) - \mathrm{sh}(a\lambda)}\right], \tag{3}$$

where $\lambda = \frac{l}{R}$, l – half the distance between the centers of two neighboring particles.

Thus, using relations (1) - (3), one can calculate the dependence of the concentration of free electrons in thermal dusty plasma on various system parameters, in particular, on the concentration (or volume fraction) of the dust particles themselves. Next, the values of n_{ed} will be calculated for various system parameters. The value of the parameter λ is directly related to the volume fraction of dust particles in the entire system $\overline{V}_d = \pi/(6\lambda^3)$. The value of ni is taken based on the condition of nondegeneracy of the electronic subsystem. The work takes $n_i = 5\cdot10^{21}$ m^{-3}. Here we note that real substances with such a concentration of free charge carriers at the considered temperatures do not exist in nature. However, this model can describe a number of effects at a qualitative level, which can manifest themselves, for example, in thermal plasma with nanoparticles of various substances.

3. Results

Calculations showed that the dependence of the concentration of free electrons in the space between dust particles on the volume fraction of dust particles has a maximum at a certain value (Fig. 1).

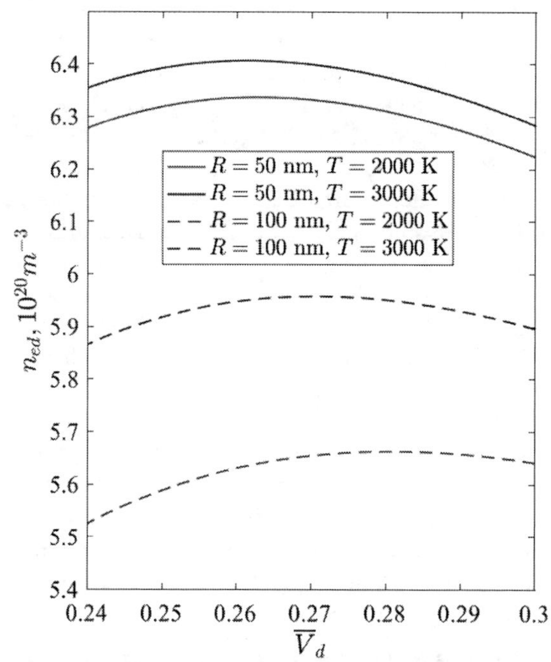

Fig. 1. Dependence of the concentration of free electrons in thermal dusty plasma on the volume fraction of dusty nanoparticles at different particle radii and system temperature.

The concentration of free electrons in the space between particles increases with an increase in the concentration of nanoparticles. However, if the concentration of particles becomes sufficiently high, then a further increase in the concentration of nanoparticles leads to a decrease in the concentration of free electrons in the plasma. This is explained by the fact that when the particles are close to each other, the emission of electrons from one particle is significantly affected by electrons from another particle. There is some dependence on the particle size of the value of the parameter $\bar{V}_d$, at which the maximum average concentration of free electrons in the space between the particles is realized. For example, for $R = 50$ nm and $T = 2000$ K, the maximum n_{de} is observed at $\bar{V}_d$=0.265, and at $R = 100$ nm and $T = 2000$ K, the same maximum falls on $\bar{V}_d$=0.28. Consequently, with a smaller particle size, their concentration at which the maximum electron yield is carried out is less than with larger particles. This effect is explained by a higher concentration gradient of free electrons near the surface of smaller particles.

4. Conclusion

Based on the previously developed analytical model, the concentration of free electrons in thermal dusty plasma is calculated depending on the volume fraction of dusty nanoparticles. It is established that this dependence has a nonmonotonic character.

5. Acknowledgements

This paper has been supported by the Kazan Federal University Strategic Academic Leadership Program.

References

[1] Pustylnik M Y, Pikalev A A, Zobnin A V, Semenov I L, Thomas H M, Petrov O F 2021 *Contributions to Plasma Physics* **61** e202100126
[2] Mitrani J M, Shneider M N, Stratton B C, and Raitses Y 2016 *Appl. Phys. Lett.* **108** 054101
[3] Rudinsky A V, Yagodnikov D A 2019 *High Temperature* **57** (5), 753-760
[4] Vishnyakov V I 2012 *Phy. Rev. E* **85** 026402.
[5] Khrapak S A, Morfill G E, Fortov V E, D'yachkov L G, Khrapak A G, and Petrov O F 2007 *Phys. Rev. Lett.* **99** 055003
[6] Fairushin I I, Dautov I G, Kashapov N F and Shamsutdinov A R 2017 *Technical Physics Letters* **43** (1) 27-30
[7] Fayrushin I and Dautov G 2013 *Journal of Physics: Conference Series* **479** 012013
[8] Fairushin I I 2020 *High Energy Chemistry* **54** (6) 477-479
[9] Fayrushin I I, Dautov I G and Kashapov N F 2017 *International Journal of Environmental Science and Technology* **14** (12) 2555-2560
[10] Davletov A E, Kurbanov F, and Mukhametkarimov Ye S 2018 *Physics of Plasmas* **25**, 120701
[11] Shigeta M, Murphy A B 2011 *J. Phys. D: Appl. Phys.* **44** 174025.
[12] Fairushin I I, Saifutdinov A I, and Sofronitskiy A O 2020 *High Energy Chemistry* **54** (2), 150–153

Justification of the choice of the optimal design of equipment for electrolyte-plasma polishing

B K Tazmeev[1], O N Sokolenko[1], V M Pogosyan[1], R N Tazmeeva[2]

[1]Kuban State Agrarian University named after I. T. Trubilin, Krasnodar, Russia
[2]Kazan Federal University, Naberezhnye Chelny Institute, Naberezhnye Chelny, Russia

tazmeevb@mail.ru

Abstract. The article proposes a method of complex comparison of equipment for electrolyte-plasma polishing of surfaces. The Harrington function is used to evaluate the equipment. The values of the function are limited by the numerical interval 0.2 - 0.8. The parameters of the equipment are selected as the evaluation parameters: maximum polishing area per cycle, productivity, power, area occupied by the installation, cost. As a result of a comprehensive assessment, the best type of equipment for electrolyte-plasma polishing, which provides a resource-saving and energy-saving mode of operation was selected. A computer program has been developed with the help of which a generalized evaluation parameter has been calculated and a comparative analysis of the equipment has been performed.

1. Introduction

The aim of the study is to substantiate the choice of the best design of equipment for electrolyte-plasma polishing using the upgraded Harrington function.

Polishing of metals and alloys by the electrolyte-plasma method is an effective technological process that saves resources used and has high environmental and economic parameters. Relatively new high-performance and environmentally friendly technology allows processing products made of stainless and carbon steels, copper and aluminum alloys, brass, zinc, titanium, silicon in aqueous solutions of salts. This technology of material processing has a sufficient number of advantages compared to traditional electrochemical polishing in acid solutions having a high concentration. In addition to saving material and financial resources and high productivity, the electrolyte-plasma polishing method has the best technical characteristics and allows getting a high quality of the processed surface: product processing speed, surface cleanliness class, absence of abrasive particles introduction, degreasing of the surface.

Many research papers have been shown the study of plasma methods of material processing [1-8].

Plasma polishing of metals and alloys can be carried out using an atmospheric pressure gas discharge with a liquid electrode [9-10].

2. Materials and methods

A generalized criterion of complex evaluation calculated using Harrington functions was adopted as a criterion for evaluating the best type of equipment for electrolyte-plasma polishing [11].

$$D_i = \sqrt{\prod_1^n d_{ij}^{kj}} \rightarrow 1.0 . \quad (1)$$

Content from this work may be used under the terms of the Creative Commons Attribution 3.0 licence. Any further distribution of this work must maintain attribution to the author(s) and the title of the work, journal citation and DOI.
Published under licence by IOP Publishing Ltd

Here *Di* – generalized parameter of integrated assessment *i*-th of the equipment; d_{ij} – desirability of each *j*-th evaluation parameter of the *i*-th equipment; *n* – number of equipment under consideration; k_j – value coefficient of each *j*-th parameter.

The parameters that determine the choice of equipment are shown in the table 1. The Harrington function is modernized in the form of equation (2) and presented as a graph (figure 1)

$$\mathrm{d}(y') = e^{-e^{-(y'-2)}}. \tag{2}$$

Here d(*y'*) – a function that determines the significance of each evaluation parameter.

The natural values of the evaluation parameters are converted into dimensionless quantities according to formulas (3) - (7).

For maximum polishing area in one cycle (S_{max}, cm^2)

$$y_1' = 0.0021x_{1,j} + 0.461. \tag{3}$$

For productivity (P_r, m^2/h)

$$y_2' = 0.068x_{2,j} + 0.05. \tag{4}$$

For equipment capacity (*P*, kW)

$$y_3' = 0.1228x_{3,j} - 2.0676. \tag{5}$$

For the area occupied by the equipment (S_{in}, m^2)

$$y_4' = -0.275x_{4,j} + 4.39. \tag{6}$$

For cost (*C*, €)

$$y_5' = -2.43x_{5,j} + 4.17. \tag{7}$$

Table 1. Estimated *j*-th parameters of devices under consideration.

Type of equipment (i)	Model of equipment	Value parameters (*j*)				
		S_{max}, cm^2	P_r, m^2/h	*P*, kW	S_{in}, m^2	*C*, €
1	*EPP-1000/2000*	40000	8.0	2000	34	43400
2	*EPP-400/500*	10000	4.2	500	18	20000
3	*PlasmaC-III*	4000	3.2	250	15	28900
4	*PlasmaC-V*	20000	6.0	800	32	50500
5	*UPP-630*	11000	4.6	630	16	22500
	Weighting factor	0.3	0.3	0.1	0.1	0.2

The criteria for evaluating devices are given in tables 1 and 2. After finding the desirability functions d(*y'*) we calculate the generalized criterion D_i.

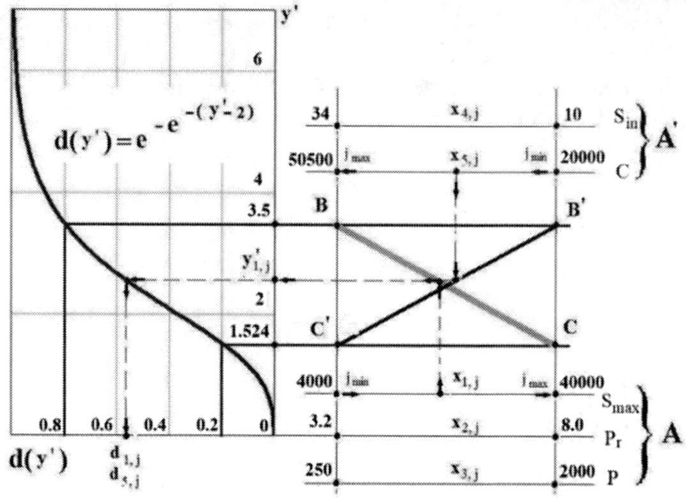

Figure 1. The upgraded d-Harrington desirability function, the scales A and A′ of the estimated *j*-th parameters and the dimensionless scale of the function d(y'). j_1 – maximum polishing area per cycle S_{max}, cm^2; j_2 - productivity, P_r, m^2/h; j_3 - installation capacity P, kW; j_4 - the area occupied by the installation S_{in}, m^2; j_5 - cost C, €.

3. Results and discussion

The results of the quality analysis of compared devices are presented in tables 2 and 3.

Table 2. Desirability function of equipment evaluation parameters

n	Model of equipment	Desirability function of evaluation parameters					Generalized criterion of integrated assessment	
		S_{max}, cm^2	P_r, m^2/h	P, kW	S_{in}, m^2	C, €	D_i	D_i^k
1	*EPP-1000/2000*	0.8	0.8	0.8	0.2	0.388	0.525	0.899
2	*EPP-400/500*	0.388	0.364	0.253	0.708	0.8	0.458	0,881
3	*PlasmaC-III*	0.2	0.2	0.2	0.8	0.708	0.331	0.785
4	*PlasmaC-V*	0.512	0.575	0.427	0.308	0.2	0.36	0.838
5	*UPP-630*	0.4	0.486	0.392	0.650	0.788	0.523	0.873

The equipment of the following models was studied: *EPP-1000/2000*, *EPP-400/500*, *PlasmaC-III*, *PlasmaC-V*, *UPP-630*. The devices differ in basic parameters. The comparison of equipment was made according to the estimated parameters shown in table 1.

As follows from the data in Table 2, the following devices correspond to the best (maximum) value of the desirability function (0.8) according to one of the parameters:

- according to the maximum polishing area in one cycle, productivity and capacity - the device of the *EPP-1000/2000* model;
- by the area occupied by the equipment - *PlasmaC-III*;
- by cost – *EPP-400/500*.

Bat the same time, the *Plasma C-III* device has a low desirability (0.2) in terms of maximum polishing area per cycle, performance and capacity. The same (low) desirability value has the *EPP-1000/2000* device in terms of the area occupied by the equipment and *PlasmaC-V* in terms of cost. These results of the research do not make it possible to form final conclusions about the preference of the particular equipment. The evaluation method using the upgraded Harrington function can be used to make a preliminary decision on the choice of equipment. Table 3 contains calculated data on the generalized evaluation criterion, taking into account the weighting factor k_j and without taking it into account.

Table 3. Generalized criteria for a comprehensive assessment of equipment.

№	Models of compared machines	Generalized estimated parameters		Taking into account the rounding
		Without taking into account the weighting factor, D_i	With taking into account the weighting factor, D_i^k	
1	*EPP-1000/2000*	0.525	0.899	0,9
2	*EPP-400/500*	0.458	0,881	0,9
3	*PlasmaC-III*	0.331	0.785	0,8
4	*PlasmaC-V*	0.36	0.838	0,8
5	*UPP-630*	0.523	0.873	0,9

According to the results of the analysis of table 2, it becomes obvious that the *UPP-630* machine does not occupy a leading position in terms of such parameters as the maximum polishing area per cycle, productivity and capacity from the point of view of desirability. But the relatively low cost and the occupied area contributed to the fact that the generalized criterion D_i of a comprehensive assessment of the *UPP-630* device differs from the *EPP-1000/2000* by only 0.002.

It should be noted that the installation parameters according to table 1 are divided into two groups as shown in figure 1. Parameter group A: maximum polishing area per cycle S_{max}, cm^2; productivity, P_r, m^2/h; installation capacity P, kW – should have as much value as possible. Parameter group A′: the area occupied by the installation S_{in}, m^2 and the cost C, € - should be as small as possible. According to the results of the calculation of the generalized desirability function D_i^k, which takes into account the degree of desirability of each of the parameters in the range from 0 to 1.0, the installations received the greatest preference: *EPP-1000/2000, EPP-400/500* and *UPP-630.*

4. Conclusions

The proposed evaluation method allows an objective approach to solving the problem of choosing the best design of equipment for electrolyte-plasma polishing using the upgraded Harrington function in the range of desirability of the j-th evaluation parameters 0.2 - 0.8. Five evaluation parameters of installations with their own scales of permissible values of these parameters have been identified. The dependences of the transmission of the natural values of the selected evaluation parameters into dimensionless ones on the scale y', which are used to calculate the desirability function in dimensionless quantities and the generalized criterion of complex evaluation, are determined. A necessary condition for an objective assessment of the quality of the compared machines is the same test conditions (temperature regime, single set of processed products, etc.). Calculations have shown that the application of the weighting factor of the compared parameters leads to smoothing the difference in the values of the generalized equipment evaluation criterion.

Journal of Physics: Conference Series **2270** (2022) 012002 doi:10.1088/1742-6596/2270/1/012002

References

[1] Seo B, Park H.-K, Kim H G, Kim W R, Park K 2021 *Surf. Coatings Technol.* **406** 126640
[2] An S et al. 2020 *Surf. Coatings Technol.* **405** 126504
[3] Belkin P N, Kusmanov S A, Parfenov E V 2020 *Appl. Surf. Sci. Adv.* **1** 100016
[4] Kusmanov S A et al. 2019 *J. Mater. Eng. Perform.* **28** 5425–5432
[5] Kusmanov S, Zhirov A, Kusmanova I, Belkin P 2019 *Surface Engineering* **35**(6) 507-511
[6] Danilov I et al. 2019 *Micromachines* **10** 10030214
[7] Diaz-Rodriguez S et al. 2019 *RSC Advances* **9**(4) 2292–2301
[8] Asri R I M et al. 2017 *Materials Science and Engineering C* **77** 1261–1274
[9] Tazmeev K K, Arslanov I M, Tazmeev B K, Tazmeev G K 2019 *J. Phys.: Conf. Ser.* **1393** 012061
[10] Tazmeev G K, Tazmeeva R N, Tazmeev B K 2020 *J. Phys.: Conf. Ser.* **1588** 012050
[11] Harrington E 1965 *Industrial Quality control* **21**(10) p 494-498

EFFECT OF SILICON FULLERENES ON CHANGES IN THE HEAT CAPACITY OF NACL ELECTROLYTES

S S Rajabova[1], T R Tilloeva[2], M M Safarov[2], M A Zaripova[2]

[1]branch of the Moscow Power Engineering Institute in Dushanbe, Dushanbe, Tajikistan
[2]Tajik technical university after named by academic M.C.Osimi, Dushanbe, Tajikistan

mahmad1@list.ru

Abstract. The paper presents the results of an experimental study of the isobaric heat capacity of solutions of electrolytes of the water and NaCl system in the temperature range 288-373K. To measure the heat capacity of electrolytes, the method of monotonous heating was used (developed by Professor M.M.Safarov). The total relative error of measuring the heat capacity of this place with a confidence probability of 95% is on average equal to 3.8%. It is established that the specific isobaric heat capacity of electrolytes decreases with increasing temperature and concentration of sodium chlorine. The addition of fullerene leads to an increase in the heat capacity of solutions of the water and sodium chlorine system. On the basis of experimental data and the law of thermodynamic similarity, approximation dependences were obtained. Using the expressions obtained, it is possible to calculate the heat capacities of electrolytes at different temperatures and concentrations of fullerene nanoparticles.

An analysis of the electronic structure of fullerenes shows the presence of π-electron systems, for which there are large values of the nonlinear susceptibility. Fullerenes indeed have nonlinear optical properties. However, due to the high symmetry of the C60 molecule, second harmonic generation is possible only when asymmetry is introduced into the system (for example, by an external electric field). From a practical point of view, the high speed (~250 ps) is attractive, which determines the suppression of the second harmonic generation. In addition, C60 fullerenes are also capable of generating the third harmonic. Literature data on the heat capacity of aqueous solutions of NaCl at different temperatures (288-323)K and the mass concentration of NaCl are given in the work (Table 1) [2].

Table 1. Specific isobaric heat capacity (Cp,J/(kg.K)) of NaCl aqueous solutions depending on temperature and atmospheric pressure [3].

Sample / T, K	№1[3]	№2	№3	№4	№5
288	3987,7	3853.8	3665,8	3519,6	3272,9
293	3975,2	3845.6	3657.5	3515.4	3272,9
313	3925,0	3799.6	3632,4	3494,5	3264,2
333	3882,2	3753,6	3603,2	3469,4	3239,5
353	3847,2	3699,3	3578,1	3415,1	3206,1
373	3791,3	3644,9	-	3314,7	3172.6

Content from this work may be used under the terms of the Creative Commons Attribution 3.0 licence. Any further distribution of this work must maintain attribution to the author(s) and the title of the work, journal citation and DOI.
Published under licence by IOP Publishing Ltd

№1-(H_2O+5%NaCl); ;№2-(H_2O+10%NaCl); №3-((H_2O+12,5%NaCl); №4-(H_2O+15%NaCl); №5-(H_2O+24,26%NaCl).

Table 2. Specific isobaric heat capacity (Cr,J/(kg.K)) of aqueous solutions of NaCl (H_2O + 5%NaCl) depending on temperature, fullerene (Si) concentration and atmospheric pressure (0.101) MPa.

Sample / T, K	№1[3]	№2	№3	№4	№5	№6
288,2	3987,7	4032,6	4077,8	4122,3	4167,3	4210,6
293,6	3975,2	4020,5	4064,4	4110,6	4155,3	4201,9
313,2	3925,0	3969,4	4014,5	4055,4	4104,4	4149,3
333,6	3883,2	3928,1	3973,2	4018,3	4063,5	4108,4
353,4	3837,2	3882,8	3927,1	3972,3	4017,7	4062,6

№1-(H_2O+5%NaCl+0,5%Si; №2-(H_2O+5%NaCl+1,0%Si; №3-((H_2O+5%NaCl+1,5%Si; №4-(H_2O+5%NaCl+2,0%Si; №5 -(H_2O+5% NaCl+ 2,5%Si; №6-(H_2O+5% NaCl+3,0%Si.

The results of the study of the heat capacity of aqueous NaCl solutions depending on temperature at atmospheric pressure by the authors of [2] and our measurements on the heat capacity of the studied solutions and nanosized silicon fullerenes embedded in them are shown in Tables 2-4 and Figure 1.

Table 3. Specific isobaric heat capacity (Cr,J/(kg.K)) of aqueous solutions of NaCl (H_2O+ 10%NaCl) depending on temperature, concentration of fullerene (Si) and atmospheric pressure (0.101) MPa.

Sample / T, K	№1	№2	№3	№4	№5	№6
288.2	3853.8	3903.6	3953.8	4003.6	4053.6	4103.5
293.6	3845.6	3895.4	3945.5	3955.7	4005.5	4055.6
313.2	3799.5	3849.9	3899.5	3949.6	3999.6	4049.5
333.6	3753.6	3803.4	3453.6	3858.4	3908.5	3958.7
353.4	3699.3	3749.8	3799.4	3849.3	3897.3	3947.8
373.4	3644.9	3694.4	3744.2	3794.3	3844.4	3894.5
393.0		3639.2	3689.6	3739.3	3786.5	3836.5
413.5		3584.3	3634.1	3684.3	3734.8	3793.3
423.4		3529.5	3579.3	3629.4	3679.5	3729.6
433.2		3474.6	3524.6	3547.6	3624.0	3674.5

No.1- (H_2O + 10% NaCl + 0.5% Si (fullerene)); №2-(H_2O+10%NaCl+1.0% Si(fullerene)); №3-((H_2O+ 10%NaCl+1.5%Si(fullerene); №4-(H_2O+10%NaCl+2.0%Si(fullerene); No. 5- (H_2O + 10% NaCl + 2.5% Si (fullerene)); No. 6-(H_2O + 10% NaCl + 3.0% Si (fullerene).

Table 4. Specific isobaric heat capacity (Cp, J/(kg.K)) of NaCl aqueous solutions (H_2O+ 15.0%NaCl) depending on temperature, fullerene (Si) concentration and atmospheric pressure (0.101) MPa.

Sample / T, K	№1[3]	№2	№3	№4	№5	№6
288,2	3519,6	3569,6	3619,8	3669,6	3719,6	3769,6
293,6	3515,4	3565,4	3615,5	3665,7	3715,5	3765,8
313,2	3494,5	3544,9	3594,5	3644,6	3694,6	3744,5
333,6	3469,4	3519,4	3569,6	3619,4	3669,5	3719,3
353,4	3415,1	3465,8	3515,3	3565,3	3615,6	3665,6
373.4	3314,7	3364,4	3414,2	3464,3	3514,4	3564,8
393,0	-	3265,2	3315,6	3365,3	3415,5	3465,7

413,5	-	3156,3	3206,1	3256,3	3306,8	3356,4
423.4	-	3057,5	3107,3	3157,4	3207,5	3257,6
433,2	-	2957,6	3007,6	3057,6	3107,0	3157,2

№1-(H2O+15%NaCl+0,5%Si(fullerene));№2-(H2O+15%NaCl+1,0%Si(fullerene))); 3((H2O+15%NaCl+1,5%Si (fullerene)); №4-(H2O+15%NaCl+2,0%Si(fullerene)); №5-(H2O+15% NaCl+2,5%Si (fullerene)); №6 - (H2O+15% NaCl+3,0%Si (fullerene)).

Further, the results of an experimental study of the heat capacity of electrolytes with silicon fullerenes embedded in them showed that the addition of silicon fullerenes in the electrolyte increases the heat capacity of solutions of the system (NaCL+H2O).

As the results showed, the specific isobaric heat capacity of electrolytes with silicon fullerenes embedded in them increases with an increase in the concentration of fullerenes over the entire temperature range based on the straight line law.

The equation of a straight line has the form:

$$(C^{*}) = F\,(n_{Si}) + D,\ \underline{J/\ (kg\cdot K)} \quad (1)$$

Table 5. Coefficients F and D of equation (4) for the studied electrolytes at temperature T1=353K and atmospheric pressure.

№	Sample	F, J/ (kg·K)	D, J/ (kg·K.)
1	(H2O+5%NaCl)+(от 0 до 2,5%Si)	90,063	3837,3
2	(H2O +10%NaCl)+(от 0 до 2,5%Si)	99,166	3699,8
3	(H2O +12,5%NaCl)+(от 0 до 2,5%Si)	110,11	3578,3
4	(H2O +15%NaCl)+(от 0 до 2,5% Si)	100.14	3415,2
5	(H2O +24,26%NaCl)+(от 0 до 2,5%Si)	130.62	3206.2

Using equation (1), taking into account the values of Table 5, it is possible to numerically calculate the specific isobaric heat capacity of colloidal electrolytes of the (NaCl + H2O) system at various temperatures, silicon fullerene concentrations and water concentrations at atmospheric pressure.

Low-temperature plasma can be used for the synthesis of various nanostructures and is well suited for the modification of various surfaces. This is shown in many works [4-18].

References

[1] Kaleeva A. A., Timerkaev B. A., Shamsutdinov R. S., Saifutdinova A. A., Shakirov B. R. Micro-arc method for the synthesis of silicon nanostructures. Journal of Physics: Conference Series 1870 (2021) 012012 IOP Publishing doi:10.1088/1742-6596/1870/1/012012 (Kaleeva A.A., Timerkaev B.A., Shamsutdinov R.S., Sayfutdinov A.I. , Shakirov B. R. Microarc method for the synthesis of silicon nanostructures. / 1st All-Russian conference with international participation. "Gas-discharge plasma and synthesis of nanostructures", "KNRTU-KAI", Kazan, 2020.-p.120-122.)

[2] Kovalevskaya N.S. Heat capacity of aqueous solutions of NaCl./N.S.

[3] Handbook on thermophysical properties of gases and liquids / N.B. Vargaftik, L.P. Filippov et al.

[4] Shamsutdinov, R. S., & Timerkaev, B. A. (2021, April). The influence of a supersonic flow of gas at glow discharge. In Journal of Physics: Conference Series (Vol. 1870, No. 1, p. 012019). IOP Publishing. DOI 10.1088/1742-6596/1870/1/012019.

[5] Saifutdinov, A. I., & Sofronitskii, A. O. (2021). Numerical Study of Breakdown and Formation Dynamics of Arc Discharge Plasma Parameters at Ultrahigh Pressures. High Energy Chemistry, 55(3), 228-232.DOI: 10.1134/S0018143921030115

[6] Shemakhin, A. Y., Zheltukhin, V. S., Shemakhin, E. Y., Terentev, T. N., & Sofronitsky, A. O. (2020, July). Experimental installation to study the RF plasma flow at low pressures with experiment data synchronization. In Journal of Physics: Conference Series (Vol. 1588, No. 1, p. 012018). IOP Publishing. DOI: 10.1088/1742-6596/1588/1/012018

[7] Saifutdinova, A. A., Sofronitskiy, A. O., Timerkaev, B. A., & Saifutdinov, A. I. (2020). Plasma-Chemical Decomposition of Hydrocarbons on the Basis of the Micro-Arc Discharge with Disc Electrodes Rotating in the Bulk of Raw Materials. Russian Physics Journal, 62(11), 2132-2136.DOI: 10.1007/s11182-020-01957-0

[8] Fairushin, I. I., Saifutdinov, A. I., & Sofronitskiy, A. O. (2020). Numerical and Experimental Studies of the Synthesis of Copper Nanoparticles in a High-Pressure Discharge. High Energy Chemistry, 54(2), 150-153. DOI: 10.1134/S0018143920020071

[9] Asadullin, T. Y., Galeev, I. G., Sofronitskiy, A. O., & Gizeev, M. M. (2019, November). Acoustic impact on electric discharge parameters during sterilization of freeze-dried products. In Journal of Physics: Conference Series (Vol. 1370, No. 1, p. 012016). IOP Publishing. DOI: 10.1088/1742-6596/1370/1/012016

[10] Fairushin, I. I., Saifutdinov, A. I., Sofronitskiy, A. O., Timerkaev, B. A., & Dautov, G. Y. (2019, October). Development of plasma reactor design for synthesis of copper nanoparticles using multi-scale simulation. In Journal of Physics: Conference Series (Vol. 1328, No. 1, p. 012088). IOP Publishing. DOI: 10.1088/1742-6596/1328/1/012088

[11] Asadullin, T. Y., Galeev, I. G., & Sofronitskiy, A. O. (2019, October). The application of pulsed discharge for sterilization of freeze-dried product. In Journal of Physics: Conference Series (Vol. 1328, No. 1, p. 012071). IOP Publishing. DOI: 10.1088/1742-6596/1328/1/012071

[12] Timerkaev, B. A., Ganieva, G. R., Kaleeva, A. A., Israfilov, Z. K., & Sofronitskii, A. O. (2019). Growing of carbon nanotubes from hydrocarbons in an arc plasma. Journal of Engineering Physics and Thermophysics, 92(5), 1248-1252. DOI: 10.1007/s10891-019-02040-3

[13] Dautov, G. Y., Kashapov, N. F., Dautov, I. G., & Sofronitskiy, A. O. (2018, July). Research of the influence of the geometry of the discharge chamber on the characteristics of the arc plasmatron. In Journal of Physics: Conference Series (Vol. 1058, No. 1). IOP Publishing. DOI: 10.1088/1742-6596/1058/1/012035

[14] Timerkaev, B. A., Andreeva, A. A., & Sofronitskiy, A. O. (2017, November). Discharge creeping along the surface in the process for producing nanomaterials. In Journal of Physics: Conference Series (Vol. 927, No. 1, p. 012068). IOP Publishing. DOI: 10.1088/1742-6596/927/1/012068

[15] Sadikov, K. G., Sofronitskiy, A. O., & Larionov, V. M. (2017, November). The effect of electrically conductive additives on the plasma pyrolysis of heavy hydrocarbons. In Journal of Physics: Conference Series (Vol. 927, No. 1, p. 012046). IOP Publishing. DOI: 10.1088/1742-6596/927/1/012046

[16] Timerkaev, B. A., Andreeva, A. A., & Sofronitskiy, A. O. (2017). Discharge creeping along the surface in the process of cleaning and strengthening of the materials surface. In Journal of Physics: Conference Series (Vol. 789, No. 1). IOP Publishing. DOI: 10.1088/1742-6596/789/1/012063

[17] Sadikov, K. G., Sofronitskiy, A. O., & Dautov, I. G. (2017). Functional plasma sprayed coatings on magnesium ceramic substrates. In Journal of Physics: Conference Series (Vol. 789, No. 1, p. 012043). IOP Publishing. DOI: 10.1088/1742-6596/789/1/012043

[18] Timerkaev, B. A., Sofronitskiy, A. O., & Andreeva, A. A. (2016). Carbon nanotubes formation in the decomposition of heavy hydrocarbons creeping along the surface of the glow discharge. In Journal of Physics: Conference Series (Vol. 669, No. 1, p. 012062). IOP Publishing. DOI: 10.1088/1742-6596/669/1/012062

PROPERTIES OF ELECTRIC DISCHARGE OF A JET ANODE AND AN ELECTROLYTIC CATHODE

M F Akhatov[1], R K Galimova[1], R R Mardanov[1], A A Nizameev[1], N A Loginov[2]

[1]Kazan National Research Technical University named after A. N. Tupolev - KAI, Kazan, Russia
[2]Kazan State Agrarian University, Kazan, Russia

ahatov.81@yandex.ru

Abstract. The paper shows some results of the definite current-voltage characteristic (CVC) of interaction of electric discharge with a jet electrolytic cathode and a steady anode at different jet lengths (lc), jet diameter (dc), electrolyte flow rate (G) for various electrolyte concentrations.

1. Introduction

For the first time, electrical discharges (ED) in water were carried out more than 200 years ago. However, the resulting powerful hydrodynamic impulses did not find practical application in those years. The effect of this type of discharge was forgotten for a long time. Later, when creating powerful high-voltage installations, they again encountered electrical discharges in liquids. The destructive effect that occurs during the electrical breakdown of dielectric liquids has formed a strong opinion about the futility of ED in a liquid. For many decades, this opinion had been persisting among scientists and electrical engineers [1].

The development of technology has again drawn attention to ED in liquids, as this type of discharge and devices developed on their basis are the cheapest and environmentally friendly [2]. In real time, active experimental studies of the electric discharge of a flow with an electrolytic cathode
and an iron anode at atmospheric and reduced pressures are being conducted, at which a weakly conductive liquid (aqueous solutions, electrolytes, process and basic water) is used as a liquid electrode. Basically, in existing studies, aqueous solutions of salts are used as electrolytes [3]. However, there are a large number of varieties of electrolytes that differ from each other by their chemical compositions, and by their physical and chemical properties.

2. Experimental setup and experimental methodology

Experimental setup with electrolytic electrodes and metal (see Fig. 1) is composed of an electrolytic bath, electric power supply systems and test-measuring devices. The power supply resource 1 supports constant voltage by conductors 2 to a bit shortage of steel plate 4. The bath's 3 function is to be a container to accumulate electrolyte. Electrode 5 supplies conductor to separation funnel 8 with electrolyte 7, fixed on the tripod 9. Electrolyte consumption is regulated by a valve 6. The hood 10 is used for suction from working zone of formed gases. The span of discharge voltage characteristics are $U_p=0\div1.5$ kV, currents $I=0.01\div2$ A.

Content from this work may be used under the terms of the Creative Commons Attribution 3.0 licence. Any further distribution of this work must maintain attribution to the author(s) and the title of the work, journal citation and DOI.
Published under licence by IOP Publishing Ltd

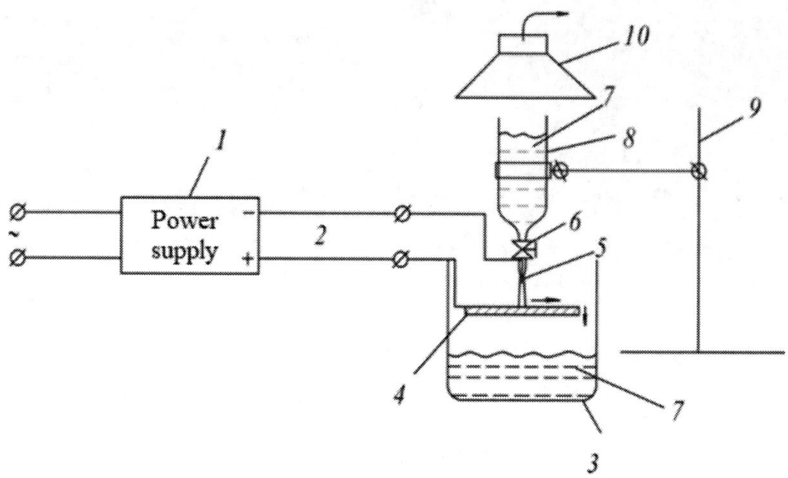

Figure 1. Functional installation.

3. Experiment

The discharge ignited between the electrolytic
cathode and the steady anode at atmospheric pressure, a solution of $NH4NO3$ (ammonium nitrate) in water of different concentrations was used as a jet electrolytic cathode, and the anode gave a plate of electrical steel.

4. Discussion of experimental findings

Some combustion discharge modes are given in Table 1.

Table 1. Combustion discharge modes

I, мА	60	200	240	280	500
U,B	840	811	780	730	500
lc, мм	100	10	120	40	20
dc, мм	2	2	2	2	2
G, г/c	1,82	1,82	1,82	1,82	1,82
Electrolyte concentration,%	26	26	20	26	20

Figures 2-4 show the CVC of an electric discharge with a jet electrolytic cathode and an iron anode for different jet lengths l, DC diameter and electrolyte flow rate G.

Figure 2. CVC of an electric discharge with a jet electrolytic cathode and a solid anode lc = 20 mm, dc = 2 mm and G = 1.82 g/s for different electrolyte concentrations.

From the review of the experimental results in Figure 2, it can be seen that for Graphs 1 and 2, at
an electrolyte concentration of 26% and 20%, with an increase in the value of the I discharge current, the current discharge voltage becomes less, i.e. CVC contains a falling process.

In case concentration decreases to 13%, the graph of the 3 CVC discharge in the area from 570 mA to 750 mA has a rectilinear character (the ED burns stably). A further current increase from 750 to 900 mA results a gradual decrease in the discharge voltage. Current increase from 900 mA to 1 A results a sharp decrease in voltage (the ED does not burn stably).

Figure 3 shows the CVC of an electric discharge with jet electrolytic cathode and solid anode, for an aqueous solution of $NH4NO3$ of different concentrations at lc = 40 mm.

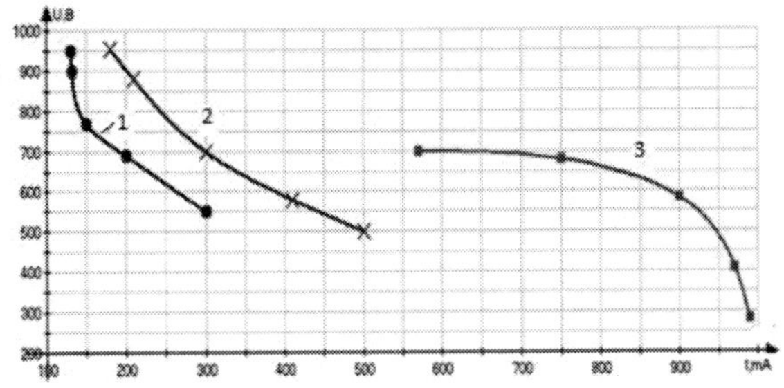

Figure 2. The CVC of the electric discharge between the jet electrolytic cathode and the anode at solid lc = 20 mm, dc = 2 mm and G = 1.82 g / s 1-26%; 2-20%; 3-13%..

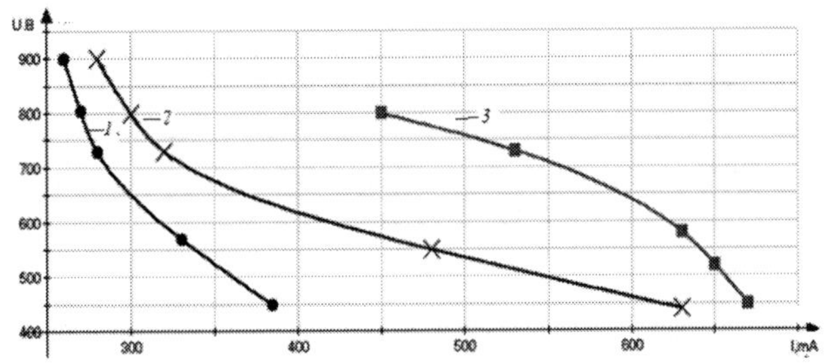

Figure 3. The CVC of an electric discharge between a jet electrolytic cathode and a solid anode at l = 40 mm, dc = 2 mm and G = 1.82 g / s 1-26%; 2-20%; 3-13%.

The figure shows that at electrolyte concentrations of 26% and 20%, the CVC is almost the same, i.e. it has a falling character. Unlike the CVC with a jet length of 20 mm, the 13% electrolyte solution does not have straightness sections. As the current increases, the voltage gradually decreases.

Figures 5 show the CVC and discharge at a jet length with L = 80 mm, dc= 1.5 mm and G= 1.82 g/s.

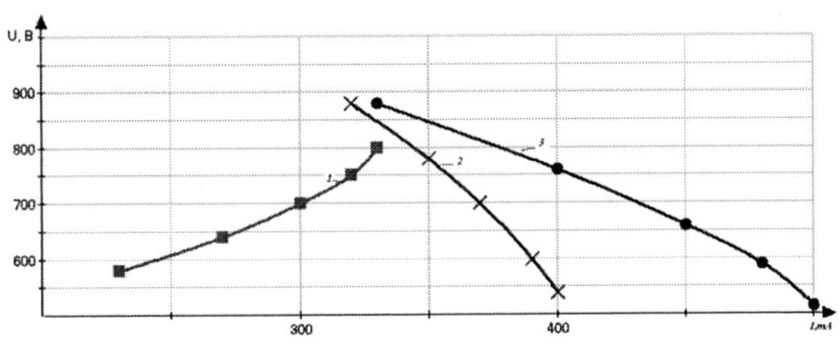

Figure 4. The CVC of electric discharge between a jet electrolytic cathode and a solid anode at l = 80 mm, dc = 1.5 mm and G = 1.82 g / s. 1-13%.; 2-20%; 3-26 %.

As can be seen, in the figures there are graphs of CVC both of an increasing nature for 13% solution (ED is not observed) and decreasing 20 and 26% solutions. This is determined by the stability of the discharge combustion. As the length of the discharge jet increases, it begins to burn unstable, and the discharge occurs not only at the jet-metal boundary, but also along the jet electrolytic cathode.

5. Conclusion

Structure of electrolyte and its concentration affect the combustion ED in jet electrolytic cathode. The shape of discharge and its CVC depend on the length of the jet. Experimental setup and experimental methodology.

The research was supported by a grant from the Russian Science Foundation (project No. 22-29-00021).

References

[1] Kayumov R.R. Experimental study of the effect of electric discharge on carbon fiber/Zakirov Zh.U., Kayumov R.R., Khazeev K.I.// Journal of Physics: Conference Series This link is disabled, 2021, 1870(1), 012007
[2] Akhatov M.F. Glycerin as an electrolytic electrode / Sh. Akhatov, G., Kayumov R.R., Valeeva R.R., Akhatov M.F.// Physical journal: conf. Series. 2019.V. 1328 Is.1,01200410
[3] Bagautdinova L.N. Low-power electric discharges with metal, dielectric and electrolytic electrodes at low frequencies and atmospheric pressure / Gaisin A.F., Sadriev R.S., Bagautdinova L.N., Nasibullin R.T., Gaisin F.M., Mastyukov S.S. // High temperature, 2020, 58(6), pp. 777-780
[4] Galimova R.K. Calculation of portable properties of some real gas mixtures at high temperatures / Taxeitov R.R., Galimova R.K., Yakupov Z.Ya. // Physical Journal: Conference Series, 2020, 1588(1), 012065
[5] Fakhrutdinova, I.T. Features of the development of an electric discharge between a jet anode and a liquid cathode / Galimzyanov I.I., Gaisin A.F., Fakhrutdinova I.T., Shakirova E.F., Akhatov M.F., Kayumov R.R. // High temperature, 2018, 56(2), pp. 296-298
[6] Tsareva, A.M. Determination of natural frequencies and vibration modes of a disk of constant thickness with a central mount / Makaeva, R.Kh ., Tsareva A.M., Karimov, A.Kh . // Russian Aeronautics, 2008, 51(1), pp. 53-59
[7] Gaisin A.F. Special features of a multichannel discharge in a porous solid cathode/Loginov N.A, Son E.E., Gaisin A.F., Gaisin F.M. //High Temperature. 2009. T. 47. № 4. C. 603-605.

The dependence of the surface roughness of aluminum nitride films on the processing methods when magnetron sputtering is used.

L Baranova[1, 2], V Strunin[1, 2], B Baysova[1, 2]

[1]Dostoevsky Omsk State University, Omsk, Russian Federation
[2]Institute of Radiophysics and Physical Electronics OSC SB RAS, Omsk, Russian Federation

baranova@omsu.ru

Abstract. The paper investigates the dependence of the surface roughness of aluminum nitride films on the modes of magnetron deposition. The paper shows the difference between the surfaces of polycrystalline aluminum nitride films on ceramized glass and silicon substrates grown under the same conditions. Atomic force microscopy was used to compare the surface roughness parameters of aluminum nitride obtained at different substrate temperatures and the power dissipation in the target. The optimal conditions for forming aluminum nitride films through magnetron sputtering have been determined.

1. Introduction

The paper considers the conditions for formation and their effect on the surface relief of aluminum nitride films obtained through magnetron sputtering of an aluminum target in argon and nitrogen atmosphere. Aluminum nitride films are widely used in the design and manufacture of microelectronic piezoelectric films of the devices for generating and selecting microwave signals [1].

In modern thin film technology, magnetron sputtering systems have become widely used due to a number of advantages. These are control of deposition processes, the possibility of obtaining materials of various properties, high quality coatings [2].

This method enables us to achieve relatively high deposition rates and obtain uniform thickness films on large area substrates. High sputtering rates allow us to reduce the contamination of sputtered films by foreign impurities. The possibility of processing in a reactive medium of dilute gas makes it possible to obtain films of nitrides, carbides, oxides and other compounds [3]. On the other hand, the magnetron method of forming aluminum nitride films with specified characteristics is a multiparameter process task of mutually consistent processes that determine the interrelation between film properties, namely, the elemental and phase composition, the texture and the structure of crystallites, surface morphology, electromechanical coupling coefficient, with such main operating parameters of magnetron sputtering systems as electrode voltage, discharge current, target current density, specific capacity, working gas composition, actuation gas pressure, substrate temperature.

One of the key characteristics of the film quality is the surface roughness, which determines the frequency and resonant properties of BAW resonators.

Content from this work may be used under the terms of the Creative Commons Attribution 3.0 licence. Any further distribution of this work must maintain attribution to the author(s) and the title of the work, journal citation and DOI.
Published under licence by IOP Publishing Ltd

The aim of the paper is to reduce the degree of surface roughness of aluminum nitride films by determining the interrelation between the process conditions of film formation and their properties, and to determine the optimal sputtering conditions.

2. The experimental procedure

A production magnetron installation for applying multicomponent coatings STEMS116-01 was used to form the films. The deposition of aluminum nitride films was carriedout on silicon and ceramized glass substrates. The pressure of the Ar:N 2gas mixture during sputtering provided a collisionless motion of aluminum atoms between the sputtered target and the substrate (a distance of 100 mm), and it was equal to 0.07 Pa. Aluminum nitride films were formed at the substrate temperatures of 300-400 0C and with the capacity of 400-900 wattssupplied to the target.

Magnetron sputtering of an aluminum target is performed by kinetic impact of positively charged Ar ions with an energy of 300-500 eV. Argon ions are generated during the processes of ionization of argon atoms by an electron impact occurringin the cathoderegionsof a glow discharge plasma, DC or AC. A mixed stream of ions and atoms enters the target of inert (Ar) and re-active (N2) gases. Chemically active atoms and ionsduring the bombardment of the target do not only participate in its sputtering, but also react with the target atoms forming compounds on the surface, in particular, AlN. Inthis case, atomization of both pure target material (Al)and its compounds (AlN) is possible. The atoms sputtered off from the target are transported to the substrate, while their interaction with plasma particles and actuation gas atoms, including the formation of chemical bonds, remains possible . The ratio of argon and nitrogen was considered as 4:5.

The surface morphology was examined through atomic force microscopy (AFM), the dependences of the surface roughness of aluminum nitride films obtained by magnetron sputteringon the temperature of the substrate and the power dissipated in the target were investigated [4].

The films were measured at the Analytical and Technological Research Center "High Technology and New Materials" (ATRC HT NM) of Novosibirsk State University. To study the morphology of the film surface and compare the values of roughness on silicon and ceramized glass substrates, the topographies of samples were measured with atomic force microscopy (AFM) on an NT-MDT Solver Next microscope. The sections of the surface were photographed with a scale of 5x5 microns, and a resolution of 512x512 points for different substrates. Fig.1. shows 2D topographies of the surface of the samples.

With the two-dimensional image of the surface relief of the films taken into consideration, it is possible to estimate the grain sizes, their uniform distribution over the surface, the uniformity of their shape and size.

The samples were flat surfaces. In the images of scanning probe microscopy (SPM), the particles of various diameters and small furrows are distinguishable; the furrows have a depth of only about a few nanometers. The surface appears to be a granular structure, the bases of the columnar structure are visible, a nano-crystalline surface has grains of 50-100 nm in size, the nanopeaks height is 8 nm.

The values of Ra of the average roughness along the profile (along the line) and Rq of the RMS roughness along the profile (along the line) were calculated in the SPM data processing program "ImageAnalysis 3.2.4 10128".

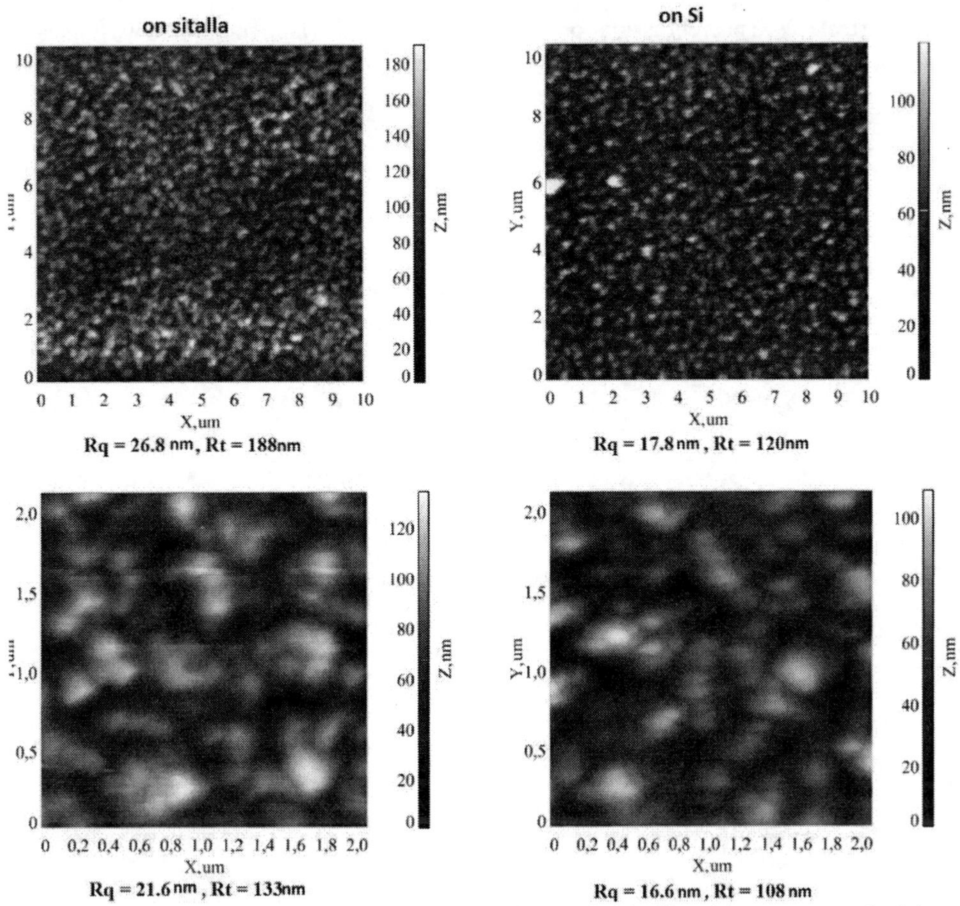

Figure 1. AFM image of the surface of the aluminum nitride film: in contrast mode (phase composition) (on the left you can see the AlN film on the metal, on the right there is the AlN film on silicon).

3. The experimental procedure

When the films grown under the same conditions (actuation gas pressure, substrate temperature, target capacity, etc.) were compared, it turned out that the roughness of the film and the dimensions of AlN polycrystallites in AFM images on a silicon substrate were smaller than those of the films on a ceramized glass substrate. This is due to the fact that the surface of the silicon substrate itself is smoother; its roughness was about 0.35 nm, the roughness of the ceramized glass substrate was 2.2 nm. The thickness of the sputtered films is about 1 microns, so the roughness of the substrate affected the roughness of the film.

Figure 2 shows the results of measurements of Ra and Rq roughness of the films depending on the substrate temperature at a target capacity of 650 watts. The minimum roughness of both Ra and Rq corresponds to a temperature of 300 0 C. As the substrate temperature increases, the roughness increases from 8 to 18 nm too.

The substrate temperature is a parameter that determines the mobility of adsorbed atoms. When the temperature of the substrate increases, the surface diffusion coefficient of the adsorbed atoms increases too, which contributes to their mobility. The adsorbed atoms on the surface of the growing film acquire sufficient mobility to regroup into grains with the lowest value of surface energy. The merging of grains is accompanied by their enlargement resulting in an increase in the roughness of the surface of the film due to the formation of deep intergranular cavities on the surface of the film.

Next, the effect of the power dissipation in the target on the surface roughness was investigated, the dependences were obtained for substrate temperatures of 300 0C and 350 0 C.

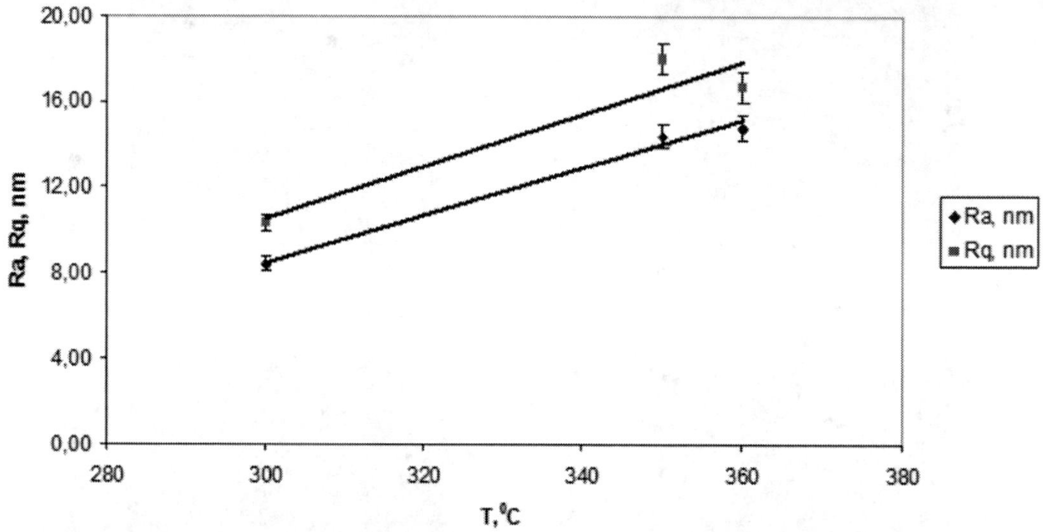

Figure 2. Dependence of the roughness of aluminum nitride films on the substrate temperature

Figure 3 shows a diagram of the dependence of roughness on the capacity at a temperature of 300 0C.

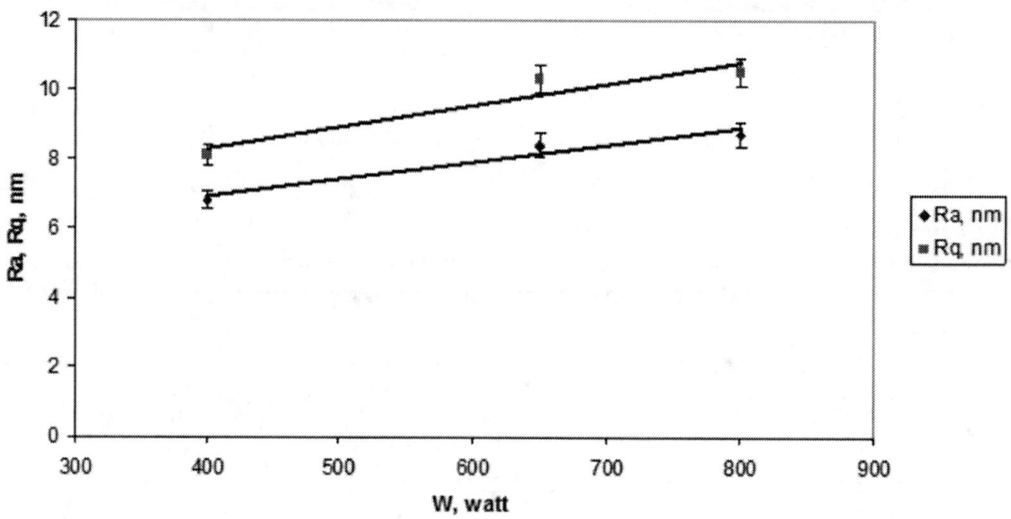

Figure 3. Dependences of the roughness of aluminum nitride films on the target capacity at a substrate temperature of 300 °C.

Figure 4 shows a diagram of the dependence of roughness on the capacity at a temperature of 350 0C.

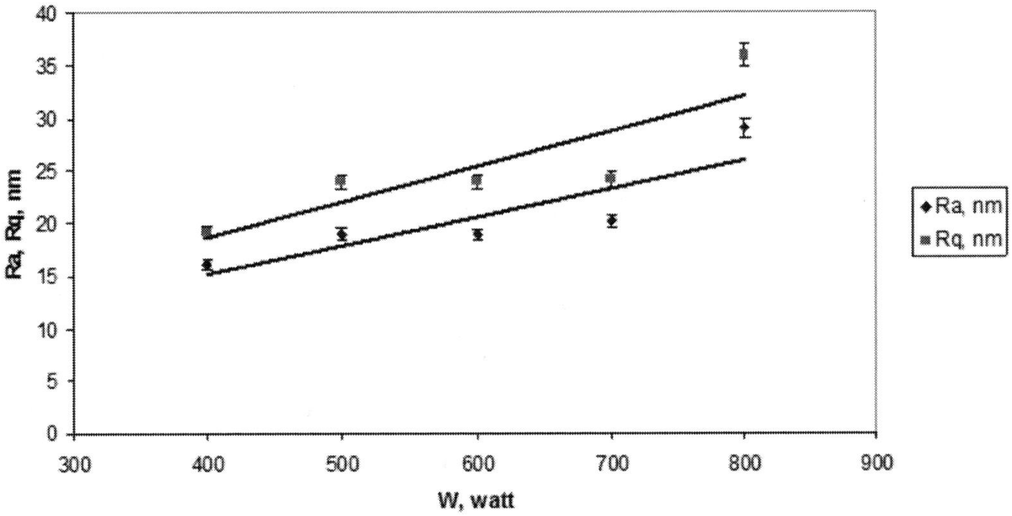

Figure 4. Dependences of the roughness of aluminum nitride films on the target capacity at a substrate temperature of 350 ^{0}C.

When the power increases, the values of Ra and Rq roughness also increase from 13 to 32 nm at a substrate temperature of 350 0 C, and from 6 to 11 nm at a substrate temperature of 300 0 C. For substrates with a higher temperature, the values of roughness at the same capacities are higher. The increase in capacity contributes to more argon ions affecting the target. As a result, there is an increase in the number of aluminum atoms moving to the substrate per unit of time. Thus, when the power increases, the deposition rate of films increases too. An increase in the deposition rate leads to a higher roughness of the film surface. The atom adsorbed on the substrate can remain stationary or move along the substrate due to the surface diffusion. An increase in the deposition rate makes the surface diffusion more difficult.

Thus, we have proven that the most optimal condition for forming aluminum nitride films with minimal surface roughness is a substrate temperature of 300 0C and a target capacity of 400 watts.

References

[1] J. Gao, G. Liu, J. Li and G. Li, Recent developments of film bulk acoustic resonators // Functional Materials Letters, 2016. – Vol. 9, No. 3. – 1630002 (10 p.)

[2] Stephan Danko, Dirk Bluhm, Valentin Bolsinger, Wladislaw Dobrygin, Oliver Schmidt and Ralf Peter Brinkmann A global model study of silane/hydrogen discharges// Plasma Sources Sci. Technol. 22 (2013) 055009 (10pp)

[3] Bosscher W., Lievens H. Advances in magnetron sputter sources // Thin Solid Films. 1999. V.351, p.15-20

[4] Kuznetsova E.A., Ambartsumov M.G., Medvedev A.E., Shevchenko I.M. Investigation of thin films of aluminum nitride with contact atomic force microscopy. // Publishing house: OOO ID "TESERA". 2018.

The work was carried out within the state assignment of Omsk Scientific Center of SB RAS (the project state registration number 121121700062-3)
The authors express their gratitude to the staff of Analytical and Technological Innovation Center "High Technologies and New Materials" (ATIC HTNM) of NSU for the assistance in carrying out the measurements.

Journal of Physics: Conference Series **2270** (2022) 012006 doi:10.1088/1742-6596/2270/1/012006

SURFACE TENSION OF COLLOID SYSTEMS OF CASTOR OIL AND BENZENE AT P=0.101 MPa

A R Rajabov[1], D S Hakimov[2], M M Safarov[3]

[1]Tajik State Pedagogical University named after S. Aini, Dushanbe, Tajikistan
[2]Institute of Energy of Tajikistan, Kushoniyon district, Tajikistan
[3]Tajik technical university after named by academic M.C.Osimi, Dushanbe, Tajikistan

mahmad1@list.ru

Abstract. The article describes the production and use of benzene and lubricants. The combination of these solutions represents a new chemical formula. Methods for measuring the surface tension coefficient of these liquids are given. The research results are presented in the form of tables and diagrams.

1. Introduction
Benzene was discovered by the English naturalist M. Faraday in 1825. The correct formula C_6H_6 was attributed to him, but the true structure of the molecule remained a mystery for a long time.

2.Experimental section
In industry, aromatic hydrocarbons are obtained from oil or from coal, during the dry distillation of which, along with tar and coke, a gas is formed, 1 m3 of which contains about 30g of benzene. Coal tar also contains benzene derivatives. Table 1 shows the surface tension coefficients of the components of the test samples (at 20°C).

Table 1. The surface tension coefficients of the components of the test samples (at 20°C)

Components	**σ, 10^{-3}N/m**
benzene	24,0
castor oil	35,4 (18°C)

Table 2. Experimental data on the measurement of the surface tension coefficient (**σ·10^3, N/m**) of the studied solutions on the concentration of benzene

n,%	№1	№2	№3	№4	№5	№6	№7	№8	№9
σ·10^3, N/m exp.	35,5	33,7	32,9	32,5	27,7	27	25,9	25,1	24
σ·10^3, N/m cal.	35,45	33,93	32,41	30,88	29,36	27,84	26,32	24,80	23,283
Δ,%	0,13	-0,69	1,489	4,957	-6	-3,137	-1,645	1,179	2,988
The total arithmetic mean error is Δ = 0.25%									

Content from this work may be used under the terms of the Creative Commons Attribution 3.0 licence. Any further distribution of this work must maintain attribution to the author(s) and the title of the work, journal citation and DOI.
Published under licence by IOP Publishing Ltd

Journal of Physics: Conference Series 2270 (2022) 012006 doi:10.1088/1742-6596/2270/1/012006

Sample № 1- (100% castor oil); Sample № 2- (87.5% castor oil + 12.5% benzene); Sample № 3- (75% castor oil + 25% benzene); Sample № 4- (62.5% castor oil + 37.5% benzene); Sample № 5- (50% castor oil + 50% benzene); Sample №6-(37.5% castor oil +62.5% benzene); Sample №7-(25% castor oil +75% benzene);Sample №8-(12.5% castor oil +87.5% benzene); Sample №9-(100% benzene.

To measure the surface tension coefficient of the studied solutions at atmospheric pressure and room temperature, we used the drop method. The total relative error in measuring the surface tension coefficient of solutions at a confidence level α=0.95 is 1.05%.

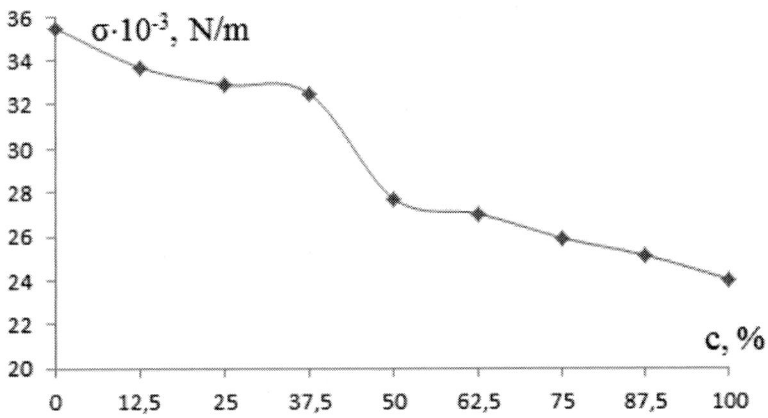

Figure 1. Diagrams of changes in the surface tension coefficient (σ 10^3, N/m) of the system (castor oil +benzene) at atmospheric pressure (P= 0.101MPa) and room temperature (T= 293K) as a function of benzene concentration. Sample №1 - (100% castor oil); Sample №2 - (87.5% castor oil + 12.5% benzene); Sample № 3 - (75% castor oil + 25% benzene); Sample № 4 - (62.5% castor oil + 37.5% benzene); Sample №. 5- (50% castor oil + 50% benzene); Sample №6-(37.5% castor oil +62.5% benzene); Sample №7-(25% castor oil +75% benzene); Sample №8-(12.5% castor oil +87.5% benzene); Sample №9-(100% benzene).

To generalize the experimental data on the measurement of the surface tension coefficients of the studied solutions, we used the following expression:

$$\frac{\sigma}{\sigma_1} = f\left(\frac{n}{n_1}\right) \tag{1}$$

where, σ, σ_1 are, respectively, the coefficients of surface tension of solutions at n and n_1: n_1= 50.0%.

3. Conclusion

According to the values of Table 1 and the graph shown in Figure 1, with an increase in the concentration of benzene, the surface tension coefficient in the solution increases linearly.

Low-temperature plasma can be used for the synthesis of various nanostructures and is well suited for the modification of various surfaces. This is shown in many works [3-23].

References

[1] Nematov A., Ismonov F.D.,Davlatov R.J., Rajabov A.R., Hakimov D.Sh. The relationship between the dynamic and thermodynamic characteristics of solu-tions of the benzene and polymer system at atmospheric pressure, T=293K.// Bulletin of the Tajik National University (Series of Natural Sciences). - Dushanbe: Sino, 2018. - №.1. -pp. 131-136..

[2] Rajabov A.R., Safarov M.M., Hakimov D.Sh., Davlatov R.Dj. The density of the castor oil system on the concentration of benzene at atmospheric pressure and room temperature.// Bulletin of the Tajik National University (Series of Natural Sciences). - Dushanbe: Sino, 2018. - №.4. -pp.76-83.

[3] Saifutdinov, A. I., & Sofronitskii, A. O. (2021). Numerical Study of Breakdown and Formation Dynamics of Arc Discharge Plasma Parameters at Ultrahigh Pressures. High Energy Chemistry, 55(3), 228-232.DOI: 10.1134/S0018143921030115

[4] Shemakhin, A. Y., Zheltukhin, V. S., Shemakhin, E. Y., Terentev, T. N., & Sofronitsky, A. O. (2020, July). Experimental installation to study the RF plasma flow at low pressures with experiment data synchronization. In Journal of Physics: Conference Series (Vol. 1588, No. 1, p. 012018). IOP Publishing. DOI: 10.1088/1742-6596/1588/1/012018

[5] Saifutdinova, A. A., Sofronitskiy, A. O., Timerkaev, B. A., & Saifutdinov, A. I. (2020). Plasma-Chemical Decomposition of Hydrocarbons on the Basis of the Micro-Arc Discharge with Disc Electrodes Rotating in the Bulk of Raw Materials. Russian Physics Journal, 62(11), 2132-2136.DOI: 10.1007/s11182-020-01957-0

[6] Fairushin, I. I., Saifutdinov, A. I., & Sofronitskiy, A. O. (2020). Numerical and Experimental Studies of the Synthesis of Copper Nanoparticles in a High-Pressure Discharge. High Energy Chemistry, 54(2), 150-153. DOI: 10.1134/S0018143920020071

[7] Asadullin, T. Y., Galeev, I. G., Sofronitskiy, A. O., & Gizeev, M. M. (2019, November). Acoustic impact on electric discharge parameters during sterilization of freeze-dried products. In Journal of Physics: Conference Series (Vol. 1370, No. 1, p. 012016). IOP Publishing. DOI: 10.1088/1742-6596/1370/1/012016

[8] Fairushin, I. I., Saifutdinov, A. I., Sofronitskiy, A. O., Timerkaev, B. A., & Dautov, G. Y. (2019, October). Development of plasma reactor design for synthesis of copper nanoparticles using multi-scale simulation. In Journal of Physics: Conference Series (Vol. 1328, No. 1, p. 012088). IOP Publishing. DOI: 10.1088/1742-6596/1328/1/012088

[9] Asadullin, T. Y., Galeev, I. G., & Sofronitskiy, A. O. (2019, October). The application of pulsed discharge for sterilization of freeze-dried product. In Journal of Physics: Conference Series (Vol. 1328, No. 1, p. 012071). IOP Publishing. DOI: 10.1088/1742-6596/1328/1/012071

[10] Timerkaev, B. A., Ganieva, G. R., Kaleeva, A. A., Israfilov, Z. K., & Sofronitskii, A. O. (2019). Growing of carbon nanotubes from hydrocarbons in an arc plasma. Journal of Engineering Physics and Thermophysics, 92(5), 1248-1252. DOI: 10.1007/s10891-019-02040-3

[11] Dautov, G. Y., Kashapov, N. F., Dautov, I. G., & Sofronitskiy, A. O. (2018, July). Research of the influence of the geometry of the discharge chamber on the characteristics of the arc plasmatron. In Journal of Physics: Conference Series (Vol. 1058, No. 1). IOP Publishing. DOI: 10.1088/1742-6596/1058/1/012035

[12] Timerkaev, B. A., Andreeva, A. A., & Sofronitskiy, A. O. (2017, November). Discharge creeping along the surface in the process for producing nanomaterials. In Journal of Physics: Conference Series (Vol. 927, No. 1, p. 012068). IOP Publishing. DOI: 10.1088/1742-6596/927/1/012068

[13] Sadikov, K. G., Sofronitskiy, A. O., & Larionov, V. M. (2017, November). The effect of electrically conductive additives on the plasma pyrolysis of heavy hydrocarbons. In Journal of Physics: Conference Series (Vol. 927, No. 1, p. 012046). IOP Publishing. DOI: 10.1088/1742-6596/927/1/012046

[14] Timerkaev, B. A., Andreeva, A. A., & Sofronitskiy, A. O. (2017). Discharge creeping along the surface in the process of cleaning and strengthening of the materials surface. In Journal of Physics: Conference Series (Vol. 789, No. 1). IOP Publishing. DOI: 10.1088/1742-6596/789/1/012063

[15] Sadikov, K. G., Sofronitskiy, A. O., & Dautov, I. G. (2017). Functional plasma sprayed coatings on magnesium ceramic substrates. In Journal of Physics: Conference Series (Vol. 789, No. 1, p. 012043). IOP Publishing. DOI: 10.1088/1742-6596/789/1/012043

[16] Timerkaev, B. A., Sofronitskiy, A. O., & Andreeva, A. A. (2016). Carbon nanotubes formation in the decomposition of heavy hydrocarbons creeping along the surface of the glow

discharge. In Journal of Physics: Conference Series (Vol. 669, No. 1, p. 012062). IOP Publishing. DOI: 10.1088/1742-6596/669/1/012062

[17] Timerkaev, B. A., & Shamsutdinov, R. S. (2020, July). The Influence of Supersonic Gas Stream on Spatial Structure of Glow Discharge. In Journal of Physics: Conference Series (Vol. 1588, No. 1, p. 012061). IOP Publishing. DOI 10.1088/1742-6596/1588/1/012061

[18] Galeev I G, Asadullin T Y 2017 Improving of stability of the volumetric glow discharge in the gas flow J. of Phys. Conf. Ser. 789(1) 012012

[19] Asadullin T Y, Galeev I G 2017 Combined method of stabilization of the glow discharge in gas flow J. of Phys. Conf. Ser. 927(1) 012003

[20] Galeev I G, Asadullin T Ya 2016 Obtaining fullerene-containing soot during combustion of gaseous hydrocarbons in an external electric field J. of Phys. Conf. Ser. 669 012016

[21] Galeev I G, Asadullin T Y 2016 Enhancing of the glow discharge stability in chamber with cathode sections coated by a discontinuous dielectric coating J. of Phys. Conf. Ser. 669(1) 012015

[22] Shakirov A S, Galeev I G and Asadullin T Ya 2020 Features of formation of fullerene-containing soot during combustion gaseous and liquid hydrocarbons J. of Phys. Conf. Ser. 1588(1) 012013

[23] Shamsutdinov, R. S., & Timerkaev, B. A. (2021, April). The influence of a supersonic flow of gas at glow discharge. In Journal of Physics: Conference Series (Vol. 1870, No. 1, p. 012019). IOP Publishing. DOI 10.1088/1742-6596/1870/1/012019.

Spectral studies of inductively coupled plasma characteristics of low pressure discharges for two configurations of vacuum chambers

A Yu Shemakhin[1], E S Samsonova[1, 2], E F Voznesenskiy[2], Y O Zhelonkin[2], E Yu Shemakhin[1], S A Fadeev[1], T N Terentev[1], R A Maksimov[2], V A Tyurin[1]

[1]Kazan Federal University, Kazan, Russian Federation
[2]Kazan National Research Technological University, Kazan, Russian Federation

shemakhin@gmail.com

Abstract. In this work, we carried out comparative studies of the electron temperature in Inductive coupled radio frequency (ICRF) for two configurations of discharge chambers – cylindrical and flat, considered as prototypes of ICP technological modules of the installation in the argon pressure range from 10 to 50 Pa. The study used optical emission spectrometry of the plasma. As a result, the dependences of the electron temperature on pressure and power were obtained. It is shown that with increasing pressure, different dynamics of the electron temperature is observed for the cylindrical and plane geometry of the inductor.

1. Introduction

Optical emission spectroscopy (OES) is a set of methods [1, 2] for determining such plasma parameters as electron density [3–6], electron temperature [3, 4, 6–10], the form of electron energy distribution functions (EEDF) [11-13]. The study of discharges by OES methods allows one to determine the main characteristics of the plasma without causing disturbances in it. This is convenient not only for fundamental research, but also for industrial applications, since access to industrial reactors is often limited, the compact implementation of the diagnostic method using an optical fiber is an additional advantage of this method. Electron temperature is one of the main parameters of a partially ionized plasma. Determination of T_e in plasma, in the processes of deposition of functional coatings and modification of materials, is an important task for predicting the properties and parameters of coatings, and for choosing the optimal operating modes of technological units.

The aim of the research is to study the dynamics of the electron temperature in ICRF discharges (13.56 MHz) for two configurations of discharge chambers - cylindrical and flat, in the argon pressure ranges from 10 to 50 Pa and the power supplied to the inductor from 0.1 to 1.0 kW.

2. Experimental setup, results and discussion

The work carried out comparative studies of two schemes for the execution of induction discharges - cylindrical (Figure 1) and flat (Figure 2), considered as prototypes of technological modules of the installation for creating an ICP. Photographs of the combustion process of an RFI discharge at various inductor geometries are shown in Figures 3 and 4. The cylindrical ICP source is a quartz water-cooled tube with an inner diameter of 27 mm and a length of 250 mm with an inductor with an inner diameter

Content from this work may be used under the terms of the Creative Commons Attribution 3.0 licence. Any further distribution of this work must maintain attribution to the author(s) and the title of the work, journal citation and DOI.
Published under licence by IOP Publishing Ltd

and height of 60 mm, 2.5 turns. The flat ICP scheme is made using a silicate glass flask with an inner diameter of 150 mm, placed on the outside of the bottom of a flat coil with an outer diameter of 80 mm, 2.5 turns.

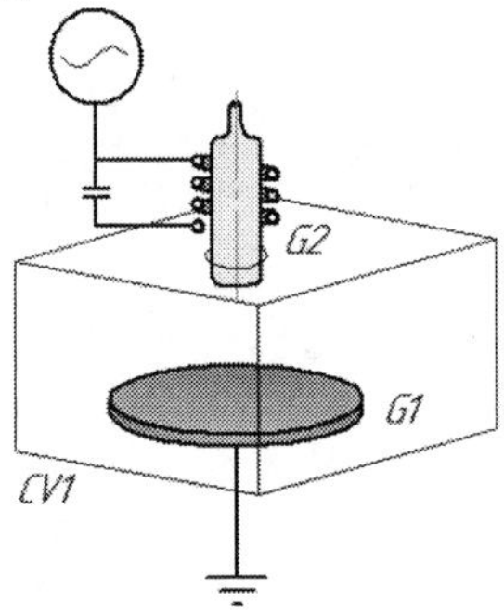

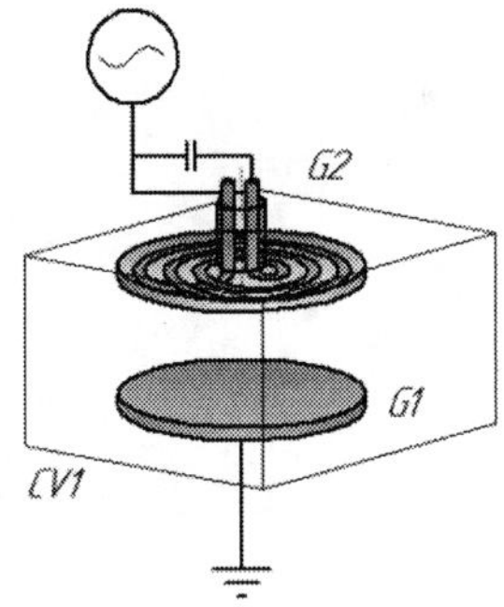

Figure 1. Diagram of a cylindrical ICP source. **Figure 2.** Schematic of a flat ICP source.

Plasma radiation is focused through collimator optics and transmitted to the optical system of the spectrometer. The spectrum decomposed into wavelengths is converted into a digital signal, which is transmitted to a PC (laptop) and processed using special software. The research uses the Ocean Optics USB4000-VIS-NIR spectrometer with Plasus PL-25-12-00 collimator optics, the spectral measurement range is 345 - 1041 nm with a resolution of 1.5 nm.

The measured spectra were normalized with respect to the exposure time, the spectra were recorded at equal distances from the radiation source towards the core of the ICP discharge glow. In the studies for both discharge circuits, uniform means of creating a vacuum were used - an assembly based on AVZ-20 and NVD-600 (DVN-150), pressure measurements - a capacitive membrane sensor MKS 627B, working gases supply - RRG-10-360 and HF power supply - AE Cesar 1330 (VCG) and AE VarioMatch 5000 (SU). The intensity of the brightest lines is higher in the case of cylindrical ICP.

The electron temperature T_e was estimated using the method of relative spectral line intensities [14]. The calculations were carried out under the assumption of a Boltzmann distribution of atoms over excited states and a Maxwellian electron energy distribution function. Ionization is determined by the Saha formula. These assumptions are fulfilled in the plasma of a ICRF discharge [15]. If the main excitation processes are determined by inelastic collisions of electrons with atoms, the ratio of the intensities of the spectral lines has the form

$$\frac{I_j}{I_l} = \frac{g_j}{g_l}\frac{A_j}{A_l}\frac{\lambda_l}{\lambda_j} e^{-\frac{E_j - E_l}{kT_e}}, \qquad (1)$$

where g_j and g_l are the statistical weights of the upper levels, A_j and A_l are the transition probabilities, λ_j and λ_l are the wavelengths, E_j and E_l are the energies of the upper levels, k is the Boltzmann constant, and T_e is the electron temperature. In the calculations, we used the intensities of two argon lines with wavelengths of 738.4 nm and 763.5 nm [16], which appear during transitions between electronic levels $3s^2 3p^5(^2P^\circ{}_{1/2})4p\ ^2[3/2]$, $3s^2 3p^5(^2P^\circ{}_{3/2})4s\ ^2[3/2]^\circ$ for λ_l = 738.4 nm and $3s^2 3p^5\ (^2P^\circ{}_{3/2})\ 4p^2[3/2]$, $3s^2 3p^5(^2P^\circ{}_{3/2})4s\ ^2[3/2]^\circ$ for λ_j = 763.5 nm. Thus, knowing the transition probabilities of the corresponding spectral lines and experimentally measuring the ratio of their intensities, we can determine the electron temperature T_e from (1) by the formula

$$kT_e = \frac{E_j - E_l}{ln\left(\frac{I_l}{I_j}\frac{g_j}{g_l}\frac{A_j}{A_l}\frac{\lambda_l}{\lambda_j}\right)}, \qquad (2)$$

where h is Planck's constant, A_j and A_l are the transition probabilities, E_j and E_l are the energies of the upper levels [17].

Figure 3. Photos of a prototype cylindrical ICP source.

Figure 4. Photo of the work of the prototype of a flat ICP source.

The statistical weights of the upper levels, the transition probabilities, the energies of the upper levels, and other corresponding constants are taken from the Atomic Spectra Database Lines Data NIST [18] and are presented in Table 1.

Table 1

λ, нм	A, с$^{-1}$	g	E, eV
738,4	$8{,}5\cdot 10^6$	5	13,30
763,5	$24{,}5\cdot 10^6$	5	13,17

For two schemes of the studied discharges, according to formula (2), the dependences of the electron temperature T_e on the pressure and power of the RF signal in the ranges of 10-50 Pa and 0.1-1.0 kW, respectively, are plotted. The curves are shown in Figures 5, 6. There is a decrease in the electron temperature with an increase in the electric power. This is typical for gas discharges at a pressure of 10 Pa and is a consequence of two-stage ionization, which, with an increase in the electron density, requires a lower electron temperature to provide the required ionization rate [19]. The measurement results are in qualitative agreement with the experimental data of other authors [19].

Figure 6 shows that with increasing pressure, a different dynamics of the electron temperature is observed for the cylindrical and flat geometry of the inductor. The structure and magnitude of the electromagnetic field depend on the geometry and distance between the inductor and the grounded electrode. In the case of a cylindrical inductor, the spectral characteristics of the jet were recorded; the decrease in the electron temperature is due to an increase in the frequency of collisions between atoms and electrons with increasing pressure and, accordingly, a more efficient transfer of energy from the electron gas to neutral particles. In the case of a flat inductor, where the distance between the electrodes is less, the spectral characteristics were recorded in the region with a higher concentration of the electromagnetic field due to the presence of the potential component of the field, inversely proportional to the distance between the electrodes. Thus, for the cases of flat and cylindrical inductors, the electron distribution function is different. In the case of a flat inductor, the electron energy exceeds the maximum of the elastic collision cross section, the electrons spend less energy in collisions with the neutral plasma component, and an increase in the electron temperature is observed.

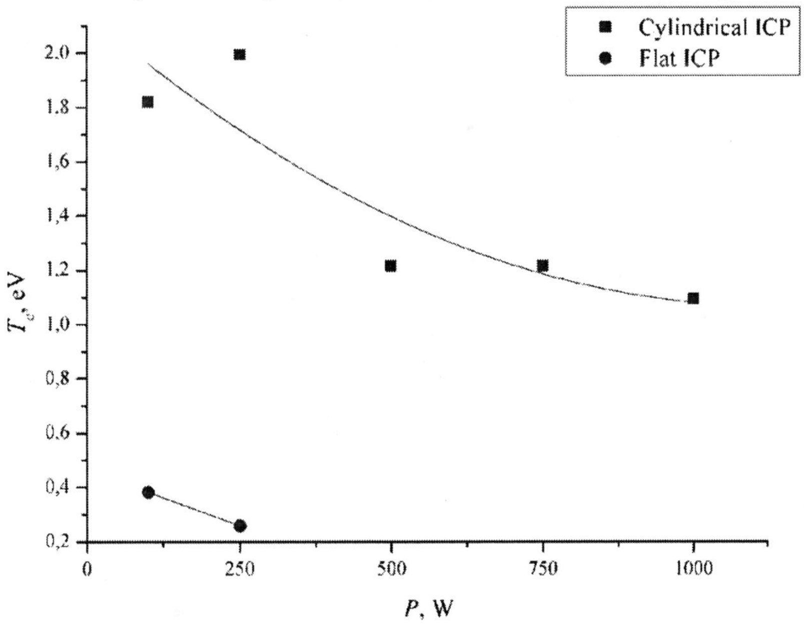

Figure 5. Dependence of T_e on RF power at a pressure of 10 Pa, curves – approximation.

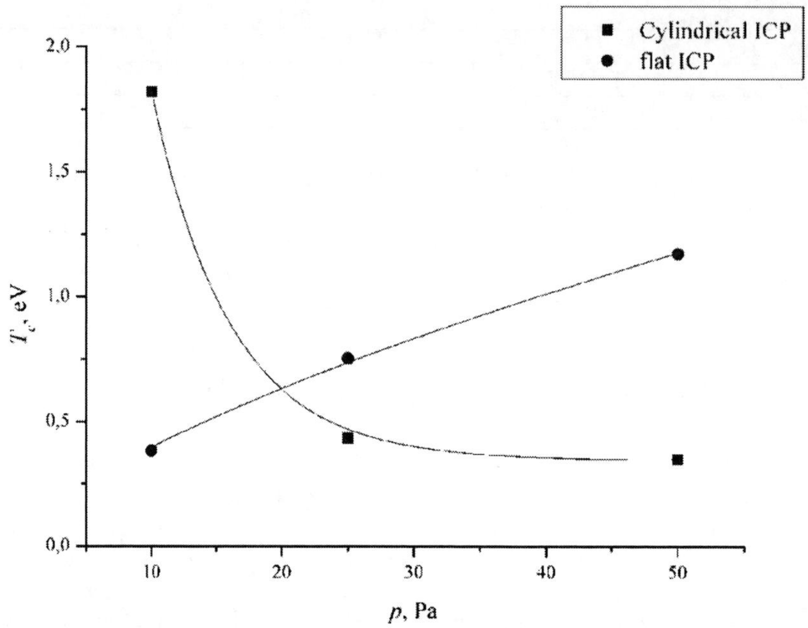

Figure 6. Dependence of T_e on pressure at RF power of 100 W, curves – approximation.

3. Conclusions

It follows from the results obtained that the cylindrical ICP has a higher electron temperature at 10 Pa, which is explained by the higher values of the magnetic field induction associated with the geometry of the inductor in comparison with the flat circuit, as well as higher values of the local pressure in the flow plasmatron. At the same time, a flat ICP source allows one to create a more voluminous plasma, which is explained by the large dimensions of the space for the discharge burning; accordingly, the diffusion and death factor of charged particles on the chamber walls is much lower.

A decrease in the electron temperature with an increase in the electric power is a consequence of two-stage ionization, which, with an increase in the electron density, requires a lower electron temperature to ensure the required ionization rate.

For a cylindrical inductor, with increasing pressure, a decrease in the electron temperature is due to an increase in the frequency of collisions between atoms and electrons. In the case of a flat inductor, where the distance between the electrodes is smaller, the spectral characteristics were recorded in the region with a larger potential component of the electromagnetic field as compared to the circuit for a cylindrical inductor. Thus, the shape of the electron distribution function, in the case of a flat inductor, differs from the case with a cylindrical circuit. The energy of electrons exceeds the maximum of the cross section for elastic collisions, electrons spend less energy in collisions with the neutral component of the plasma, and an increase in the electron temperature is observed.

From the point of view of the efficiency of using the input power into an ICP discharge, a cylindrical scheme is preferable for processing bulk materials passing along a plasma and for processes associated with gas-phase changes, for example, as a source of dissociation of molecules and creating a flux of atomic radicals for CVD processes or reactive deposition of thin films. It is also worth noting that the cylindrical ICP scheme is a more efficient source of heating the working gas, which can be used in the processes of ion-assisted heat treatment of local areas of various materials. The flat scheme of excitation of a gas discharge is widely used in the processes of plasma-chemical etching of elements of microelectronics and optics, due to the possibility of obtaining dense plasma uniformly distributed over the area. The results obtained for various methods of excitation of an inductive discharge make it possible to take into account the peculiarities of a particular scheme of execution when they are applicable in various technological problems.

Acknowledgments
The research was carried out with the financial support of the Russian Science Foundation (project No. 19-71-10055).

References

[1] Fantz U 2006 *Plasma Sources Sci. Technol.* **15** S137–47
[2] Samukawa S et al 2012 *J. Phys. D: Appl. Phys.* **45** 253001
[3] Iordanova S and Koleva I 2007 *Spectrochim. Acta B* **62** 344–56
[4] Crintea D L, Czarnetzki U, Iordanova S, Koleva I and Luggenholscher D 2009 *J. Phys. D: Appl. Phys.* **42** 045208
[5] Zhu X M, Pu Y K, Balcon N and Boswell R 2009 *J. Phys. D: Appl. Phys.* **42** 142003
[6] Zhu X-M, Chen W-C, Li J and Pu Y-K 2009 *J. Phys. D: Appl. Phys.* **42** 025203
[7] Zhu X-M, Chen W-C and Pu Y-K 2008 *J. Phys. D: Appl. Phys.* **41** 105212
[8] Boffard J B, Jung RO, Lin C C, Aneskavich L E and Wendt A E 2012 *J. Phys. D: Appl. Phys.* **45** 045201
[9] Malyshev M V and Donnelly V M 1997 *J. Vac. Sci. Technol. A* **15** 550-8
[10] Donnelly V M 2004 *J. Phys. D: Appl. Phys.* **37** R217-36
[11] Zhu Li-Guo, Chen Wen-Cong, Zhu Xi-Ming, Pu Yi-Kang and Li Ze-Ren 2009 *Review of Scientific Instruments* **80** 023105
[12] Zaidel A N, Shrader E Ya 1976 *Vacuum spectroscopy and its applications* (Moscow: Nauka) pp. 352–362. [in rus]
[13] Zhu Xi-Ming, Pu Yi-Kang, Celik Yusuf, Siepa Sarah, Schungel Edmund, Luggenholscher Dirk and Czarnetzki Uwe 2012 *Plasma Sources Sci. Technol.* **21** 024003.
[14] Knyazev B A 2003 *Low-temperature plasma and gas discharge.* (Novosibirsk: Novosibirsk State University) p 290 [in rus]
[15] Raizer Yu P, Schneider M N, Yatsenek N A 1995 *Radio-frequency capacitive discharge* (Moscow: Nauka) p 320[in rus]
[16] Siepa Sarah, Danko Stephan, Tsankov Tsanko V, Mussenbrock Thomas and Czarnetzki Uwe 2014 *J. Phys. D: Appl. Phys.* **47** 445201.
[17] Qayyum Abdul, Murtaza G, Zakaullah M, Ahmad Riaz 2003 *International Journal of Modern Physics B* **17** (14) 2749-2759
[18] NIST Atomic Spectra Database www.nist.gov
[19] Godyak V A, Piejak R B and Alexandrovich B M 2002 *Plasma Sources Sci. Technol.* **11** 525.

NANODIAMONDS FROM FUEL

B A Timerkaev[1], R B Shakirov[1], A A Zalyalieva[1], R K Gevorgyan[1], R R Shaikhattarov[1]

[1]KNRTU-KAI, Kazan, Russia

btimerkaev@gmail.com

Abstract. In this paper, we present the results of the synthesis of nanodiamonds from carbon nanostructures and diamondoids in an arc discharge. The electrodes used in the experiments were a molybdenum cathode and a copper anode coated with carbon nanostructures such as carbon nanotubes and diamondoids. The formation of carbon nanostructures on the surface of a copper electrode was carried out in an electric arc sunk into fuel oil. At the next stage, these carbon nanostructures served as a source of atomic carbon and diamondoids from the surface of the arc discharge anode. The use of fuel oil as a feedstock for creating the electrode material probably led to the retention of diamondoids contained in fuel oil during two vaporizations. It was found that, in addition to carbon nanotubes, nanodiamonds also formed on the cathode surface.

1. Introduction

Diamond is one of the most important and sought-after materials, on the one hand, as the hardest material on Earth, on the other hand, as an exceptional piece of jewelry. Therefore, the search for simple methods of artificial synthesis of diamonds has not stopped for almost a century and a half. There are two ways to create diamonds. The first method is a method of converting graphite into diamond under high pressure and high temperature. To do this, it is necessary to force carbon atoms to overcome the quantum potential barrier and rearrange into the crystal structure of diamond. In this way, diamond crumbs are created and sufficiently large diamonds are grown. The second method is the synthesis of diamonds by the CVD method in the presence of catalysts and diamond nuclei. Now a new way of creating nanodiamonds is emerging. For instance. The works [1, 2] reported on the synthesis of nanodiamonds under high pressure and temperature from adamantanes (diamonds) (chemical formula (CH4)4(CH2)6). A sensational article [3] reports on the synthesis of nanodiamonds from diamondoids. Diamondoids are the smallest (typically 0.5 to 2 nm) form of hydrogen-terminated carbon cells that can be superimposed on the diamond lattice. According to the spatial arrangement of carbon atoms, such a molecule is similar to one, two or three cells of the diamond crystal lattice. Composed entirely of sp3 hybridized bonds, diamondoids combine some of the exceptional properties of bulk diamond and hydrocarbon molecules, such as rigidity, thermal stability, and uniformity at the atomic level. The authors of this work used diamondoids as promising precursors for laser-induced synthesis of diamonds under high pressure at high temperature. The lowest pressure and temperature (P-T) at which diamond was obtained were 12 GPa (at ~2000 K) and 20 GPa (at ~900 K), respectively. At a pressure of 20 GPa, the transformation of diamondoid into diamond occurred in 19 μs. Molecular dynamics modeling showed that after dehydrogenation, the remaining diamond-like carbon cells rearrange into diamond-like structures at high

Content from this work may be used under the terms of the Creative Commons Attribution 3.0 licence. Any further distribution of this work must maintain attribution to the author(s) and the title of the work, journal citation and DOI.
Published under licence by IOP Publishing Ltd

pressure and temperature. The results of plasma-chemical synthesis of carbon nanostructures from hydrocarbons are given in [9–10]. In [11–13], methods for increasing the stability of electric discharges in the processes of synthesis of various nanostructures were studied.

Despite the success achieved in the synthesis of large diamonds and nanodiamonds by the above methods, the synthesis process requires compliance with very stringent conditions. As reported in [4], the synthesis of even microscopic diamonds can take several days.

The synthesis of nanostructures can be successfully carried out only if the temperature regime is strictly observed. That is, the synthesis temperature cannot exceed the melting temperature of the nanostructure. Therefore, information about the distribution of temperature fields in the region where the synthesis of one or another nanostructure is supposed to be carried out should be known in advance. Each type of nanostructure has its own synthesis temperature. The traditional technique for the synthesis of carbon nanotubes involves the evaporation of graphite at a temperature of 4000K and their deposition on the cathode. In this case, there should be a section on the cathode with a temperature of about 1800K. This is exactly the temperature at which the synthesis of nanotubes occurs. It would be much better if the evaporation temperature of carbon atoms also corresponded to the same temperature. Therefore, for the implementation of the synthesis of carbon nanostructures, it is advisable to use easily evaporating structures. In this work, we used an anode consisting of fullerene soot, nanotubes, and diamondoids as an anode.

It was shown in [5] that two regions can be distinguished in a carbon arc, defined as the core of the arc and the periphery of the arc. These areas have very different temperatures and densities of carbon formations. At the colder periphery of the arc, diatomic carbon molecules (C2) predominate, which are in the minority in the core of the hot arc. These differences are due to the extremely uneven distribution of the arc current, which is mainly conducted through the arc core filled with carbon atoms and ions. Thus, the ablation of the graphite anode is driven by the core of the arc, while the formation of carbon molecules occurs at the colder periphery of the arc.

It is reported in [6] that the growth kinetics of nanostructures in plasma is affected by Coulomb forces between charged particles. In a stationary plasma, the unipolar charge of particles leads to a slowdown in growth and to a limitation in the particle size. In a non-stationary atmospheric pressure arc discharge, a bipolar charge distribution is formed. Large formations are positively charged, while atoms are negatively charged. In this case, the Coulomb forces accelerate the formation of particle growth on a millisecond time scale.

2. Materials and methods

2.1. Plasma-chemical method

In this work, in contrast to HPHT, CVD methods and methods for the synthesis of microdiamonds by arc evaporation of graphite, a plasma-chemical method for the synthesis of nanodiamonds in arc plasma by evaporation of carbon nanostructures, such as carbon black containing nanotubes and diamondoids, is proposed.

In the first stage, a cathode assembly consisting of carbon black containing nanotubes and diamondoids was synthesized in an arc plasma with electrodes embedded in fuel oil. In the second stage, this cathode unit was used as the anode of the electric arc discharge.

The effectiveness of an arc discharge with electrodes recessed into fuel oil for the synthesis of carbon nanotubes is described in detail in [7–8]. The advantage of this method for the synthesis of carbon nanomaterials is the fact that electrodes are placed directly in the raw material. Therefore, both the temperature of the electrodes and the temperature of the plasma itself are moderate. On the surface of the electrodes, the temperature distribution turns out to be such that there are corresponding regions that are optimal for the synthesis of certain carbon nanoobjects. Copper rods 5 mm in diameter were used as electrodes in this work. After this stage of the experiment, nanotubes and carbon deposits formed on the surface of the copper cathode. Next, this copper electrode with carbon deposits was placed in a vacuum chamber and served as an anode in the nanodiamond synthesis facility. The synthesis was carried out in an argon medium at a pressure of 500 Torr with an arc current of about 30 A. The experiment lasted for about 30–40 s. During this time, the carbon build-up from the anode surface was transferred to the surface of the molybdenum cathode.

2.2. SEM imaging

Carbon nanoformations on the surface of the molybdenum cathode were studied using optical and electron microscopes.

3. Results and discussion

Nanoformations in form represent a carbon buildup on the cathode surface. The middle part of this outgrowth contains carbon nanotubes to a greater extent, while the periphery contains nanodiamonds.

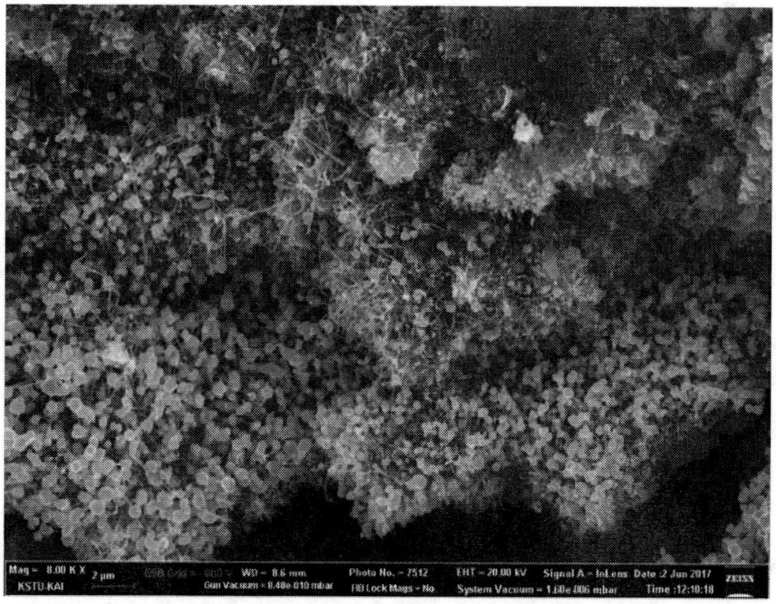

Figure 1. An electron micrograph of a carbonaceous build-up on the cathode. Magnification 8000 times.

On Fig. 1 shows an electron micrograph of deposits on the cathode. As can be seen from the figure, these deposits are nanodiamonds with dimensions of about 300–500 nm. Nanodiamonds are quite densely located relative to each other, and fluctuations in their sizes are insignificant. Carbon nanotubes are also visible there. The lengths of nanotubes vary within 2-4 µm, and the diameters are about 50 nm.

Figure 2 shows photographs of smaller nanodiamonds. Their sizes are more than 10 times smaller than the sizes of diamonds shown in Fig.1. Characteristically, nanotubes are not observed in this region.

The following figure (Fig. 3) shows an electron micrograph of even smaller nanodiamonds in the form of a diamond placer. The diamond placer is located on the surface of the graphene sheet.

In Fig.2, nanodiamonds have approximately the same dimensions of about 40-50 nm. The appearance of practically all nanodiamonds is the same. Some of them are pyramidal in shape.

The nanodiamonds shown in Fig. 3 have more regular shapes. Their sizes are less than 20 nm.

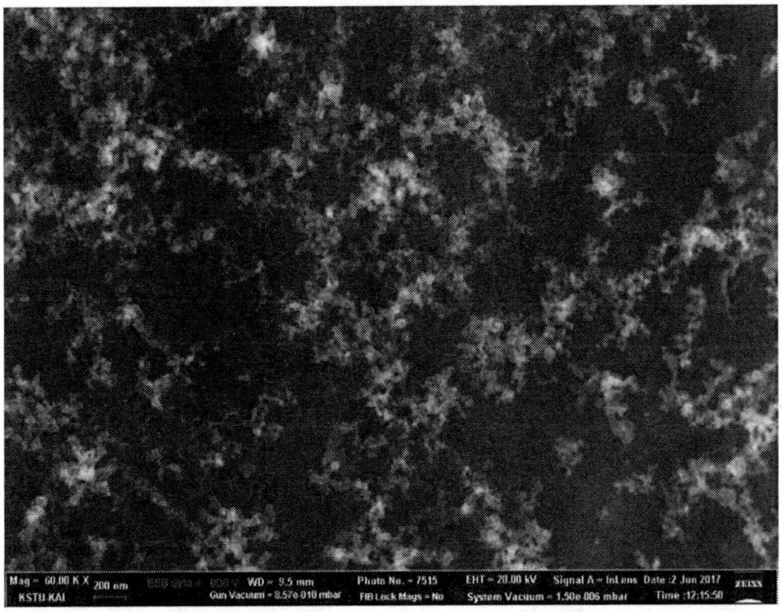

Figure 2. An electron micrograph of a carbonaceous build-up on the cathode. Magnification 60000 times.

Figure 3. Electron microscopic image of a diamond placer. Magnification 10000 times.

4. Conclusion

Thus, the plasma-chemical synthesis of nanodiamonds in an electric arc plasma was carried out by evaporating carbon nanostructures containing carbon nanotubes and diamondoids, followed by their deposition on the surface of a molybdenum cathode. The anode material was one of the electrodes of an electric arc embedded in fuel oil.

References

[1] E. A. Ekimov, O. S. Kudryavtsev, N. E. Mordvinova, O. I. Lebedev, I. I. Vlasov, High-pressure synthesis of nanodiamonds from adamantane: Myth or reality? *Chem. Nano. Mat.* 4, 269–273 (2018).

[2] E. A. Ekimov, K. M. Kondrina, N. E. Mordvinova, O. I. Lebedev, D. G. Pasternak, I. I. Vlasov, High-pressure, high-temperature synthesis of nanodiamond from Adamantane. *Inorg. Mater.* 55, 437–442 (2019).

[3] Sulgiye Park, Wnetim I.Abate ,Jin Liu , Chenxu Wang, at.all/ Facile diamond synthesis from lower diamondoids. SCIENCE ADVANCES•21 Feb 2020•Vol 6, Issue 8•DOI: 10.1126/sciadv.aay9405

[4] Palyanov Yu.N., Kupriyanov I.N., Borzdov Yu.M., Surovtsev N.V. // Scientific Reports. 2015. V. 5. P. 14789. https://doi.org/10.1038/srep14789

[5] Vekselman, V., Feurer, M., Huang, T., Stratton, B., and Raitses, Y. Complex structure of the carbon arc discharge for synthesis of nanotubes. United States: N. p., 2017. Web. doi:10.1088/1361-6595/aa7158.

[6] V. Vekselman, Y. Raitses, M. N. Shneider. Growth of nanoparticles in dynamic plasma. PHYSICAL REVIEW E 99, 063205 (2019). https://doi.org/10.1103/PhysRevE.99.063205

[7] Timerkaev, B.A., Ganieva, G.R., Kaleeva, A.A., Israfilov, Z.K., Sofronitskii, A.O. Growing of Carbon Nanotubes from Hydrocarbons in an Arc Plasma. Journal of Engineering Physics and Thermophysics, 2019, 92(5), P. 1248–1252. DOI:10.1007/s10891-019-02040-3.

[8] Ganieva G. R., Ziganshin D. I., Aukhadeev M. M., Timerkaev B. A. //J. Eng. Phys. Thermophys. 2014. V. 87. № 1. P. 699. DOI10.1007/s10891-014-1062-9

[9] Galeev I G, Asadullin T Ya 2016 Obtaining fullerene-containing soot during combustion of gaseous hydrocarbons in an external electric field *J. of Phys. Conf. Ser.* 669 012016

[10] Shakirov A S, Galeev I G and Asadullin T Ya 2020 Features of formation of fullerene-containing soot during combustion gaseous and liquid hydrocarbons *J. of Phys. Conf. Ser.* 1588(1) 012013

[11] Galeev I G, Asadullin T Y 2017 Improving of stability of the volumetric glow discharge in the gas flow J. of Phys. Conf. Ser. 789(1) 012012

[12] Asadullin T Y, Galeev I G 2017 Combined method of stabilization of the glow discharge in gas flow J. of Phys. Conf. Ser. 927(1) 012003

[13] Galeev I G, Asadullin T Y 2016 Enhancing of the glow discharge stability in chamber with cathode sections coated by a discontinuous dielectric coating J. of Phys. Conf. Ser. 669(1) 012015

Study of the Effect of Plasma Treatment on the Mechanical Properties of Mineral Fiber Threads

M V Antonova[1], A R Ibatullina[1], A S Parsanov[1], I V Krasina[1]
[1]Kazan National Research Technological University, Kazan, Russia

parsanov1982@yandex.ru

Abstract. The article describes methods for modifying glass and basalt fibers and threads, and composite materials from them. Special attention is paid to plasma methods. Twisted glass thread and basalt twisted thread are investigated in the work. Plasma treatment was used to modify the threads. The type of plasma is nonequilibrium, low-temperature; the type of discharge is high-frequency capacitive. The results of research of thread properties, such as morphological changes and breaking load of threads. The plasma-treated and non-treated samples of threads were compared. The modes of plasma modification of threads in which the increase of breaking load strength of glass threads up to 23 % and of basalt threads up to 10 % is achieved have been established.

1. Introduction

Mineral fibers find application in various industries, such as textile production, aircraft and shipbuilding, electronics and mechanical engineering. The development of these industries requires innovative materials that are significantly superior to traditional materials in performance and strength characteristics. Composite materials can meet these requirements.

For creation of modern composites, basalt fibers, which have good adhesion to various binders, are used nowadays. Composites based on basalt fibers are widely used in many industries.

For obtaining composite materials, along with basalt fibers, glass fibers are also used. One of the methods of obtaining high-strength glass fibers is its impregnation with epoxy binder.

Strength, good adhesion to binders, fire resistance and much more - all these properties are given to fibers by modification using different methods, including plasma modification. The purpose of such modification is to effectively change the surface properties of the material without worsening the physical and mechanical characteristics.

Researches on modification of basalt and glass fibers are conducted by scientists of different countries of the world [1-8].

In recent years, both industry and the scientific community have focused their attention on the development of stable composites made of basalt fibers. Polymer-based composites reinforced with basalt fiber have found application in airplanes, automobiles, ships, in construction, in petrochemical pipelines, and in wind turbine blades. Composites based on basalt fibers have excellent mechanical properties and constructional ability. But poor compatibility between the basalt fiber and the polymer matrix leads to defects and drawbacks in the composites. These drawbacks limit the development of composites based on basalt fiber reinforced polymers. The authors of [5] work have developed various methods of physical or chemical treatment of the basalt fiber surface to improve the interaction at the

Content from this work may be used under the terms of the Creative Commons Attribution 3.0 licence. Any further distribution of this work must maintain attribution to the author(s) and the title of the work, journal citation and DOI.
Published under licence by IOP Publishing Ltd

Journal of Physics: Conference Series **2270** (2022) 012009 doi:10.1088/1742-6596/2270/1/012009

interface between the fiber and the resins. As a result, a composite with improved strength properties was obtained.

Nonthermal atmospheric pressure plasma is used more and more often for modification of polymer surface. The authors [6] used argon, oxygen, hydrogen and hydrogen-nitrogen plasma at atmospheric pressure for the surface modification of basalt fibers. The treatment was carried out with the aim of detecting changes and regulating such fibre parameters as surface-active groups content, chemical resistance, roughness etc. The results showed the processes involved in etching and subsequent attachment of nitrogen and oxygen-containing polar groups, the experiments also showed a significant increase in chemical stability and high adhesion. During the study, the authors found that etching with oxygen was mostly the result of bombardment by ions, has caused a decrease in molecular weight and roughness. In this case, chemical changes on the surface of the basalt fiber with the participation of nitrogen and hydrogen are the result of neutral particles affecting the surface of the basalt fiber previously subjected to plasma activation [6].

Plasma modification methods are also used for glass fibers. For example, the authors of work [7] used tetravinylsilane polymerized in plasma to modify the surface of glass fibers. Plasma modification of glass fibers made it possible to significantly increase the interfacial shear strength as compared to untreated fibers. In tests of the interfacial shear strength of the coatings, the optimized plasma coating showed an increase in shear strength up to 26% compared to the baseline coating.

In study [8], plasma-enhanced chemical vapor deposition (PECVD) was used to modify the surface of glass fibers and preserve their volumetric properties. The fibers were used as reinforcement in a glass fiber/polyester composite. By optimizing the plasma modification parameters, the adhesion between the glass fibers and the polymer matrix was improved, as shown by a 14% increase in the shear strength of the glass fiber polymer composite.

In this work the effect of plasma treatment on the mechanical properties of basalt and glass fiber filaments was considered.

2. Materials and methods

Twisted alumina borosilicate glass thread with paraffin emulsion oiling agent; basalt twisted thread Basalt 10-68Z40-KV-12 were chosen as objects of research.

Plasma treatment was used to modify threads and tapes. Type of plasma is nonequilibrium, low-temperature; type of discharge is high-frequency capacitive.

Plasma treatment of threads was conducted using a plasma vacuum unit (capacitive discharge), an argon/propane-butane mixture in the ratio of 70/30 and 100% propane-butane was used to create the plasma medium. Treatment mode: plasma gas flow $G = 0.04$ grams per second, generator frequency $f = 13.56$ MHz, pressure in the chamber $P = 26$ Pa, treatment time $t = 5$ min, with variation of input voltage parameters at the anode $U_a = 4$ to 7 kV. The current strength at the anode was fixed during the treatment and ranged from 0.5 A to 0.85 A.

Determination of breaking load of glass and basalt threads was carried out on a testing machine Shimadzu AGS-X (Japan production). Research was conducted according to State Standard 6943.10-79 Glass textile products. Method for determination of breaking load and elongation at rupture. Microstructure of glass and basalt fibers was studied by confocal laser microscopy on Olympus LEXT 4000 microscope (Japan production).

3. Results and discussion

Based on the results of previous works in the field of studying the of cold plasma on the characteristics of different yarns and threads [9-13], in this work we present studies of the impact of plasma treatment parameters on the properties of mineral fiber threads.

To study plasma impact to the change of structural characteristics of glass and basalt filaments, their surface was studied by scanning electron microscopy. At micro scanning, the general state of thread surface before and after plasma treatment was studied.

During the study, it was found that samples of fibers that have been plasma treated have a more homogeneous fiber structure and less rough surface compared to untreated samples. Such effect probably arises due to equal distribution of oiling emulsions on fiber surface, filling of cracks and chips.

The change in mechanical characteristics of glass and basalt threads was investigated further. Mechanical characteristics of threads are often crucial for creation of composite materials, so their decrease is not recommended.

The results of researches of influence of plasma treatment on strength properties of glass and basalt threads are presented in figures 1 and 2.

For glass and basalt threads, three modes (by voltage at the anode) of treatment in gas discharges were chosen: in the first mode Ua = 4 kV; in the second mode Ua = 5 kV; in the third mode Ua = 6 kV; in all three modes, the plasma exposure time on the samples was 5 minutes. Filament samples were treated in plasma consisting of argon and propane-butane gases (percentage 70/30, respectively) and 100% propane-butane.

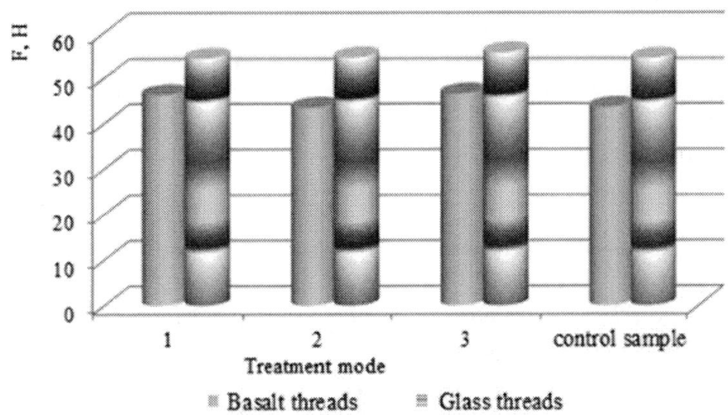

Figure 1. Dependence of yarn tensile strength from plasma treatment modes, propane-butane plasma gas

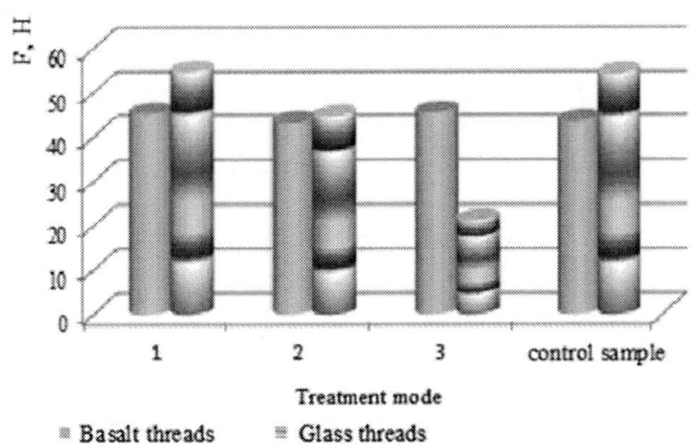

Figure 2. Dependence of yarn tensile strength from plasma treatment modes, argon/propane-butane (ratio 70/30, respectively) plasma gas

As can be seen from the graphs in Figures 1 and 2, plasma treatment in a mixture of argon/propane-butane (70/30) and 100% propane-butane at 4 kV voltage increases the breaking load of the glass fibres by up to 23%, against a standard sample. At low temperatures, the adsorption effect of air water on the glass fibres is eliminated, increasing their strength. In addition, the filling of voids and cracks with paraffin emulsion also improves the strength of the yarns. However, with an increase of voltage at the anode up to 6 kV in argon/propane-butane gas plasma, a radical degradation of the strength of glass filaments is observed. In the case of 100% propane-butane plasma, the strength of the modified samples stays higher compared to the control samples.

The results of treatment of basalt filaments have shown that plasma modification in argon/propane-butane (70/30) gas mixture and in pure propane-butane at 6 kV voltage increases their breaking load strength.

The experiments showed that the relative elongation of glass and basalt threads remains in the range of values of the control sample (Table 1).

Table 1. Dependence of relative elongation of glass and basalt threads on plasma treatment modes

№	Plasma treatment mode	Relative elongation, %			
		Control sample	1	2	3
	Glass threads				
1	Argon/propane-butane (70/30)	0,23	0,16	0,16	0,12
2	Propane-butane		0,19	0,18	0,18
	Basalt threads				
1	Argon/propane-butane (70/30)	0,17	0,19	0,18	0,18
2	Propane-butane		0,21	0,18	0,19

4. Conclusions

The paper gives an overview of researches in the field of modification of glass threads and basalt threads. The modern tendencies in world science in the field of modification of mineral fibres have been described.

The influence of plasma modification on mechanical properties and change of morphology of glass thread and basalt thread surface has been studied. The scanning electron microscope (SEM) method has established, that plasma treatment in medium of argon/propane-butane (70/30) and pure propane-butane gases contributes to fiber surface cleaning, and also to keeping its integrity.

It was found that plasma treatment in argon/propane-butane (70/30) gas mixture and in pure propane-butane at 4 kV increases the breaking strength of glass fibres up to 23%, and of basalt fibres up to 10% on average at 6 kV.

The relative elongation of glass and basalt threads after plasma treatment does not change considerably.

Therefore, the study showed that modification in cold plasma based on active, inert, polymerizing gases and based on mixtures of some of them is a perspective method to increase the strength characteristics of glass and basalt threads, as well as the transformation of their surfaces.

References

[1] Manylov M.S., Gutnikov S.I., Lipatov Y.V., Pokhodok K.V., Filimonov D.S., Lazoryak B.I. Crystallization of basalt continuous fibers in oxidizing atmosphere // Physics and Chemistry of Glass. 2012. T. 38. № 4. P. 565-573.

[2] N.M. Demina, K.P. Tsvetkova, B.K. Gromkov Researches of penetrating power of basalt yarn by wetting agents when modifying size 4C Textile Industry Technology № 2 (350) 2014. P. 72-74

[3] Zhang, XF at all. Study on Modification of Basalt Fibers Advanced Textile Materials, PTS 1-3// Volume 332-334.- 2011.- P.2028-2031 doi10.4028/www.scientific.net/AMR.332-334.2028

[4] Alibi, Z, Habibi, N, Zadhoush, A Surface Modification of Basalt Fibers by Nanostructured Silica Aerogel. / Fibers and Polymers. Volume19. Issue 9 – 2018.- Pp.1843-1849 doi10.1007/s12221-018-7710-5

[5] Jia, H (Jia, Hang) at all. Interface Modification Strategy of Basalt Fiber Reinforced Resin Matrix Composites Progress in Chemistry Volume 32. Issue9 Pp.1307-1315 DOI10.7536/PC200222

[6] G.J.Wangab at all. Surface modification and characterizations of basalt fibers with non-thermal plasma Surface and Coatings Technology Volume 201, Issue 15, 2007, P. 6565-6568 https://doi.org/10.1016/j.surfcoat.2006.09.069

[7] Cech, V at all. Enhanced interfacial adhesion of glass fibers by tetravinylsilane plasma modification Composites part a-applied science and manufacturing Volume58. 2014. P.84-89 doi 10.1016/j.compositesa.2013.12.003

[8] Cech, V at all. Continuous surface modification of glass fibers in a roll-to-roll plasma-enhanced CVD reactor for glass fiber/polyester composites Composites part a-applied science and manufacturing Volume121. 2019. Pp.244-253. doi 10.1016/j.compositesa.2019.03.036

[9] Kultyshev A.S., Study of influence of low-temperature plasma treatment on strength characteristics of threads from basalt fibers / A.S. Kultyshev [et al.] // Bulletin of technological university. 2017. VOL.20, No.10 P.79-81

[10] Abdullina V.H., Plasma methods of polyolefin fibre surface activation / V.H. Abdullina, R.S. Davletbaev // Izvestia of Samara Scientific Centre of the Russian Academy of Sciences, Vol.12, № 4(3), 2010. Pp.656-659

[11] M V Antonova et al 2018 J. Phys.: Conf. Ser. 1058 012003

[12] M V Antonova et al 2020 J. Phys.: Conf. Ser. 1588 012003

[13] A R Ibatullina et al 2020 J. Phys.: Conf. Ser., 1588(1) 012020

THE EFFECT OF JET ELECTRIC DISCHARGE ON THE STRENGTH CHARACTERISTICS OF THE SURFACE

M F Akhatov[1], R R Kayumov[1], R R Mardanov[1], I M Loginova[2]

[1]Kazan National Research Technical University named after A. N. Tupolev - KAI, Kazan, Russia
[2]Kazan State Agrarian University, Kazan, Russia

ahatov.81@yandex.ru

Abstract. In this paper, the results of measuring the density of the skin of metallic tissues using an installation for assessing the strength of damaged layers (sclerometer) MV-11m, a study of the treatment of tissues with an electrolytic discharge, are revealed.

1. Introduction.

The development of technology poses to physics the task of studying processes with high concentrations of energy, high pressures and temperatures. An electric discharge in a liquid is one of these processes. As a result of the rapid release of capacitor energy, high temperature and pressure occur in the discharge channel. Presumably, the electron temperature is higher than the temperature of heavy particles. These qualities of electrical discharges in liquids are widely used in the development of new technological processes for processing various materials and in the creation of new means of energy conversion. The prospects of using such discharges for a number of technological processes, such as obtaining oxide powder, cleaning and polishing the surface of metals [1], for heating metals [2], as well as for initiating chemical reactions in electrolytes, have been established.

2. The main part

Measurements of the surface strength of metal products were carried out using the MV-11m installation for assessing the strength of surface layers (sclerometer) [3]. In particular, the installation provides automatic registration of scratching diagrams with a diamond indenter of surface layers of materials and coatings with determination of the fracture characteristic - hardness during scratching.

The measurement procedure is following: the test sample is placed on a lifting table and clamped to the lower surface of the beam. By rotating the screw, the indenter is buried by a predetermined amount. After the installation is turned on, appropriate measurements are carried out during the operation of the sclerometer, and a graph of the dependence of the force applied to the indenter on its movement is plotted in real time [4]. The figures show the results of the study, graphs of the dependence of the force applied to the indenter on its movement before and after processing for copper (Fig.1) and steel (Fig.2). In all cases, the diamond cone of the indenter was buried to a depth of 20mkm.

Content from this work may be used under the terms of the Creative Commons Attribution 3.0 licence. Any further distribution of this work must maintain attribution to the author(s) and the title of the work, journal citation and DOI.
Published under licence by IOP Publishing Ltd

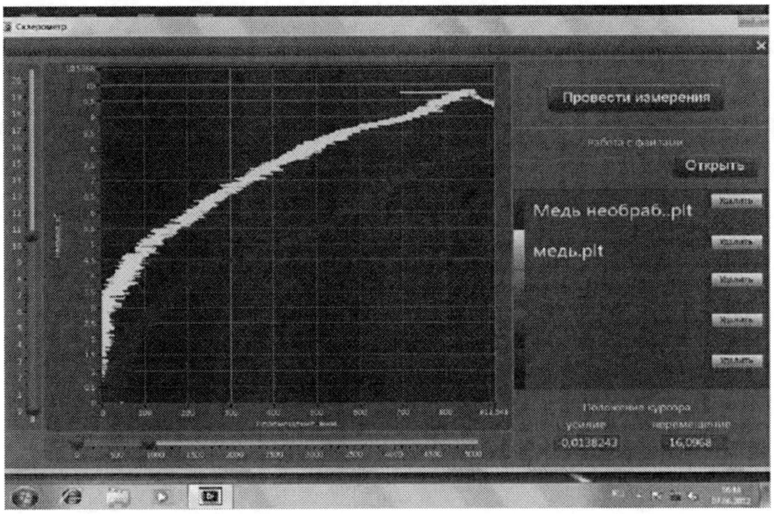

Figure 1. Graph of the dependence of the force applied to the indenter on its movement for copper before and after processing

As can be seen from the figure, when examining untreated copper, the movement of the indenter begins already with an effort of less than 1 kg or less than 10N. The second graph shows that the movement of the diamond indenter begins at a force of 3.5 kg or approximately at a load of 35N. According to this it can be concluded that when the surface of copper is treated with an electrolytic discharge, the surface hardening of the product is performed [5].

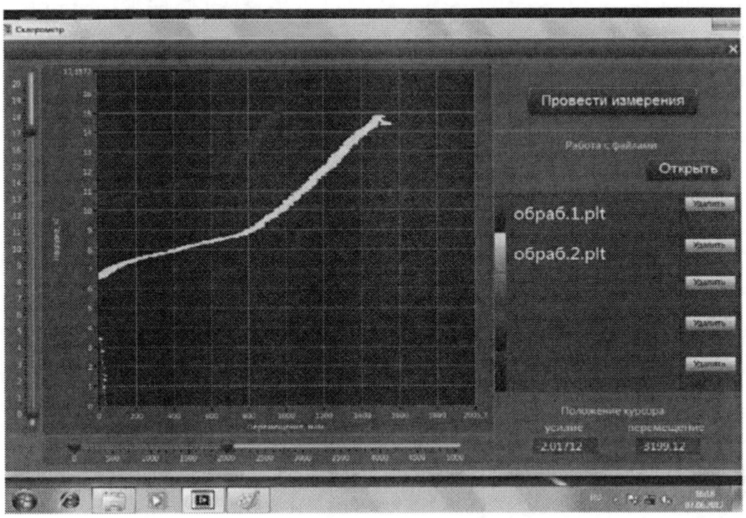

Figure 2. Graph of the dependence of the force applied to the indenter on its movement for steel before and after processing

2.1. Figure 2 shows the results of the study of the surface of the steel plate before and after processing. Here you can see that the movement of the indenter begins with a force of 7 kg or 70N. After processing a steel product with a discharge, its strength characteristics have noticeably improved, the movement of the diamond indenter along the plate begins only with a force of more than 10 kg or already with a load of more than 100N.

3. Conclusion

As a result, measurements of the hardening the samples with various surface pretreatment show an increase in its strength characteristics several times. Thus, electrolytic discharge treatment affects the strength of the processed product, which is another advantage of this discharge.

The research was supported by a grant from the Russian Science Foundation (project No. 22-29-00021).

References

[1] Kayumov R.R. Experimental study of the effect of electric discharge on carbon fiber/Zakirov Zh.U., Kayumov R.R., Khazeev K.I.// Journal of Physics: Conference Series This link is disabled, 2021, 1870(1), 012007

[2] Akhatov M.F. Glycerin as an electrolytic electrode / Sh. Akhatov, G., Kayumov R.R., Valeeva R.R., Akhatov M.F.// Physical journal: conf. Series. 2019.V. 1328 Is.1,01200410

[3] Bagautdinova L.N. Low-power electric discharges with metal, dielectric and electrolytic electrodes at low frequencies and atmospheric pressure / Gaisin A.F., Sadriev R.S., Bagautdinova L.N., Nasibullin R.T., Gaisin F.M., Mastyukov S.S. // High temperature, 2020, 58(6), pp. 777-780

[4] Galimova R.K. Calculation of portable properties of some real gas mixtures at high temperatures / Taxeitov R.R., Galimova R.K., Yakupov Z.Ya. // Physical Journal: Conference Series, 2020, 1588(1), 012065

[5] Fakhrutdinova, I.T. Features of the development of an electric discharge between a jet anode and a liquid cathode / Galimzyanov I.I., Gaisin A.F., Fakhrutdinova I.T., Shakirova E.F., Akhatov M.F., Kayumov R.R. // High temperature, 2018, 56(2), pp. 296-298

[6] Tsareva, A.M. Determination of natural frequencies and vibration modes of a disk of constant thickness with a central mount / Makaeva, R.Kh ., Tsareva A.M., Karimov, A.Kh . // Russian Aeronautics, 2008, 51(1), pp. 53-59

[7] Gaisin A.F. Special features of a multichannel discharge in a porous solid cathode/Loginov N.A, Son E.E., Gaisin A.F., Gaisin F.M. //High Temperature. 2009. T. 47. № 4. C. 603-605. A subsection

HYSTERESIS PHENOMENA AND THE EFFECT OF REORIENTATION IN A POLYMER-LIQUID CRYSTAL SYSTEM UNDER THE INFLUENCE OF LASER RADIATION AND UNIAXIAL DEFORMATION

B I Makhsudov[1], I K Fayzulloev[1], M K Egamov[2]

[1]Tajik National University, Dushanbe, Tajikistan
[2]Khujand Scientific Center of NAST, Khujand, Tajikistan

fayzulloevinoyatullo@gmail.com

Abstract. The orientational structure in a polymer-liquid crystal thin film induced by uniaxial mechanical deformation has been studied. It is shown that the dependence of the deflection angle of the director of a nematic liquid crystal on the direction of the tensile force has a hysteresis region in which the orientation structure spontaneously changes from a state with positive torsion of the director to a state with negative torsion. It is established that the effect of modification of the hysteresis loop leads to an expansion of the hysteresis region and a decrease in the field at which orientation bistability appears.

1. Theory

Recently, the search and research of new types of information display devices based on liquid crystals (LC) has been rapidly developing. An important role for the functioning of such devices is played by the phenomenon of reorientation of the predominant direction of molecules (director) in the cell under the influence of external fields (mechanical, electric, magnetic, light, acoustic) [1-4]. The reorientation of the director under the influence of external fields is accompanied by a change in a number of physical parameters that are caused by the absorption or transmission of light radiation [5]. The possibilities of using LC in photonics and optoelectronics become much wider if semiconductor injection lasers based on heterostructures are used as light radiation. To control photon flows, the best option is considered to be the use of polymer-liquid crystal composites (PJCC) [6]. These are thin transparent films containing microscopic capsules of a nematic liquid crystal in volume. This work is devoted to the solution of just such a question.

A nematic liquid crystal (NLC) is a uniaxial liquid. The macroscopic state of the NLC is described by specifying at each point a unit vector n, called the director. In the equilibrium state, the NLC is homogeneous, while inhomogeneous distributions of the director arise in the case of various elastic deformations of the LC. The introduction of the NLC capsule into the composition of the binder (polymer matrix) is carried out by dispersion in various ways (polymerization, cooling of the solution, evaporation of the solvent). These two components are incompatible, so external factors are considered separately. Observation of the structure of the NLC capsule with a polarization microscope makes it possible to obtain bright images occurring under the action of external fields (Fig. 1).

Content from this work may be used under the terms of the Creative Commons Attribution 3.0 licence. Any further distribution of this work must maintain attribution to the author(s) and the title of the work, journal citation and DOI.
Published under licence by IOP Publishing Ltd

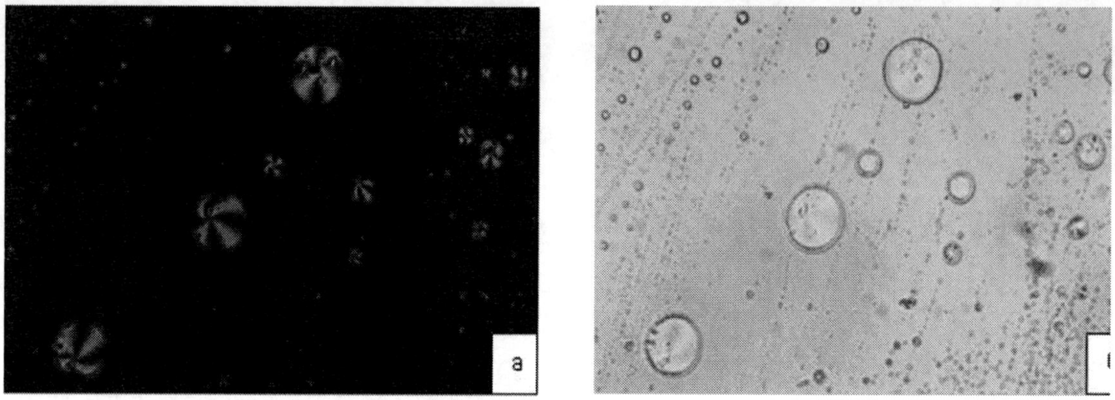

Figure 1. Micrographs of PJCC films in the geometry of crossed polarizers (a) and with the analyzer turned off (b)

As a rule, the configuration of the director is determined by solving the system of Euler-Lagrange equations [7]. However, if there is a phase transition of the first kind in the system under study, difficulties arise when solving the Euler-Lagrange equations. Therefore, instead of a numerical solution of this equation, an approach based on minimizing the free energy functional is usually used. In this paper, we have focused on the latter option.

For the production of composite films, we used a nematic liquid crystal (NLC) 4-n-heptyl-4'-cyanobiphenyl (7CB) having the following sequence of phase transitions: Cr (-70oC) - LC (28.5oC) – I (42oC). Refractive indices of the studied 7CB: $n_{\|}=n_{e,max}=1.725$ and $n_{\perp}=n_0=1.534$ at a temperature T=25oC and a wavelength λ= 0.633 micrometer. The polymer binder was polyvinyl butyral (PVB) with n_p=1,492 [8]. PVB is a soluble polymer characterized by resistance to organic solvents and heat resistance, low gas permeability, good adhesive properties, it provides tangential adhesion to molecules of alkyl cyanobiphenyls derivatives. NLC 7CB was dispersed in a polymer matrix by emulsion technology [9] during evaporation of a common solvent – ethyl alcohol. The prepared mixture was applied to the surface of the glass substrate and the solvent was evaporated in the temperature control mode. The thickness of the samples after drying was 25-30 micrometer, and the capsule size was 7CB 10-15 micrometer. The PJCC films were positioned perpendicular to the direction of the laser radiation passing through the polarizer (Fig. 2) [6].

An experimental setup was used, the peculiarity of which is that an injection laser based on stressed heteronostructures is used as a photon source [12]. This class of lasers has advantages related to miniaturization, speed, a wide range of lasers with different wavelength ranges, the ability to work in a wide range of capacities and durability. Of particular importance is the durability of injection lasers, which last up to a million or more hours in continuous operation. By selecting the appropriate heterostructures emitting different wavelengths of radiation, it is possible to investigate the dependence of nonlinear optical processes in these composites on the power and wavelength of laser radiation.

Figure 2 shows a block diagram of an experimental setup that allows us to study the nonlinear optical properties of the PJCC. The device contains a semiconductor injection laser based on InGaAsP/AlGaAs heterostructures operating in the red spectral range, an optical laser beam collimator for producing parallel laser beams, a polymer-liquid crystal composite based on a nematic liquid crystal (NLC) 4-n-heptyl-4'-cyanobiphenyl (7CB), a photodetector for recording the intensity of the laser beam. The entire experimental setup is placed in a special box, which allows you to adjust the ambient temperature.

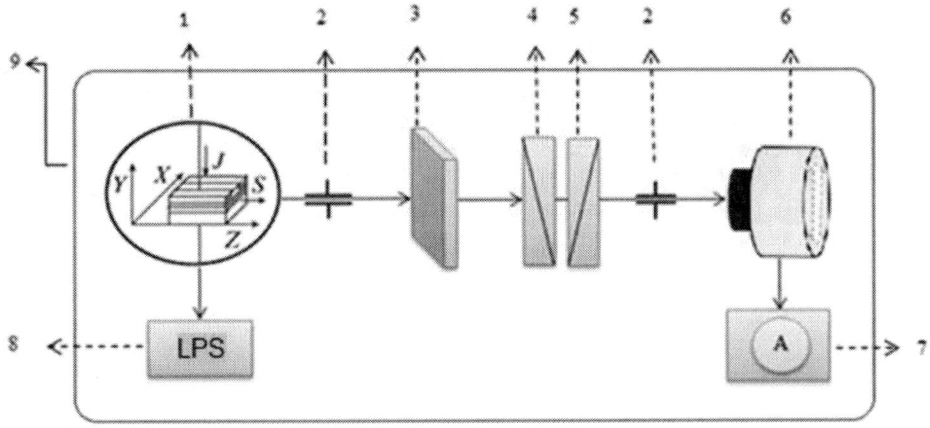

Figure 2. The scheme of an experimental setup for the study of optical properties of composites based on PJCC. 1-semiconductor injection laser based on InGaAsP/AlGaAs heterostructures, 2-optical laser beam collimator, 3-polymer-liquid crystal composite based on nematic liquid crystal (NLC) 4-n-heptyl-4'-cyanobiphenyl (7CB), 4,5-analyzer, 6-photodetector, 7-computer, 8-adjustable injection laser power supply, 9-thermostat

2. Results and their discussion

Figure 3 shows the dependence of the transmitted optical power of a semiconductor injection laser based on InGaAsP/AlGaAs heterostructures operating in the red spectral range on the magnitude of the relative deformation of the PJCC based on NLC 7CB for the orthogonal component of polarization in the forward and reverse direction of deformation. It can be seen that the phenomenon of optical hysteresis of the power of the transmitted laser radiation is observed at uniaxial deformation. It should be noted that this phenomenon was discovered for the first time for composites based on nematic liquid crystals. Starting from a certain value of the uniaxial stretching value, the dependence of the intensity of the transmitted laser beam will be constant and, starting from this point, there is a change in the direction of uniaxial deformation. As can be seen from Fig. 3, the curves in the forward and reverse directions do not coincide, a hysteresis loop of the laser radiation intensity is observed. The observed effect is associated with relaxation processes of orientation of NLC molecules. Note that when an electric field is applied to the nematic films, the phenomenon of a hysteresis loop of laser radiation intensity is not observed.

It can be seen from Fig.3 that at a fixed value of T with an increase in Δl/l0, a sharp increase in the rotation angle of the director first occurs, corresponding to an increase in the orientation distortions of the director inside the capsule.However, during the reverse process (i.e. removal of the tensile load), the nature of the curves changes qualitatively. The behavior of the rotation angle of the director inside the NLC 7CB capsule when removing the tensile load near $\pi/2$ is described by the formula:

As can be seen from the formula, when removing the tensile load, the angular coefficient dφ/dl is negative, and the angle φ0 is an unambiguous function of Δl/l0. When stretched, the angular coefficient becomes positive and a symmetrical hysteresis loop appears on the curve T=f(Δl/l0), corresponding to the orientation bistability.

With an increase in the angle of rotation of the director, an equilibrium orientation transition, accompanied by a jump in the orientation of the director from a state with positive torsion (Fig. 3, the upper branch of the curve) to a state with negative torsion (Fig. 3, the lower branch), occurs at $\varphi = \pi/2$.The transition of the system, in this case, to a state with opposite torsion, in our opinion, is explained by the fact that the direction of the director in both cases is physically equivalent. In this case, the boundary conditions are symmetric, therefore, the free energies of states with the orientation of the mechanical field $\varphi0=\pi/2$ for positive ($\varphi0>0$) and negative ($\varphi0<0$) torsion of the director are the same.

The increase in the director's rotation angle corresponding to the behavior along the upper metastable curve in Fig. 3, i.e., an increase in the director's orientation angle in the center of the capsule, leads to an increase in elastic energy. The behavior along the lower curve in Fig. 3, when the condition φ0>π/2 is met, leads to a decrease in elastic energy, since the angle φ0 in absolute magnitude becomes smaller, since the angle between the director and the direction of the mechanical field decreases.

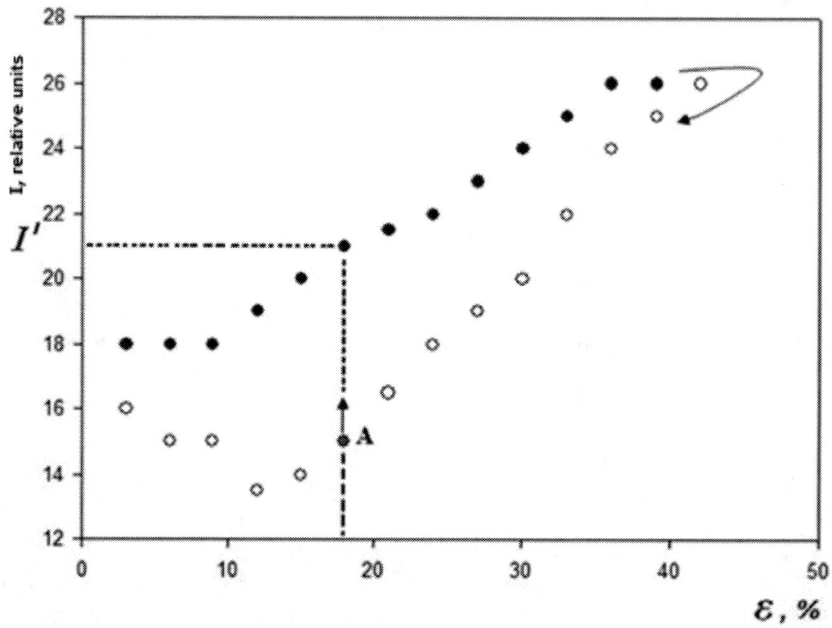

Figure 3. Dependence of the transmitted optical power of a semiconductor injection laser based on InGaAsP/AlGaAs heterostructures on the magnitude of the relative deformation of the PJCC based on NLC 7CB for the orthogonal component of polarization in the forward ( ) and reverse (o) direction of deformation

Further experimental results show that the power of the transmitted laser radiation at a fixed value of the magnitude of the inverse relative deformation gradually increases.

Figure 4 shows the results of measuring the temperature dependence of the growth kinetics of the transmitted optical power of a semiconductor injection laser based on InGaAsP/AlGaAs heterostructures on the delay time at point A in Fig. 3.

The results of calculations of the experimental data obtained show that in the absence of an external field (external tensile force), the deflection angles of the director are zero, which corresponds to the original undisturbed structure. With an increase in the tensile load, the angle of rotation of the director φ0 in the center of the capsule and the angle of rotation of the director at the interface ψ0 increase and asymptotically tend to the direction of the external field. These data are more clearly shown in Fig. 4. It can be seen that the dependence of the rotation angle of the director φ0 in the center of the capsule reaches saturation at lower field values than the dependence φ0 =f(Δl/l0); at the same time, the lower the coupling energy σ, the greater the difference between the curves for φ0 and ψ0, since the polymer matrix is more malleable to the action of the mechanical field than the LC system. In addition, the intensity of III becomes equal to what it had when the condition T<Ti is met, at the same time decreases to the value of III at T<Ti. This means that the plane of inclination of the NLC 7CB molecules inside the droplet is oriented along the direction of stretching.

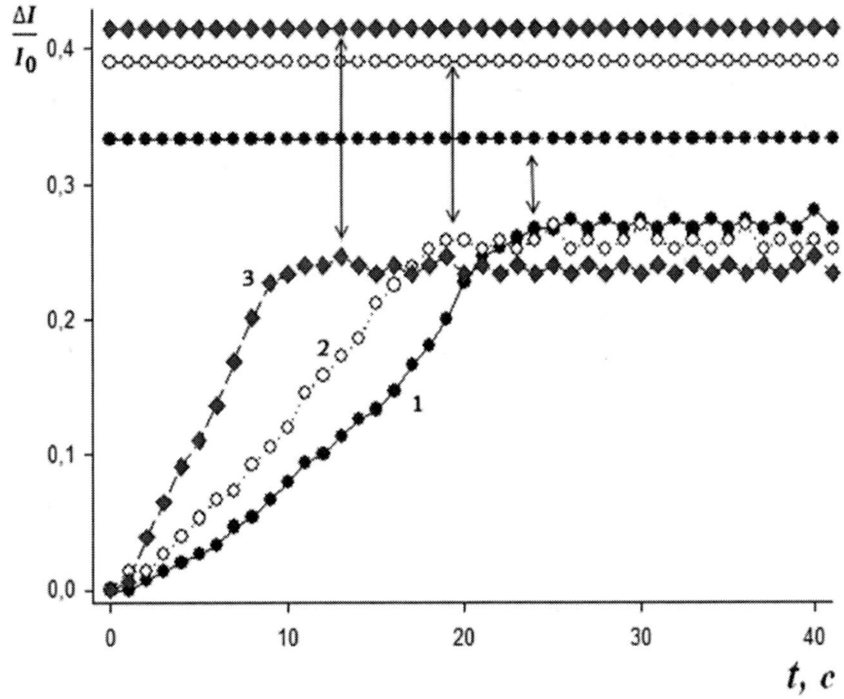

Figure 4. Temperature dependence of the growth kinetics of the transmitted optical power of a semiconductor injection laser based on InGaAsP/AlGaAs heterostructures on the delay time at point A in Fig.3.

305K (1); 309K (2); 313K (3)

3. Interpretation of scientific results

Spontaneous polarization of P0 in the nematic layer is perpendicular to the plane of inclination of long molecules [3]. The reorientation of the molecules causes the rotation of the plane of inclination of the molecules and, accordingly, the rotation of spontaneous polarization in neighboring layers. If the thickness of the PJCC sample in the direction of the normal to the composite plane is less than the period of the domains of the crystal lattice, or a mechanical field sufficient to spin the axis of the molecule is applied to the film, the NLC 7CB has a macroscopic spontaneous polarization of P0. In a mechanical field, the polar axis of the NLC 7CB should be oriented along the field (i.e. the axis of strain of stretching), and the plane of inclination of the molecules in the near-surface layers should be perpendicular to the direction of stretching. This is due to the high sensitivity of the NLC 7CB to external influences indicated in the introduction of this work. It should be noted that the effect we found is qualitatively different from the one previously known in NLC 7CB with a reorientation of 180o with respect to the direction of spontaneous polarization [10,11]. This reorientation is not specific to thin films and is observed both in films and in bulk composites, with the plane of inclination of the molecules remaining perpendicular to the directions of spontaneous polarization and the stretching field. In our case, there is a reversal of the structure by 180o with respect to the direction of the mechanical field.

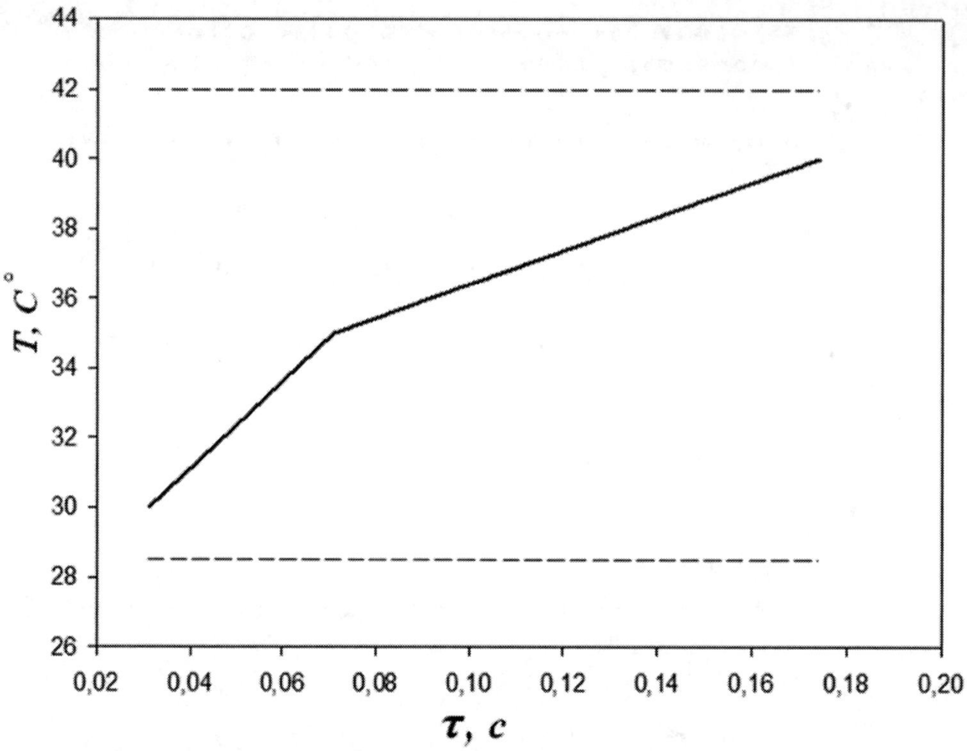

Figure 5. Temperature dependence of the characteristic time of reverse relaxation of the molecules of nematic PJCC, (--) – the limit of the isotropy of NLC

Changes in the characteristic time of the reverse relaxation of NLC molecules of 7CB with an increase in temperature (Fig. 5) characterizes the temperature change in the angle of inclination of the molecules in the capsule volume. The surface ordering is most strongly manifested above the temperature of the phase transition from the crystalline to the nematic state. In this temperature range, the angle of inclination significantly depends on the distance of the domain layer to the capsule surface. In our opinion, a change in the angle of disorientation of molecules in the domains of the 7CB film can cause spontaneous polarization, detected according to Fig. 3.

4. Conclusion

Our comparative analysis showed that flexoelectric polarization may occur in a plane perpendicular to P0, noticeably exceeding the spontaneous polarization of P0 in magnitude, which is associated with a significantly large bending of the director in the 7CB film. With the S-shaped bending of the director, the polarization in the upper and lower parts of the film are directed in different directions. For thin films, the angle of inclination of the NLC 7CB molecules in the center of the capsule may become sufficiently small or higher than the phase transition temperature. In this case, the upper and lower parts of the capsule inside the polymer matrix can be oriented independently, and it will be possible to switch to the S-shaped orientation of the director. The effects we have discovered can be used to record and store information as static memory or bistable state switches.

Low-temperature plasma can be used for the synthesis of various nanostructures and is well suited for the modification of various surfaces. This is shown in many works [13-32].

References

[1] Egamov M.Kh., Gerasimov V.P., Krakhalev M.N., Prishchepa O.O., ZyryanovV.Ya., Loiko V.A. Polarizing properties of a stretched film of a polymer-dispersed liquid crystal with a surfactant dopant. //Journal of Optical Technology, 2014, v.81, №7. – p.414-417

[2] Prishchepa O.O., Krakhalev M.N., Sutormin V.S., ZyryanovV.Ya., Rudyak V.Y. Electrically turning periodic structures in cholesteric layer with conical-planar boundary conditions. //Scientific Reports. 2021, v.11, № 1. – p. 217-223

[3] Loiko V.A., Konkolovich A.V., Miskevich A. A, Krakhalev M.N., Prishchepa O.O., ZyryanovV.Ya. Small-angle scattering and radiation polarization by stretched polymer film with nematic liquid crystal droplets having a single-domain structure. //Optics and spectroscopy. 2020, v.128, № 3. –p.331-338

[4] Pozhidaev E.P., Mudretsov D.F., Shoshin V.M., Bobylev Yu.P., Zhukov A.A., Veselago V.G. Polaroid shutter of infrared radiation based on the scattering effect in a nematic liquid crystal. //Letters in ZhTF. 2013, vol.39, issue 11. - p.45-51

[5] Loiko V.A., Konkolovich A.V., Zyryanov V.Ya. Polarization of light by a polymer film containing elongated liquid crystal droplets with inhomogeneous surface adhesion //Optics and Spectroscopy, 2017, vol.122, No. 6. - pp.1016-1028

[6] Makhsudov B.I., Fayzulloev I.H., Egamov M.H. Influence of radiation power on optical properties of polymer-liquid crystal composites //Bulletin of the Tajik National University. Series of Natural Sciences. 2019, No. 3. - pp.83-89

[7] Stewart I.W. The Static and Dynamic Continuum Theory of Liquid Crystals: A Mathematical Introduction, Liquid Crystals Book Ser., Taylor and Francis, London. 2004

[8] Rosenberg M.E. Polymers based on polyvinyl acetate. Leningrad.: Chemistry. 1983. - 174 p.

[9] Drzaic P.S. Liquid Crystal Dispersions. Singapore: World Scientific, 1995. – 430 p.

[10] Takuya Ohzono, Kaoru Katoh, Hiroyuki Minamikawa, Mohand O. Saed, Eugene M. Terenjev.Internal constrains and arrested relaxation in main-chain nematic elastomers. //Nature Communications. 2021, 9, 2531-2541 [3] More references

[11] White T.J., Broer D.J. Programmable and adaptive mechanics with liquid crystal polymer networks and elastomers. //Nat. Mater. 2015, № 12. – p.1087-1098

[12] Makhsudov B.I. Influence of thermal neutrons on the radiative characteristics of INGAAS/ALGAAS heterolasers. -Quantum Electronics, 2015, vol. 45, No. 3, pp. 216-217.

[13] Saifutdinov, A. I., & Sofronitskii, A. O. (2021). Numerical Study of Breakdown and Formation Dynamics of Arc Discharge Plasma Parameters at Ultrahigh Pressures. High Energy Chemistry, 55(3), 228-232.DOI: 10.1134/S0018143921030115

[14] Shemakhin, A. Y., Zheltukhin, V. S., Shemakhin, E. Y., Terentev, T. N., & Sofronitsky, A. O. (2020, July). Experimental installation to study the RF plasma flow at low pressures with experiment data synchronization. In Journal of Physics: Conference Series (Vol. 1588, No. 1, p. 012018). IOP Publishing. DOI: 10.1088/1742-6596/1588/1/012018

[15] Saifutdinova, A. A., Sofronitskiy, A. O., Timerkaev, B. A., & Saifutdinov, A. I. (2020). Plasma-Chemical Decomposition of Hydrocarbons on the Basis of the Micro-Arc Discharge with Disc Electrodes Rotating in the Bulk of Raw Materials. Russian Physics Journal, 62(11), 2132-2136.DOI: 10.1007/s11182-020-01957-0

[16] Fairushin, I. I., Saifutdinov, A. I., & Sofronitskiy, A. O. (2020). Numerical and Experimental Studies of the Synthesis of Copper Nanoparticles in a High-Pressure Discharge. High Energy Chemistry, 54(2), 150-153. DOI: 10.1134/S0018143920020071

[17] Asadullin, T. Y., Galeev, I. G., Sofronitskiy, A. O., & Gizeev, M. M. (2019, November). Acoustic impact on electric discharge parameters during sterilization of freeze-dried products. In Journal of Physics: Conference Series (Vol. 1370, No. 1, p. 012016). IOP Publishing. DOI: 10.1088/1742-6596/1370/1/012016

[18] Fairushin, I. I., Saifutdinov, A. I., Sofronitskiy, A. O., Timerkaev, B. A., & Dautov, G. Y. (2019, October). Development of plasma reactor design for synthesis of copper nanoparticles

using multi-scale simulation. In Journal of Physics: Conference Series (Vol. 1328, No. 1, p. 012088). IOP Publishing. DOI: 10.1088/1742-6596/1328/1/012088

[19] Asadullin, T. Y., Galeev, I. G., & Sofronitskiy, A. O. (2019, October). The application of pulsed discharge for sterilization of freeze-dried product. In Journal of Physics: Conference Series (Vol. 1328, No. 1, p. 012071). IOP Publishing. DOI: 10.1088/1742-6596/1328/1/012071

[20] Timerkaev, B. A., Ganieva, G. R., Kaleeva, A. A., Israfilov, Z. K., & Sofronitskii, A. O. (2019). Growing of carbon nanotubes from hydrocarbons in an arc plasma. Journal of Engineering Physics and Thermophysics, 92(5), 1248-1252. DOI: 10.1007/s10891-019-02040-3

[21] Dautov, G. Y., Kashapov, N. F., Dautov, I. G., & Sofronitskiy, A. O. (2018, July). Research of the influence of the geometry of the discharge chamber on the characteristics of the arc plasmatron. In Journal of Physics: Conference Series (Vol. 1058, No. 1). IOP Publishing. DOI: 10.1088/1742-6596/1058/1/012035

[22] Timerkaev, B. A., Andreeva, A. A., & Sofronitskiy, A. O. (2017, November). Discharge creeping along the surface in the process for producing nanomaterials. In Journal of Physics: Conference Series (Vol. 927, No. 1, p. 012068). IOP Publishing. DOI: 10.1088/1742-6596/927/1/012068

[23] Sadikov, K. G., Sofronitskiy, A. O., & Larionov, V. M. (2017, November). The effect of electrically conductive additives on the plasma pyrolysis of heavy hydrocarbons. In Journal of Physics: Conference Series (Vol. 927, No. 1, p. 012046). IOP Publishing. DOI: 10.1088/1742-6596/927/1/012046

[24] Timerkaev, B. A., Andreeva, A. A., & Sofronitskiy, A. O. (2017). Discharge creeping along the surface in the process of cleaning and strengthening of the materials surface. In Journal of Physics: Conference Series (Vol. 789, No. 1). IOP Publishing. DOI: 10.1088/1742-6596/789/1/012063

[25] Sadikov, K. G., Sofronitskiy, A. O., & Dautov, I. G. (2017). Functional plasma sprayed coatings on magnesium ceramic substrates. In Journal of Physics: Conference Series (Vol. 789, No. 1, p. 012043). IOP Publishing. DOI: 10.1088/1742-6596/789/1/012043

[26] Timerkaev, B. A., Sofronitskiy, A. O., & Andreeva, A. A. (2016). Carbon nanotubes formation in the decomposition of heavy hydrocarbons creeping along the surface of the glow discharge. In Journal of Physics: Conference Series (Vol. 669, No. 1, p. 012062). IOP Publishing. DOI: 10.1088/1742-6596/669/1/012062

[27] Timerkaev, B. A., & Shamsutdinov, R. S. (2020, July). The Influence of Supersonic Gas Stream on Spatial Structure of Glow Discharge. In Journal of Physics: Conference Series (Vol. 1588, No. 1, p. 012061). IOP Publishing. DOI 10.1088/1742-6596/1588/1/012061

[28] Galeev I G, Asadullin T Y 2017 Improving of stability of the volumetric glow discharge in the gas flow J. of Phys. Conf. Ser. 789(1) 012012

[29] Asadullin T Y, Galeev I G 2017 Combined method of stabilization of the glow discharge in gas flow J. of Phys. Conf. Ser. 927(1) 012003

[30] Galeev I G, Asadullin T Ya 2016 Obtaining fullerene-containing soot during combustion of gaseous hydrocarbons in an external electric field J. of Phys. Conf. Ser. 669 012016

[31] Galeev I G, Asadullin T Y 2016 Enhancing of the glow discharge stability in chamber with cathode sections coated by a discontinuous dielectric coating J. of Phys. Conf. Ser. 669(1) 012015

[32] Shakirov A S, Galeev I G and Asadullin T Ya 2020 Features of formation of fullerene-containing soot during combustion gaseous and liquid hydrocarbons J. of Phys. Conf. Ser. 1588(1) 012013

NON-THERMAL PLASMA INSTALLATION AS A PRE-TREATMENT METHOD OF BARLEY SEEDS (Hordeum vulgare L.)

D Petrukhina, S Gorbatov, O Tkhorik, I Medzhidov, V Tikhonov, V Shishko, I Ivanov, A Tikhonov, N Gluchshenko, V Kharlamov

Russian Institute of Radiology and Agroecology, Obninsk, Russia

gorbatovsa004@gmail.com

Abstract. Microbial organisms are key pathogens for plants, animals, and humans. The elimination of pathogenic microbes is an essential topic for researchers. Non-thermal atmospheric plasma (NTAP) can potentially inactivate pathogenic microorganisms. This paper draws the results of the biocidal effect study in pre-treatment of barley seeds by NTAP. The effect was observed in 7-day-old seedlings. The results of reducing the total microbial number (by 36.7% after 5 minutes of exposure) were obtained. Thus, treatment with non-thermal plasma can reduce the microbiological contamination of agricultural plant seeds.

Keywords: non-thermal plasma, seed disinfection, total microbial number, agricultural plant seeds, barley.

1. Introduction

In the agro-industrial complex, an essential task is to increase yields and improve the storability of grown raw materials and plant foods. According to the International Organization for Food and Agriculture (FAO at the UN), grain losses during storage are at least 10% and increase when the storage conditions for temperature and humidity deviate from optimal. Losses are mainly associated with pest damage and microbiological spoilage. In addition, with long-term storage of grain in adverse conditions, the ability to germinate is disturbed, and quality decreases.

One of the currently developing physical methods of pre-sowing treatment agricultural products is non-thermal atmospheric pressure plasma (NTAP) [1]. Pre-sowing seed treatment with NTAP can reduce microbiological contamination [2]. Previous studies have shown that the treatment of barley seeds with NTAP does not significantly affect germination capacity and the length of sprouts and roots. However, it reduces the infection with phytopathogenic fungi [3]. In addition, domestic studies [1, 4, 5] demonstrated the bactericidal effect of NTAP on several microorganisms: yeast, gram-negative, and gram-positive bacteria, and their spore forms. It is also worth noting the formation of free radicals due to plasma chemical reactions, which are strong oxidants [6, 7]. They cause membrane degradation and DNA disruption [1, 8].

Content from this work may be used under the terms of the Creative Commons Attribution 3.0 licence. Any further distribution of this work must maintain attribution to the author(s) and the title of the work, journal citation and DOI.
Published under licence by IOP Publishing Ltd

This work aims - to access the biocidal effect of NTAP on the surface microbiota of barley seeds and morphometric parameters of 7-day-old seedlings.

2. Materials and Methods

In the experiments, we used the barley seeds (Hordeum vulgare L.) of the Vladimir sort, harvested in 2018. Each investigated case of the experiment included 150 seeds. The seeds were pre-treated with NTAP, which uses a microwave discharge of a coaxial configuration. Argon was used as a plasma-forming gas. [3, 9]. The scheme of the experimental installation is shown in Figure 1.

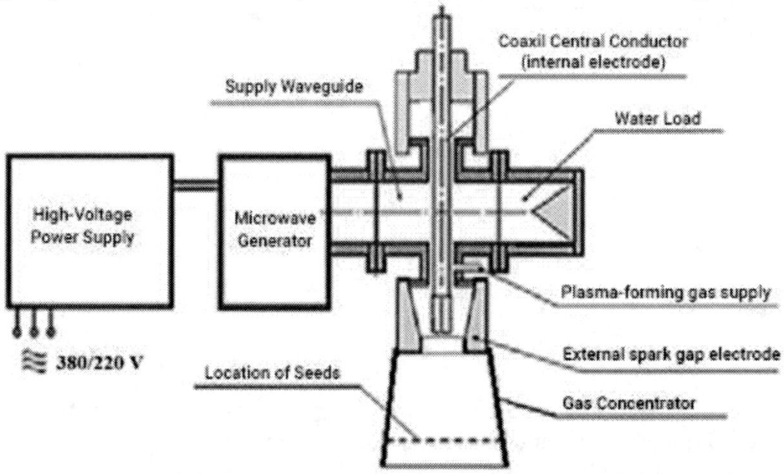

Figure 1. The scheme of the experimental installation

A 20 cm long gas concentrator was additionally installed to reduce the flow temperature and maintain the concentration of active particles. The concentrator in the form of a truncated cone was made of stainless steel 0.4 mm thick.

Seeds were pre-treated with NTAP at three levels of time: 1, 5 and 10 minutes. The distance from the nozzle to the seeds was 13 cm. Argon consumption was 5 l/min, and the power of the microwave generator was 1.2 kW. The gas flow temperature at the seed placement level was measured using a thermal imager and was about 37 C°.

Further, the control and treated seed groups were germinated on wet filtered papers in a thermostat at 20-21 C° in accordance with Russian National Standard 12038-847. Each group contained 50 seeds. On the third day, the seed germination energy was determined. On the seventh day, laboratory germination, growth strength, sprout and root length, and wet and dry mass were determined. Then a phytosanitary examination of barley seeds was carried out. The degree of damage and the prevalence of diseases by *Helminthosporium*, *Fusarium*, and *Penicillium* were determined in accordance with Russian National Standard 12044-93.

3. Results and Discussion

The spring barley of the Vladimir variety has high adaptability to various cultivation conditions resistant to drought and soil acidity. The variety is moderately resistant to loose smut, highly susceptible to helminthosporiasis, moderately resistant to leaf rust and powdery mildew. Germination energy characterizes the ability of seeds to give uniform and even seedlings in the field, which guarantees good evenness and survival of plants. This indicator shows the percentage of germinated seeds of the total number of seeds planted in the experiment. Our experiments studied the effectiveness of presowing treatment with APNTP of spring barley seeds, which was determined by changes in morphological parameters and phyto-expertise. Following exposure to APNTP on the 7th day of germination, the length of the sprout and root did not change statistically significantly (Table 1). Our experiments decreased germination energy of barley seeds was observed during 5 and 10-minute exposure to plasma. Next, the percentage of seeds that gave seedlings under standard conditions was determined - laboratory germination. Following exposure to APNTP, laboratory germination did not change statistically significantly. It was also found that pre-sowing plasma treatment of seeds before germination does not affect the growing strength of 7-day-old seedlings.

Table 1. Parameters of 7-day-old barley seedlings after exposure to APNTP on seeds before germination

Parameter	**Exposure time**			
	Control	**1 min**	**5 min**	**10 min**
Sprout length	96 ± 4,00	93 ± 5,00	93 ± 4,00	92 ± 5,00
Root length	132 ± 4,00	133 ± 4,00	131 ± 4,00	125 ± 4,00
Germination energy	96 ± 0,00	93 ± 3,06	89 ± 4,16	91 ± 2,31
Laboratory germination	100 ± 0,00	97 ± 6,00	99 ± 2,00	99 ± 2,00
Seed vigour	97 ± 2,00	95 ± 4,00	96 ± 4,00	97 ± 2,00

The percentage of water, wet and dry weight per 1 plant was determined (Table 2) for a more detailed study of the effect of plasma on the initial growth processes of spring barley. It was found that the percentage of water content in 7-day-old barley seedlings after exposure to APNTP on seeds before germination does not change statistically significantly. The study results showed that the fresh weight of seedlings after irradiation at 10 minutes decreased, but the dry weight did not change.Other paragraphs are indented (BodytextIndented style).

Table 2. Wet and dry weight of 7-day-old barley seedlings after exposure to APNTP on seeds before germination

Exposure time	**Weight per 1 plant, g**		**% of water in seedlings**
	Wet	**Dry**	
Control	0,29 ± 0,02	0,0425 ± 0,0041	85,40 ± 1,42
1 min	0,28 ± 0,01	0,0425 ± 0,0008	84,69 ± 0,46
5 min	0,29 ± 0,01	0,0441 ± 0,0023	84,79 ± 0,33
10 min	0,26 ± 0,00	0,0379 ± 0,0054	85,29 ± 2,07

Next, the effect of APNTP on the phytosanitary state of barley seeds was studied. It was established that the seeds were infected by *Penicillium*, neither in control nor in the irradiated samples. The degree of damage by *Fusarium*, *Helmintostorium*, *Aspergillus* did not change statistically significantly (Table 3). The table presents the results of assessing the effect of plasma treatment on the prevalence of seed infections in spring barley. After exposure to APNTP on barley seeds, the prevalence of Fusarium root rot and fungi of the genera *Helmintostorium*, *Aspergillus* did not change statistically significantly.

Table 3. The degree of damage and prevalence of helminthosporiasis (Helmintosporium spp.) in 7-day-old barley seedlings after exposure to ANTP on seeds before germination

Exposure time	**Damage degree by helminthosporiasis, %**	**Prevalence of the disease, %**
Control	25,00 ± 5,00	75,00 ± 12,00
1 min	24,00 ± 11,00	68,00 ± 19,00
5 min	25,00 ± 9,00	73,00 ± 24,00
10 min	22,00 ± 5,00	70,00 ± 7,00

Thus, the obtained results of the study showed the absence of significant violations of the sowing qualities of barley seeds during presowing treatment with APNTP. The studied seed samples had a low initial infection with diseases (except for helminthosporiasis); therefore, no pronounced disinfecting effect was observed. However, the biocidal effect of argon ANTP was shown when exposed to barley seeds, determined by a decrease in the total microbial number of the surface microbiota (by 36.7% when exposed for 5 min) (Table 4).

Table 4. Dependence of the number of CFU QMA&OAMO of barley microbiota on the duration of exposure to APNTP on seeds.

Exposure time	1 day Mean ± standard deviation×10^4 CUF/g	2 day Mean ± standard deviation×10^4 CUF/g
Control	8,61 ± 0,51 (100 %)	14,48 ± 0,70 (100 %)
1 min	6,08 ± 0,59 (70,6 %)*	10,56 ± 0,80 (72,9 %)*
5 min	7,21 ± 0,73 (83,7 %)*	9,16 ± 0,67 (63,3 %)*
10 min	11,43 ± 4,99 (132,8%)	13,63 ± 4,12 (94,1%)

* — significant differences ($p \leq 0.05$) from control according to the Mann-Whitney U test.

As a result of studies on the impact of APNTP on barley seeds for 1 to 10 minutes, there was no change in their main morphometric parameters. However, exposure for 10 minutes reduced fresh weight in 7-day-old barley seedlings. decrease in seed germination energy was observed after 5 and 10 min of plasma exposure, but laboratory germination did not change significantly. A decrease in seed germination energy was observed after 5 and 10 min of plasma exposure, but laboratory germination did not change significantly. The damage degree and the prevalence of seed diseases by various fungi were similar to control samples. A biocidal effect was shown, expressed in a decrease in the total microbial number of the surface microbiota of seeds.

References

[1] Kobzev E N et al 2014. Prospects of the use of low temperature plasma for betacontamination food products. *Engineering and technology* **5** 60-63

[2] Mravlje J et al 2021. Development of cold plasma technologies for surface decontamination of seed fungal pathogens: present status and perspectives. *J. of Fungi* V. 7 **8** 650

[3] Park Y et al 2018. The biological effects of surface dielectric barrier discharge on seed germination and plant growth with barley. *Plasma Processes and Polymers* **15 (2)** 1600056

[4] Gomboeva S V et al 2017. Effects of low-temperature plasma on plant products. *Food Processing: Techniques and Technology* V. 46 **3** 129-134

[5] Kolesnikov V A et al 2016. Research of the inactivation of pathogens in water under exposure to low temperature plasma. *Hygiene and Sanitation* **95(6)** 588-592 (In Russ.)

[6] Piskarev I M et al 2018. Sources of gas-discharge plasma: effect of the absorbed dose and active particle composition on physicochemical transformations in biological substrates. *Modern Technologies in Medicine* **10(2)** 90-100

[7] Ivanova I P et al 2012. The analysis of active products of spark discharge plasma radiation determining biological effects in tissues. *Modern Technologies in Medicine* **2** 20-30

[8] Trofimova S V et al 2013. Structural and biochemical changes in prokaryotic cells under the influence of low-temperature plasma radiation. *Medical Almanac* **3 (27)** 87-88. Перевод через гугл-переводчик

[9] Ivanov I A et al 2019. Microwave complex for obtaining low-temperature plasma at atmospheric pressure. *J. of Phys.: Conf. Series.* V. 1393 **1** 012042

Application of the Harrington function in substantiating the choice of the best equipment for plasma processing of materials

B K Tazmeev[1], V V Tsybulevsky[1], R N Tazmeeva[2], G K Tazmeev[2]

[1]Kuban State Agrarian University named after I. T. Trubilin, Krasnodar, Russia
[2]Kazan Federal University, Naberezhnye Chelny Institute, Naberezhnye Chelny, Russia

tazmeevb@mail.ru

Abstract. The possibility of using the Harrington desirability function in choosing the most suitable equipment for plasma processing of materials is confirmed. The search for the optimal variant was carried out by calculating the maximum value of the generalized index of desirability. Four types of equipment for plasma cutting of metal are considered. A comparative analysis was carried out on the following indicators: the cost of equipment and technological cutting speeds for low carbon steel, stainless steel, and aluminum. An example of choosing the best equipment from an alternative series is considered. A system of four scales of partial evaluation indicators was used. Graphical and analytical evaluation methods are presented. The value of the generalized desirability index equal to 0.65 corresponds to the optimal variant of equipment for plasma cutting of metal.

1. Introduction

At present, to accelerate scientific and technological progress, new innovative approaches to increasing the technological level of production processes and creating machines with the best technical and economic indicators play an important role. But for scientifically based decision making on the choice of technological equipment, objective and reliable research methods are needed. It should be noted that growth in the productivity of new equipment is not always accompanied by adequate reduction in production costs. There is a need to develop new evaluation methods when choosing a final decision when choosing a particular machine or technology. In the study of high-tech equipment, a comprehensive assessment of the determining parameters should be carried out and only according to this indicator should preference be given to the best option. The purpose of the article is to show the possibility of using the Harrington desirability function [1] for objective decision making on the example of a comparative analysis of the parameters of metal plasma cutting. The advantage of the proposed approach is a comprehensive assessment of the advantages of the selected machine according to a complex (generalized) indicator. At the same time, the technological, economic, environmental and other parameters of the equipment should be considered in a comprehensive way.

2. Materials and methods

To make a decision about the choice of the best technique option from an alternative series, a complex assessment method is used according to the generalized index D_i, which takes into account the influence of many private assessment indicators with their own units of measurement. In the special literature [2],

Content from this work may be used under the terms of the Creative Commons Attribution 3.0 licence. Any further distribution of this work must maintain attribution to the author(s) and the title of the work, journal citation and DOI.
Published under licence by IOP Publishing Ltd

there are known methods for complex assessment of systems with many parameters that have different dimensions or without them at all using the desirability function [3], the values of which are in the range of zero to one (figure 1)

$$d_{ij} = \exp\{-\exp[-y']\}, \tag{1}$$

where d_{ij} – is the desirability function for the i option and the j evaluation indicator; y' – is a dimensionless quantity, as a rule, linearly related to the optimization parameter.

When performing calculations, it is customary to consider the lower level of desirability (not very desirable), equal to 0.2, and the upper level (excellent), equal to 0.8, to be acceptable. There is a known method for converting estimated dimensional parameters X_{ij} into dimensionless scale values y' using reference lines [4-8] and an analytical method, in the case of a large number of parameters.

Dimensionless evaluation indicators must be positive, so formula (1) should be reduced to the form (2)

$$d_{ij} = \exp\{-\exp[-(y'-2)]\}. \tag{2}$$

Consider a graphical method for converting dimensional parameters X_{ij} into a dimensionless desirability value d. To do this, it is necessary to plot the function, formula (2) - figure 2. On the graph, points A and B mark the limit values corresponding to the desirability of 0.2 and 0.8. On the scale y' they correspond to the values y'_A and y'_B. To translate the dimensional parameters of a "direct" relationship (the larger the parameter value, the higher the desirability - for example, technological cutting speed, unit productivity, etc.), it is necessary to build an additional scale X_1 of the same direction under the scale y' – index S. Its scale will be determined after the minimum and maximum values of this parameter are fixed on this scale - the points x^S_{min} and x^S_{max}, which determine the value of the desirability function 0.2 and 0.8, respectively.

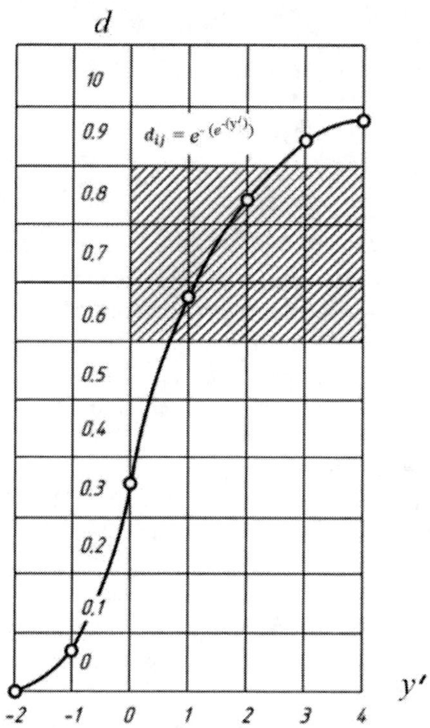

Figure 1. Harrington function graph.

Knowing the scale range, it is easy to put on it the current value of this parameter - the X_c. point. Having restored the perpendicular from this point to the intersection with the curve of the graph, it is necessary to draw a straight line through the intersection point C^S and on the scale d find the desired value of the dimensionless desirability d_c^S.

When translating the dimensional parameters of the "inverse" relationship (the larger the parameter, the lower the desirability; for example, the cost of equipment, the cost of consumed electricity, etc.), an additional X_2 scale is built inversely to the X_1, index O.

The method to convert a dimensional parameter into a dimensionless quantity d_c has a similar algorithm.

As can be seen, the graphical method is clear, simple, and laborious.

The analytical method of converting the numerical value of any parameter of any dimension into a dimensionless desirability value allows one to solve the problem without constructing a graph of the function.

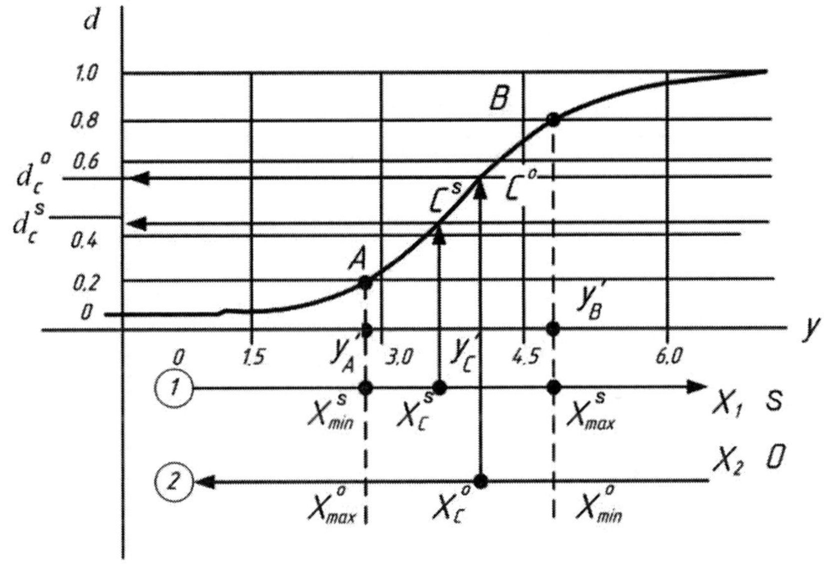

Figure 2. Desirability function graph d_{ij}.

To do this, it is necessary to determine the scale of conversion of the dimensional parameter to the scale x of the graph - figure 1. It can be seen from the figure that it is equal to

$$M = \left| \frac{x_{\max} - x_{\min}}{x_A - x_B} \right|, \tag{3}$$

where M – translation scale; x_{max}, x_{min} – limiting dimensional values of the translated parameter.

To convert the actual value of the dimensional parameter into a code value corresponding to the y' scale of figure 1, we need to use the equations

$$x_C^S = x_A + \frac{x_{Ci} - x_{i\min}}{M}. \tag{4}$$

For the "reverse" dependence

$$x_C^o = x_A + \frac{x_{i\max} - x_{ci}}{M}.. \tag{5}$$

Here xc^S, xc^o – code value of the dimensional parameter in the scale range, X; x_{Ci} – the current value of the i-dimension parameter.

By substituting the code value of the dimensional parameter into formula (4) or (5), one can determine the dimensionless numerical indicator of its effectiveness.

The generalized indicator of the efficiency of an individual machine or a system of machines can be determined by the following formula

$$D_i = \sqrt{\prod_1^n d_{ij}} \to 1.0, \quad (6)$$

where D_i – is a generalized performance indicator of each i option; d_{ij} – a private indicator of the desirability of the j-indicator of the assessment according to the i-variant; n – number of indicators j.

3. Results and discussion

Table 1 shows the results of the calculations that were performed to select the optimal plasma cutting model..

Table 1. Particular and generalized indexes of a comprehensive assessment of the compared devices.

№	Device Model	Private j estimated indicators				Generalized index D_i
		Equipment cost, C_o, €	Technological cutting speed, v_1, mm/min	Technological cutting speed, v_2, mm/min	Technological cutting speed, v_3, mm/min	
1	Powermax45	2 402	400	540	745	0,41
2	Powermax65	2 774	1140	920	1330	0,25
3	Powermax85	3 934	1600	1400	1930	0,65
4	Powermax105	5 550	2060	1860	2450	0,60

The results of the calculation of the complex assessment indicator Di (table 1) showed the following: the best desirability (0.65) corresponds to device number 3. This was facilitated by the relatively low cost of equipment and high technological indicators.

Figure 3 shows a graph of desirability functions and their application to evaluate the devices compared and select the best option.

The points B and B', C and C' indicate the limiting values of the parameters studied of machines corresponding to the desirability 0.2 and 0.8, which on the scale y' correspond to the values y'_B and y'_C. To translate the dimensional parameters of the "direct" dependence S (the higher the parameter value, the higher the desirability, for example, the productivity of the installation, etc.), it is necessary to build an additional scale X_l under the scale y' with the same direction S (figure 2). Its scale will be determined after the maximum and minimum value of this parameter are fixed on this scale, satisfying the desirability of 0.2 and 0.8, respectively.

Knowing the scale of the range group A, it is not difficult to postpone the current value of this parameter to them. From these points x_j the perpendiculars to the graph of the function $d(y')$ were reconstructed using the straight BC or $B'C'$ and the values of d_{ij} were found.

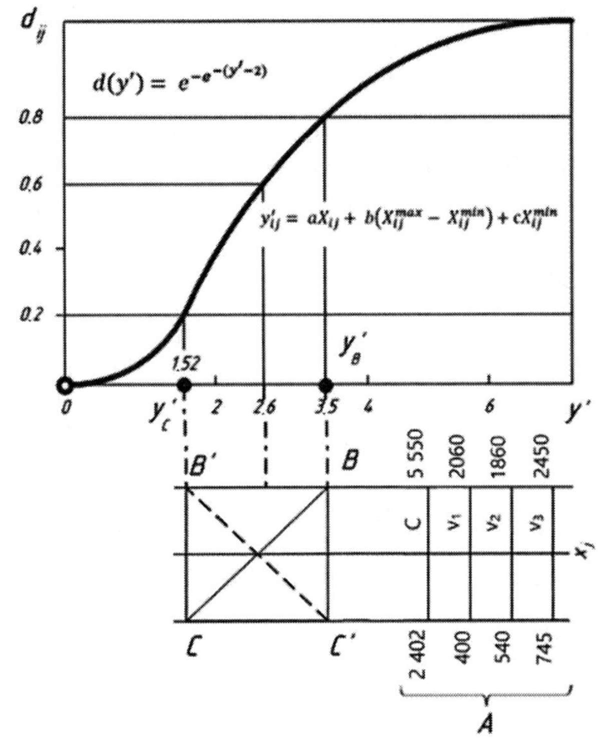

Figure 3. Harrington's upgraded desirability function with forward and backward scale *B'C'*.

When translating the dimensional parameters of the "inverse" dependence (the larger the parameter, the lower the desirability; for example, installation cost, electricity consumption, etc.), an additional scale 2 is built in the opposite direction to the scale y', index O (figure 2).

The scale range and the method for converting a dimensional parameter into a dimensionless quantity d have a similar algorithm.

Thus, the proposed method for a comprehensive assessment of equipment, systems, or technologies is as follows:

1) using well-known methods, it is necessary to determine the largest possible number of factors affecting the system and establish the boundaries (limits) of their changes (for example: productivity, labor costs, fuel, energy, metal consumption, unit cost, serviceability, etc.);

2) after establishing (by calculation, measurements, according to reference data, etc.) the actual values of the parameters (factors), determine the performance indicators (desirability) of each parameter in a graphical or analytical way;

3) determine the result criterion D_i;

4) analyze the D_i values and make a conclusion on the most optimal option. The highest value of D_i, approaching one, determines the best option.

4. Conclusions

As a result of the research, the Harrington function was found to be an effective means of analysis when choosing the most preferred equipment for plasma cutting of metal. The use of a system of four partial evaluation indicators is substantiated when choosing the best equipment from alternative options according to the resulting indicator D_i. Reliable dependencies were obtained for converting values from natural to dimensionless (scale y'). The desirability d_i of each particular indicator and the generalized criterion D_i are calculated. As a result of the application of the Harrington function, it was found that the

Journal of Physics: Conference Series **2270** (2022) 012013 doi:10.1088/1742-6596/2270/1/012013

highest value of the generalized indicator D_i corresponds to the POWERMAX 85 plasma cutting machine.

References

[1] Harrington E 1965 *Industrial Quality control* **21**(10) p 494-498

[2] Novik F S, Arsov A B 1980 Optimization of metal technology processes by experimental planning methods (Moscow: Mashinostroenie) p 304

[3] Tsybulevsky V V, Maslov G G, Tazmeev B K 2021 *IOP Conference Series: Earth and Environmental Science* **848**(1) 012166

[4] Maslov G G, Didmanidze O N, Tsybulevsky V V 2006 Kompleksnoye proyektirovaniye mekhanizirovannykh proizvodstvennykh protsessov v rasteniyevodstve (Moscow: UMTS Triada) p 255

[5] Tsybulevsky V V, Maslov G G 2021 IOP Conference Series: Earth and Environmental Science **677**(3) 032097

[6] Nagirny Yu P 1990 *Mekhanizatsiya i elektrifikatsiya sotsialisticheskogo sel'skogo khozyaystva* No 5 6

[7] Adler Y P, Makarova Y V, Granovskiy Y V 1976 *Planirovaniye eksperimenta pri poiske optimal'nykh resheniy* (Moscow: Nauka) p 208

[8] Kantorovich L V, Goretko A B 1972 *Optimal'nyye resheniya v ekonomike* (Moscow: Nauka) p 232

Experimental installation for study the low-power ICRF plasma at low pressures with equipment for experiment data synchronization

T N Terentev[1], V S Zheltukhin[1, 2], A Yu Shemakhin[1], I K Nekrasov[1], E S Samsonova[1, 2], E Yu Shemakhin[1], I E Tsokalo[3], V A Tyurin[1]

[1]Kazan Federal University, Kazan, Russian Federation
[2]Kazan National Research Technological University, Kazan, Russian Federation
[3]Medical Academy named after S.I. Georgievsky of V.I. Vernadsky Crimean Federal University, Simferopol, Russian Federation

shemakhin@mail.com

Abstract. An installation for numerical and experimental studies of low-pressure radio frequency plasma for surface modification of functional materials with equipment for data synchronization are presented. The equipment for data synchronization as well as intermediate results for plasma generation are showed.

1. Introduction

There are two main methods of cold RF plasma generation: capacitive coupled plasma (CCP) and inductive coupled plasma (ICP) [1]. In ICP electromagnetic energy is transmitted from a RF generator to plasma in a wide range of frequencies from 1 to 50 MHz, and gas pressure in range from 10 upto 1000 mTorr; the optimal frequency range is 0.05-13.56 MHz [2]. One of the advantages of this kind of plasmas is its temperature ("cold" plasma), so material surface is not heated to critical temperatures at processes ofplasma modification of different materials [3, 4]. The installation will let us to verifyresults of mathematical simulation of nonlinear processes occurring in the low pressure RF plasma [5, 6], as well as processes of interaction with a sample. Special attention is paid to synchronization of incoming data such as gas flow rate, gas pressure in a vacuum chamber, temperature, and electron energy distribution function (EEDF).

2. Another section of your paper

The main elements of the installation are RF generator, vacuum chamber, ESPION Probe Interface Unit (EPIU), Langmuir Probe, flowmeter and its control unit, thermocouple, manometer, rotary pump. The block diagram of the installation one can see on Fig.1.

RF generator was assembled on IRFP250N transistors. The frequency of RF generation is 1.76 MHz. This frequency was chosen in accordance with the article [2]. As follows from experimental data, the minimum value of the rotary pump in the vacuum chamber is 0.6 Torr. The vacuum chamber and the RF generator were placed in a steel box to shield the radiation and to avoid the penetration of electromagnetic radiation into the diagnostic environment. Other elements such as the flowmeter, EPIU, manometer (which is an analog device) were placed outside the box to avoid the influence of

Content from this work may be used under the terms of the Creative Commons Attribution 3.0 licence. Any further distribution of this work must maintain attribution to the author(s) and the title of the work, journal citation and DOI.
Published under licence by IOP Publishing Ltd

electromagnetic radiation. The power of the generator is approximately 30 W. The generator was designed so that the RF frequency is variable from 1.76 to 40 MHz. An example of plasma generation without argon flow is shown in Fig. 2(a).

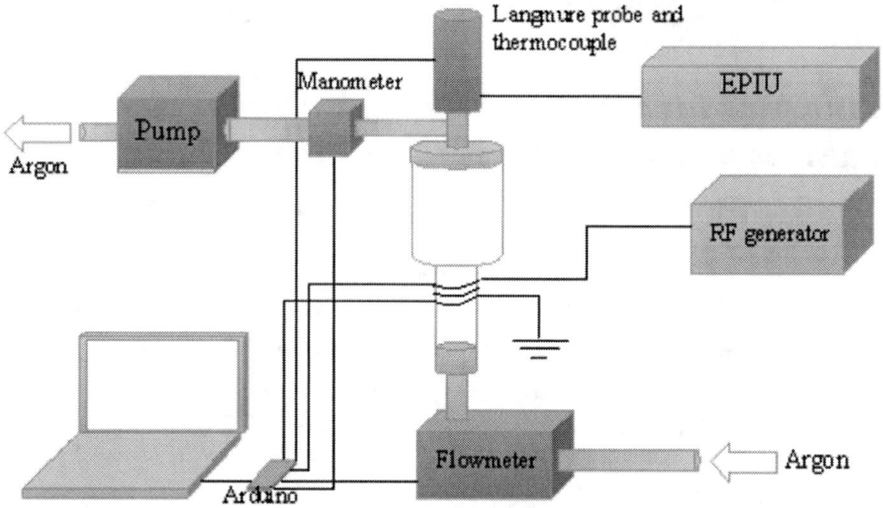

Figure 1. Installation block diagram.

The device MKS M100B having a maximum gas flow of 20 sccm was selected as the flowmeter. It is designed for regulation the flow of argon into the vacuum chamber. The minimum step of flow variation is 0.2 sccm. The device is analogue type; it is configured with a connector type D 15 pin. It was decided to realize the project of flowmeter control based ona small, complete, and breadboard-friendly boardAurduino Nano (AN). The main feature of AN for using in installation is a plotter on a serial connection in the program Aurduino IDE. AN has analog and digital pins. However, the analog pins can only receive a signal, and digital pins can't conduct current more than 50 mA. Thus, the plotter on one of the analogue pins plot the signal in real time.

M100B transmits the value of argon flow from 0 to 5 V. Thus, the plotter on one of the analogue pins in real time can plot the received signal. Flowmeter transmits the value of argon flow from 0 to 5 V by 8 pins. This signal is plotted in real time in the plotter. Then the data from the COM-port (in the plotter window) are copied into *.xlsx* format, and then the program converts the ordinate axis from volts to sccm. There is a direct proportional relationship between voltage and flow value. Adjustment by the argon flow is performed by the DC voltage block from 0 to 5 V. Flow regulation with AN is difficult because only digital pins work in OUTPUT mode and it has current limitation. At the maximum gas flow value, the M100B consumes 350 mA by the flow control pin. An example of plasma generation with argon flow can be seen in Fig. 2(b).

EPIU receives the signal from Langmuir Probe and plots the EEDF graph. Test current-voltage characteristics for 50k resistor in ESPsoft program interface are shown in the figure. 3.

The plasma setup consists of a discharge tube and a vacuum chamber. At pressures more than 1.5 Torr plasma is not generated in the plasmatron because of weak power of generator. Therefore, the task was to select plasmatron materials (more precisely, flanges that are connected to the quartz discharge tubein combination with vacuum sealant) that will withstand low pressure. It was found that PLA plastic in combination with vacuum sealant rebuff low pressure for a long time and allows to use the pump life to the maximum. Therefore, the lower, intermediate and upper flanges of the vacuum chamber were made of PLA plastic. The melting point of PLA plastic is 403 K and that of silicone sealant is 513 K. For this reason, these materials do not have direct contact with plasma in the space between the discharge tube and the vacuum chamber due to the indentation from the end of the plasma torch to the attachment point of the main vacuum chamber.

Figure 2. Plasma generation without (a) and with (b) argon flow in plasma equipment.

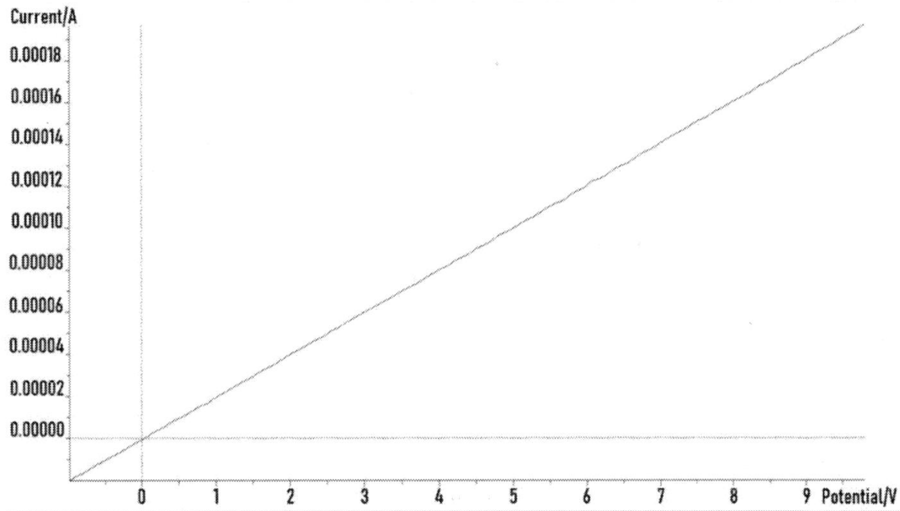

Figure 3. Current-voltage characteristics for 50k resistor in ESPsoft.

3. Equipment for data synchronization system

The data synchronization system includes a flowmeter, manometer, thermocouple, oscilloscope and EPIU. EEDF charts with EPIU are displayed on a separate screen for more convenient data processing. The interface is showed on the figure. 4.

Data from the flowmeter and manometer are fed to the AN, high voltage probes and thermocouple signals on the Arduino UNO. Through the serial COM-port the graphs are built in real time. Data from the secondary winding of the RF generator output transformer are taken off the high voltage probe, which has a built-in attenuator at 30 dB in amplitude. The signal is received by Arduino UNO, which transmits the data to a PC. Measurements are displayed in a special program that acts as an oscilloscope.

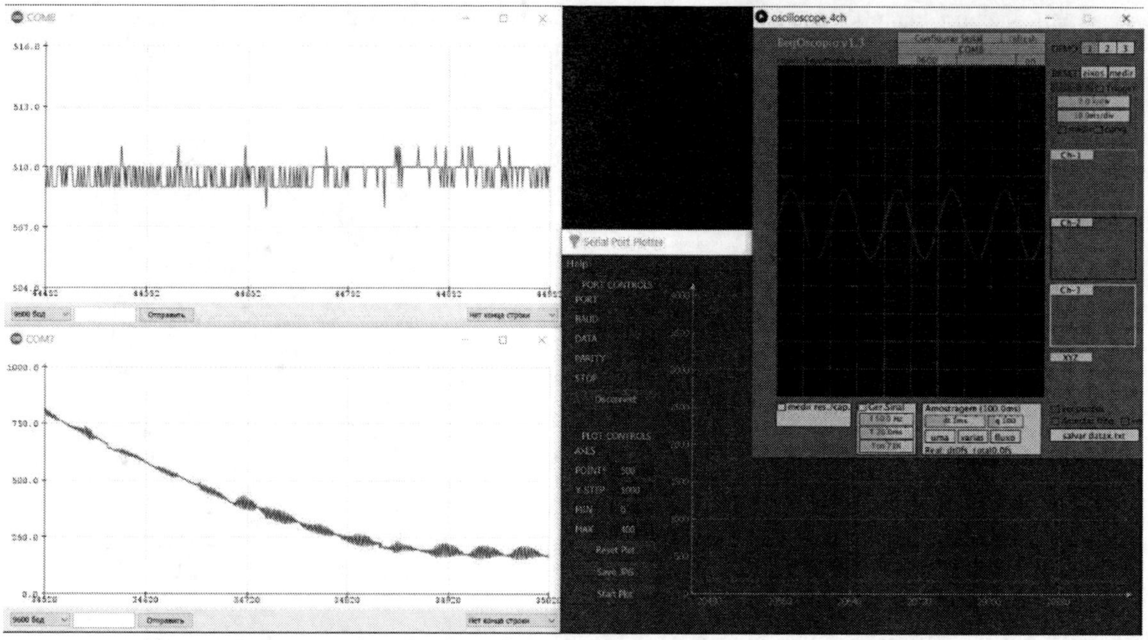

Figure 4. Data synchronization of temperature, pressure and argon flow rate.

Before recording EEDF, it is necessary to ensure the accuracy of the set and measured parameters such as temperature, pressure and gas flow. Only then EEDF measurements can begin. To ensure that temperature and pressure data are not mixed with data prior to recording the EEDF, a sketch is updated when starting the EEDF measurements and all available temperature and pressure data in the cache is reset with the sketch.

Acknowledgment
The study was funded by Russian Science Foundation, project № 19-71-10055.

References

[1] H. Conrads, M. Schmidt, Plasma generation and plasma sources, Plasma Sources Sci. Technol. 9 (2000) 441 – 454.
[2] Lieberman M. and Lichtenberg A. 1994 Principles of Plasma Discharges and Materials Processing (New York: Wiley)
[3] GrillA.1994 Cold Plasma in Materials Fabrication: from Fundamentals to Applications, IEEE Press, New York.
[4] AbdullinI.Sh, Zheltukhin V.S., Sagbiyev I.R., and Shayekhov M.F. 2007 Modification of nanolayers in radio-frequency discharge at low pressure [Modifikaciya Nanosloev v vysokochastotnoj Plazme Ponizhennogo Davleniya – in Russian] (Kazan: Kazan Technol. Univ. Press [Izdatel'stvo Kazanskogo Tekhnologicheskogo Universiteta])
[5]Badriev, I.B., Chebakova, V.Y., Zheltukhin, V.S.(2017) Capacitive coupled RF discharge: Modelling at the local statement of the problem,J. Phys. Conf. Ser.**789**(1) 0120048
[6]Shemakhin, A. Y., andZheltukhin, V. S., Mathematical Modelling of RF Plasma Flow at Low Pressures with 3D Electromagnetic Field, Advances in Materials Sci. and Eng. (2019)7120217

INVESTIGATION OF THERMAL CONDUCTIVITY OF NANOSTRUCTURES USING A MATHEMATICAL MODEL

H.S. Juraev[1], G.H. Juraeva[1]

[1]Tajik National University, Dushanbe, Tajikistan

hayrullo_58@mail.ru

Abstract. The transfer of heat through nanostructures differs significantly from the corresponding processes inside macroscopic bodies. Various research methods, both theoretical and experimental, are applied to such objects. This approach to nanostructures gives great advantages. The hyperbolic equations we have obtained describe the wave process of the thermal signal flow and the properties of heat transfer in nanostructures across a thin layer, in which the simplest task is to determine the thermal conductivity.

1. The first section in your paper

The study of nanostructures from the point of view of heat transfer in them is important because of the rapid development of semiconductor lasers that increase radiation properties, and the presence of various structures in almost all fields of science, technology and others.

First of all, dimensional effects that are absent in macroscopic solids attract attention. The dimensional effects can be classical in the range from 1 nm to 100 nm and quantum, which are realized at scales comparable to the de Broglie wavelength.

Active studies of nanostructures have been started relatively recently [1-6], although the statistical theory of thermal conductivity of solids was founded in the works of Peierls and Clemens [7, 8].

Nanostructures (multilayer nanofilms) are artificially formed layered structures consisting of a large number (up to tens of thousands) of alternating nanolayers of various materials. The list of materials used to create multilayer nanostructures is quite wide – it can be pure metals, alloys, semiconductor materials, materials with dielectric properties in various combinations. The thickness of the layers of multilayer nanostructures, as a rule, is in the range from 1 to 100 nm. It is worth noting that the greatest interest is the study of multilayer nanostructures with layers of thickness from units to several tens of nanometers. This is due to the fact that it is in this range that the degree of influence of the dimensional factors of nanolayers on the structure and properties of materials acquires the greatest importance. Thus, the ultra-small thickness of the layers in combination with a large number of interfaces between the layers and the alternation of layers of different materials leads to the appearance of multilayer nanostructures with unique structural features that are not present in the materials that make up the nanostructures in massive samples [9].

A one-dimensional unsteady process of heat propagation in a thermally insulated ordered multilayer nanostructure consisting of several alternating layers of AlAs and GaAs materials is considered.

The heat transfer process (heat signal) occurs in the AlAs layer, and the positive signal simultaneously with the heat flow is the main effect on the GaAs layer. During the pulse, a positive

Content from this work may be used under the terms of the Creative Commons Attribution 3.0 licence. Any further distribution of this work must maintain attribution to the author(s) and the title of the work, journal citation and DOI.
Published under licence by IOP Publishing Ltd

thermal signal on the surface is subjected to such a dose of energy to realize the decomposition of the upper layer of the material and achieve a voltage of the order of several megabars on the surface. In this case, the pulse goes to the lateral surface covered with AlAs depending on the time.

The mathematical model describing these processes in the simplest isotropic case consists of the energy conservation equation for the subsystem layer and the equation for the relaxation of the heat flow for the subsystem layer.

At the beginning of the process, when the flow of the thermal signal is just beginning to grow, the term with the derivative of the flow is very small, and the traditional approximation works. Therefore, the mathematical model is described in the form of a nonlinear heat equation:

$$\frac{\partial T}{\partial t}=\frac{k_1k_2}{k_3T^{5/2}}\frac{\partial}{\partial x}\left(T^{5/2}\frac{\partial T}{\partial x}\right), \tag{1}$$

where $k_1=\frac{k_0T_0^2t_*}{\tau_0P_0}$, $k_2=\frac{P_0}{\rho C_\upsilon T_0t_*}$, $k_3=\frac{t_*}{\tau_0T_0^{3/2}}$ - coefficient of thermal conductivity; C_υ - the heat capacity of the substance; τ_0 - relaxation time of the heat flow; t - time; x - coordinate; $T=T(x,t)$ - temperature; P_0 - maximum flow of the heat signal.

By introducing an integral function

$$W(T)=\int_{T_0}^{T}(T^{5/2})dT=\frac{2}{7}(T^{7/2}-T_0^{7/2}) \tag{2}$$

equation (1) can be reduced to the form:

$$\frac{\partial W}{\partial t}=\beta\frac{\partial^2 W}{\partial r^2}, \tag{3}$$

where $\beta=\frac{k_1k_2}{k_3}\left(\frac{\text{м}^3\cdot\text{К}}{\text{Дж}}\right)$.

In the low temperature region, the heat transfer process (heat signal) in the AlAs layer decreases sharply with a decrease in temperature, the main effect is on the GaAs layer, as a result of which the free path increases sharply and reaches a value comparable to the size of the GaAs layer. Since the walls of the layer have a low reflection coefficient, a further decrease in temperature does not lead to an increase in the free path length, since in this case it is determined precisely by the size of the GaAs layer. The value of the thermal conductivity coefficient for temperature is small, so the dependence of the thermal conductivity coefficient $\beta\left(\frac{\text{м}^3\cdot\text{К}}{\text{Дж}}\right)$ on $W(T)$ will be insignificant. The temperature dependence of the thermal conductivity of the layer in this temperature range is determined by the dependence of the thermal capacity AlAs (GaAs) on temperature. At low temperatures, the heat capacity is proportional to $T^{5/2}$, and the coefficient of thermal conductivity is also proportional to $T^{5/2}$.

Given the above, the following assumption will be quite true:

$W(T)=W(x,t)$.

Note that equation (3) with respect to the function $W(T)$ is linear. Since the coefficient of thermal conductivity $\lambda(T)=T^{5/2}$ depends on temperature, and the coefficient $\beta=\frac{k_1k_2}{k_3}\left(\frac{\text{м}^3\cdot\text{К}}{\text{Дж}}\right)$ weakly depends on the function $W(T)$, for an AlAs layer with a low coefficient of thermal conductivity $T^{5/2}$, the dependence of β on $W(T)$ can be neglected.

Our goal is to determine the heat transfer in the layered structure at a subsequent time by the method of artificial hyperbolization [1], that is, we will look for a solution to the hyperbolic equation of thermal conductivity:

$$\alpha\frac{\partial^2 W}{\partial t^2}+\frac{\partial W}{\partial t}=\beta\frac{\partial^2 W}{\partial r^2}, \tag{4}$$

where α - is the regularization parameter (relaxation time of the heat flow, (ns)).

In the case of a plane geometry, the heat transfer in a layered structure, relative to the function $W(x,t)$ according to [10], has the form:

$$W_\alpha(x,t)=\sum_{k=1}^{N}W_{0k}\exp(-\omega_k(\alpha)t)*$$
$$*\left[1+\alpha\frac{\omega_k(\alpha)}{\sqrt{1-4\alpha\beta\eta_k^2}}\left(\exp\left(-\frac{\sqrt{1-4\alpha\beta\eta_k^2}}{\alpha}t\right)-1\right)\right]\sin(\eta_k x), \tag{5}$$

where $\omega_k(\alpha)=\dfrac{2\beta\eta_k^2}{1+\sqrt{1-4\alpha\beta\eta_k^2}}$, η_k – the coefficient of dielectric permittivity (see [5]).

This relation (2) allows us to find the temperature distribution in nanostructures with a known physical characteristic of materials.

Then

$$T_\alpha(x,t)=\left[(T_{\alpha 0}(x))^{7/2}+3{,}5W_\alpha(x,t)\right]^{2/7}. \tag{6}$$

The heat flow functions $q_\alpha(x,t)$ and $T_\alpha(x,t)$ are connected in layers as follows:

$$q_\alpha(x,t)=-(T_\alpha(x,t))^{5/2}\left.\frac{\partial T_\alpha}{\partial x}\right|_{x=x_n}. \tag{7}$$

Thus, from the expressions (4) and (6) obtained, it is possible to determine the values of the temperature and heat flux of the substance in the layers. If the temperature and heat flow of the substance are known, the desired thermophysical characteristics (thermal conductivity - λ, specific heat capacity - c_ρ, density - ρ) on the surface of a substance, it can be determined by solving the nonlinear equation (1) in the form of expressions (4) and (6).

In the course of the work, numerical calculations were carried out according to the algorithm constructed according to the regularization scheme in the form of expressions (5) and (6). The study was carried out in a thermally insulated ordered multilayer nanostructure consisting of 400 alternating layers of AlAs and GaAs materials.

As a reference, the thermophysical characteristics of the materials under consideration are given (table 1) [11].

Table 1. Thermophysical characteristics of materials.

Thermophysical characteristics	Material	
	AlAs	GaAS
C, J/kg*K	424	327
p, kg/m^3	3730	5320
v_s, m/s	6258	5242
p_k, m2*K/W	0,625*10^{-9}	

The results are presented in the form of graphs (Fig. 1.a) and b)) depending on the material properties of the layers, their thickness and the conductivity of the interfaces between the layers. Figure 1 (a) and 1 (b) show small and large temperature jumps at the boundaries of the layers due to the different magnitude of the singularity.

From the numerical calculation results obtained, it follows that with increasing layer thickness, the phenomenon of heat transfer slows down. So, in multilayer nanostructures, the thickness of the layers has a value of 10-100 nm. The calculation was obtained at different layer thicknesses for the case of the same thickness of the AlAs and GaAs layers. In the calculation, it was also found that the ratio between the heat transfer rates is preserved over time at different layer thicknesses. Therefore, naturally, in the vicinity of the boundary of monotonous instability of the problem with respect to long-wave thermal signals, the equation connecting k_1, k_2 and k_3, using the method of artificial hyperbolization, is a stable calculation for the thermal conductivity of nanostructures.

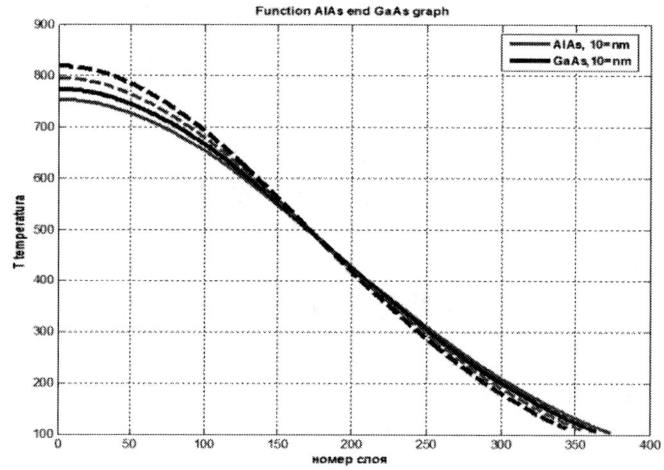

a)

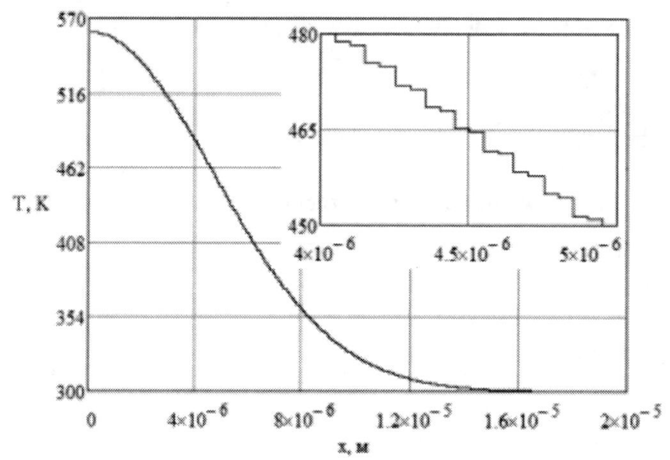

b)

Figure 1. a) Characteristic temperature distribution over layers of AlAs and GaAs nanostructure at thickness h=10 nm and time t=15*10-6 c; b) results of work [11]

Based on the results obtained and the presented graphic material, it can be said that the stability, the time to enter the mode and the temperature control error of the nanostructure are influenced by the temperature boundary characteristics, the temperature range and the type of material, as well as the regularization parameter (relaxation time). Moreover, it should be noted that each of the above factors has its own degree of influence on the parameters of solving the problem for the nonlinear wave equation of thermal conductivity, and this method of solving by regularization is stable. Low-temperature plasma can be used for the synthesis of various nanostructures and is well suited for the modification of various surfaces. This is shown in many works [12-26].

References

[1] Dmitriev A.S. Thermal processes in nanostructures. -M.: Publishing house of MEI, 2012.-302 p.
[2] Fisher T.S. Thermal Energy at the Nanoscale. - Singapore: World Scientific, 2013. - 171 p.
[3] Cahill D.G., Ford W.K., & Goodson K.E., Mahan G.D., Madjumar A., Maris H.J., Merlin R. Nanoscale Thermal Transport // J. Appl. Phys., 2003, V. 93. № 2. - P. 793-802.
[4] Cahill D.G., Braun P.V., & Chen G., Clarke D.R., Fan S., Goodson K.E., Keblinski P., at el. Nanoscale Thermal Transport II. // Appl. Phys. Rev, 2014, V. 1. № 1. - P. 011305. [3] More references
[5] Dmitriev A.S. Introduction to nanoteplophysics. - M.: BINOM, 2015. - 792 p. [3] More references
[6] Khvesyuk V.I., Scriabin A.S. Thermal conductivity of nanostructures // Thermophysics of High Temperatures, 2017, volume 55, No. 3. - pp. 447-471.
[7] Peierls R.E. Zur kinetischen Theorie der Warmelitung in Kristallen // Ann. Phys. 1929, V. 395. - P. 1055-1066. [3] More references
[8] Klemens P. Thermal Conductivity and Lattice Vibrational Modes // Solid State Phys., 1958, V. 7. - P. 1-23.
[9] Bazaleeva, K.O. Structural changes in multilayer nanofilms Ti/Al / K.O. Bazaleeva, V.S. Kraposhin, P.A. Tsygankov et al. // Materials Science. - 2008. - No. 8. - pp. 35-39.
[10] Juraev, H.S. Phenomena of energy and mass transfer in condensed media: mathematical modeling, optimization, practical applications. /H.S. Juraev //-Dushanbe: ER-graph. -2021. - 236 p.
[11] Vorobyov, D.A. Method of calculation of nonstationary heating of nanostructures / D.A. Vorobyov, V.I. Khvesyuk // Science and Education (Electronic Scientific and Technical Journal) of Bauman Moscow State Technical University. 2013. -No.9. -pp.541-550. http:// tcchnomag.bmstu.ru/doc/617255.html
[12] Shamsutdinov, R. S., & Timerkaev, B. A. (2021, April). The influence of a supersonic flow of gas at glow discharge. In Journal of Physics: Conference Series (Vol. 1870, No. 1, p. 012019). IOP Publishing. DOI 10.1088/1742-6596/1870/1/012019.
[13] Saifutdinov, A. I., & Sofronitskii, A. O. (2021). Numerical Study of Breakdown and Formation Dynamics of Arc Discharge Plasma Parameters at Ultrahigh Pressures. High Energy Chemistry, 55(3), 228-232.DOI: 10.1134/S0018143921030115
[14] Shemakhin, A. Y., Zheltukhin, V. S., Shemakhin, E. Y., Terentev, T. N., & Sofronitsky, A. O. (2020, July). Experimental installation to study the RF plasma flow at low pressures with experiment data synchronization. In Journal of Physics: Conference Series (Vol. 1588, No. 1, p. 012018). IOP Publishing. DOI: 10.1088/1742-6596/1588/1/012018
[15] Saifutdinova, A. A., Sofronitskiy, A. O., Timerkaev, B. A., & Saifutdinov, A. I. (2020). Plasma-Chemical Decomposition of Hydrocarbons on the Basis of the Micro-Arc Discharge with Disc Electrodes Rotating in the Bulk of Raw Materials. Russian Physics Journal, 62(11), 2132-2136.DOI: 10.1007/s11182-020-01957-0
[16] Fairushin, I. I., Saifutdinov, A. I., & Sofronitskiy, A. O. (2020). Numerical and Experimental Studies of the Synthesis of Copper Nanoparticles in a High-Pressure Discharge. High Energy Chemistry, 54(2), 150-153. DOI: 10.1134/S0018143920020071

[17] Asadullin, T. Y., Galeev, I. G., Sofronitskiy, A. O., & Gizeev, M. M. (2019, November). Acoustic impact on electric discharge parameters during sterilization of freeze-dried products. In Journal of Physics: Conference Series (Vol. 1370, No. 1, p. 012016). IOP Publishing. DOI: 10.1088/1742-6596/1370/1/012016

[18] Fairushin, I. I., Saifutdinov, A. I., Sofronitskiy, A. O., Timerkaev, B. A., & Dautov, G. Y. (2019, October). Development of plasma reactor design for synthesis of copper nanoparticles using multi-scale simulation. In Journal of Physics: Conference Series (Vol. 1328, No. 1, p. 012088). IOP Publishing. DOI: 10.1088/1742-6596/1328/1/012088

[19] Asadullin, T. Y., Galeev, I. G., & Sofronitskiy, A. O. (2019, October). The application of pulsed discharge for sterilization of freeze-dried product. In Journal of Physics: Conference Series (Vol. 1328, No. 1, p. 012071). IOP Publishing. DOI: 10.1088/1742-6596/1328/1/012071

[20] Timerkaev, B. A., Ganieva, G. R., Kaleeva, A. A., Israfilov, Z. K., & Sofronitskii, A. O. (2019). Growing of carbon nanotubes from hydrocarbons in an arc plasma. Journal of Engineering Physics and Thermophysics, 92(5), 1248-1252. DOI: 10.1007/s10891-019-02040-3

[21] Dautov, G. Y., Kashapov, N. F., Dautov, I. G., & Sofronitskiy, A. O. (2018, July). Research of the influence of the geometry of the discharge chamber on the characteristics of the arc plasmatron. In Journal of Physics: Conference Series (Vol. 1058, No. 1). IOP Publishing. DOI: 10.1088/1742-6596/1058/1/012035

[22] Timerkaev, B. A., Andreeva, A. A., & Sofronitskiy, A. O. (2017, November). Discharge creeping along the surface in the process for producing nanomaterials. In Journal of Physics: Conference Series (Vol. 927, No. 1, p. 012068). IOP Publishing. DOI: 10.1088/1742-6596/927/1/012068

[23] Sadikov, K. G., Sofronitskiy, A. O., & Larionov, V. M. (2017, November). The effect of electrically conductive additives on the plasma pyrolysis of heavy hydrocarbons. In Journal of Physics: Conference Series (Vol. 927, No. 1, p. 012046). IOP Publishing. DOI: 10.1088/1742-6596/927/1/012046

[24] Timerkaev, B. A., Andreeva, A. A., & Sofronitskiy, A. O. (2017). Discharge creeping along the surface in the process of cleaning and strengthening of the materials surface. In Journal of Physics: Conference Series (Vol. 789, No. 1). IOP Publishing. DOI: 10.1088/1742-6596/789/1/012063

[25] Sadikov, K. G., Sofronitskiy, A. O., & Dautov, I. G. (2017). Functional plasma sprayed coatings on magnesium ceramic substrates. In Journal of Physics: Conference Series (Vol. 789, No. 1, p. 012043). IOP Publishing. DOI: 10.1088/1742-6596/789/1/012043

[26] Timerkaev, B. A., Sofronitskiy, A. O., & Andreeva, A. A. (2016). Carbon nanotubes formation in the decomposition of heavy hydrocarbons creeping along the surface of the glow discharge. In Journal of Physics: Conference Series (Vol. 669, No. 1, p. 012062). IOP Publishing. DOI: 10.1088/1742-6596/669/1/012062

Journal of Physics: Conference Series **2270** (2022) 012016 doi:10.1088/1742-6596/2270/1/012016

One-dimensional continuum finite-difference model of the dynamics of a dusty medium in aerodynamic, electric and gravitational fields

D A Tukmakov[1]

[1]Kazan Scientific Center of Russian Academy of Sciences, 2/31 Lobachevsky Str., Kazan, 420111, RF

tukmakovda@imm.knc.ru

Abstract. This paper presents a mathematical model of the dynamics of a gas suspension under the action of an aerodynamic field, an electric field, and a gravitational field. The continuum model is used to describe the dynamics of the disperse component. The intercomponent momentum exchange included the aerodynamic drag force, the dynamic Archimedes force, and the added mass force. The Coulomb force acting on dispersed particles was taken into account. The model assumes the solution of the equations of conservation of "average density", momentum and energy for the dispersed phase. The electric field was described by the Poisson equation. The equations of the mathematical model were supplemented with boundary conditions. The equations of mechanics were integrated by explicit finite-difference method of McCormack.

1.Introduction

Studies of the dynamics of inhomogeneous media are applicable in many areas [1-10]. Including in plasma physics [3]. The heterogeneity of the studied media is one of the elements of scientific novelty in modern studies of hydrodynamics. As a branch of physics, hydrodynamics has a significant reliance on mathematical mods - methods of mathematical physics. In [3], the dynamics of monodisperse quartz particles in plasma is numerically modeled. In [5], electrohydrodynamic processes in the atmosphere are mathematically described taking into account the inhomogeneous structure of clouds. To study processes involving heterogeneous effects, it is necessary to use various branches of science [11]. The results presented in this study are theoretical in nature and allow bringing together such physical and mechanical concepts as viscosity, compressibility of the medium, intercomponent interaction and electric field.

2. Mathematical model

The dynamics of the carrier medium is described by the one-dimensional Navier-Stokes equation. The right-hand sides of the equations contain the terms responsible for the intercomponent momentum exchange $-F(u,u_1)$ and intercomponent heat transfer $-Q(T,T_1)$ [7,9,10]:

$$\frac{\partial \rho}{\partial t}+\frac{\partial \rho u}{\partial x}=0 \tag{1}$$

$$\frac{\partial(\rho\ u)}{\partial t}+\frac{\partial}{\partial x}\left(\rho\ u^2+p-\tau\right)=-F+\alpha\frac{\partial p}{\partial x} \tag{2}$$

Content from this work may be used under the terms of the Creative Commons Attribution 3.0 licence. Any further distribution of this work must maintain attribution to the author(s) and the title of the work, journal citation and DOI.
Published under licence by IOP Publishing Ltd

$$\frac{\partial(e)}{\partial t}+\frac{\partial}{\partial x}\left([e+p-\tau]u-\lambda\frac{\partial T}{\partial x}\right)=-Q-|F|(u-u_1)+\alpha\frac{\partial(pu)}{\partial x} \tag{3}$$

$$\frac{\partial \rho_1}{\partial t}+\frac{\partial(\rho_1 u_1)}{\partial x}=0 \tag{4}$$

$$\frac{\partial(\rho_1 u_1)}{\partial t}+\frac{\partial}{\partial x}(\rho_1 u_1^2)=F-F_C-g\rho_1-\alpha\frac{\partial p}{\partial x} \tag{5}$$

$$\frac{\partial(e_1)}{\partial t}+\frac{\partial}{\partial x}(e_1 u_1)=Q \tag{6}$$

$$\frac{\partial^2\varphi}{\partial x^2}=\rho_1 q_0 \tag{7}$$

Equation (1) is the gas mass continuity equation, equation (2) is the Navier-Stokes equation describing the continuity of the viscous gas momentum, equation (3) is the gas energy conservation equation. Equation (4) is the equation of continuity of the "average density"–ρ_1 of the carrier medium. The average density is the product of the physical density of the material, a constant $-\rho_{10}$=const and the volumetric content of the dispersed component $\alpha=\alpha(x,t)$, which is a function of space and time variables. Equation (5) is the equation of continuity of the momentum of the dispersed phase, equation (6) is the equation of conservation of the energy of the dispersed component. Equation (5) takes into account both the intercomponent momentum exchange $-F$, and the Coulomb force acting on particles $-F_C$, as well as the gravitational settling of particles, the term: $-\rho_1 g$. Equation (7) is the Poisson equation describing the electric field created both by the electric charge of dispersed particles and by external electric potentials. Intercomponent momentum exchange F involves several forces: $F=F_{AR}+F_A+F_{AM}$; where F_{AR} is the force of aerodynamic drag (8), F_A is the dynamic force of Archimedes (9), F_{AM} is the force of added masses (10) [1]:

$$F_{AR}=\frac{3\alpha}{4d}C_{d_i}\rho\,|u-u_1|(u-u_1) \tag{8}$$

$$F_A=\alpha\rho\left(\frac{\partial u}{\partial t}+u\frac{\partial u}{\partial x}\right) \tag{9}$$

$$F_{AM}=0.5\alpha\rho\left(\frac{\partial u}{\partial t}+u\frac{\partial u}{\partial x}-\frac{\partial u_1}{\partial t}-u_1\frac{\partial u_1}{\partial x}\right) \tag{10}$$

$$Q=6\alpha Nu_1\lambda(T-T_1)/d^2 \tag{11}$$

$$Nu_1=2\exp(-M_1)+0.459Re_1^{0.55}Pr^{0.33},\ C_d=C_d^0\eta(\alpha)\psi(M_1),\ C_d=\frac{24}{Re_1}+\frac{4}{Re_1^{0.5}}+0.4,$$

$$\psi(M_1)=1+exp\left(\frac{-0.427}{M_1^{0.63}}\right),\ \eta(\alpha)=(1-\alpha)^{-2.5},\ M_1=|u-u_1|/c_0,\ Re_1=\rho|u-u_1|d/\mu, \tag{12}$$

$$Pr=\gamma C_p\mu/\lambda,\ \tau=\frac{4}{3}\mu\frac{\partial u}{\partial x}$$

The drag coefficient of a single particle C_d is calculated taking into account the compressibility of the medium (factor ψ) corrected for multiplicity of particles (factor η) (C_d is the drag coefficient of a single spherical particle). Nu_1 is the relative Nuselt number, Re_1 and M_1 are the relative Reynolds and Mach numbers, $\tau(u)$ –function of viscous gas stresses. The electric field is described in the following way [12]:

$$divE = \frac{\rho_e}{\varepsilon\varepsilon_0},\ E = -\bar{\nabla}\varphi,\ \Delta^2\varphi = -\frac{\rho_e}{\varepsilon\varepsilon_0},\ \rho_e = \alpha\rho_{10}\cdot q_0 = \rho_1\cdot q_0,\ \varepsilon_0 = \frac{10^{-9}}{36\pi}F/m \quad (13)$$

where q_0 is the specific charge per unit mass of the solid fraction, φ is the electric field potential. E is the electric field strength, ε is the relative dielectric constant of air, ε_0 is the absolute dielectric constant of air. The Coulomb force acting on the particles of the dispersed component of the gas suspension is determined by formula (14):

$$F_C = -\rho_1\frac{\partial\varphi}{\partial x} \quad (14)$$

The equations of the mathematical model (1)-(14) were supplemented with boundary conditions. The velocities of the mixture components at the boundaries of the computational domain were set equal to zero. For the rest of the dynamic functions, the homogeneous Neumann boundary conditions were set. For the electric field potential, values were set at opposite ends of the computational domain [13,14]: $u(t,1)=0, u_1(t,1)=0,\ u(t,N)=0, u_1(t,N)=0,\ \rho(t,1)=\rho(t,2),\ \rho_1(t,1)=\rho_1(t,2),\ \rho(t,N)=$ $=\rho(t,N-1),\ \rho_1(t,N)=\rho_1(t,N-1),\ e(t,1)=e(t,2), e_1(t,1)=e_1(t,2),\ e(t,N)=e(t,N-1),\ e_1(t,N)=$ $=e_1(t,N-1),\ p(t,1)=p(t,2), p(t,N-1)=p(t,N),\ \varphi(1,t)=\varphi_1, \varphi(N,t)=\varphi_2$

3. Solution methods

Let us consider the application of the algorithm of the explicit McCormack method using the example of a scalar nonlinear partial differential equation (15) [13]:

$$\frac{\partial f}{\partial t} + \frac{\partial a(f)}{\partial x} = c(f) \quad (15)$$

The algorithm of McCormack's explicit finite-difference method for nonlinear equation (15) has the form (16) - (17):

$$f_i^* = f_i^n - \frac{\Delta t}{\Delta x}\left(a_{i+1}^n - a_j^n\right) + \Delta t c_i^n \quad (16)$$

$$f_i^{n+1} = 0{,}5(f_i^n + f_i^*) - 0{,}5\frac{\Delta t}{\Delta x}\left(a_i^n - a_{i-1}^n\right) + 0.5\Delta t c_i^n \quad (17)$$

Here Δx is the step in the spatial direction, Δt is the step in time. The Poisson equation for the potential of the electric field was solved by the finite difference method - the method of establishment applied on the gas-dynamic computational grid. The scheme of the establishment method is as follows [15]. For a one-dimensional inhomogeneous second-order differential equation (18):

$$\frac{df}{dx^2} = h(x) \quad (18)$$

To implement the method for equation (18), a non-stationary expressions is written (19), (20):

$$\frac{\partial m}{\partial t} = -\frac{\partial^2 m}{\partial x^2} - h(x) \quad (19)$$

$$\lim_{t\to\infty} m(t,x) = f(x) \quad (20)$$

The function $m(t,x)$ is found from the solution of a finite-difference equation:

$$\frac{m^{n+1,i} - m^{n,i}}{\Delta t} = -\frac{m^{n,i+2} - 2m^{n,i+1} + m^{n,i}}{\Delta x^2} + h(i\Delta x) \quad (21)$$

The function f is found from the finite-difference equations (21). The solution of the stationary linear differential equation (7) of type (18) was carried out each time after the transition from the nth to the n + 1st time layer, so the function $\varphi(x,t)$ is a non-stationary function –depends on time.
To overcome some of the shortcomings of the numerical method used, a nonlinear grid correction algorithm was used, which will reduce the local maxima and increase the local minima of the grid function [14]. The correction algorithm was performed sequentially along all nodes. Let us consider

an algorithm for correcting the solution using the example of the function *f*. If the conditions are met $\left(\delta f_{i-1/2}\cdot\delta f_{i+1/2}\right)<0$ or $\left(\delta f_{i+1/2}\cdot\delta f_{i+3/2}\right)<0$ then the algorithm of the correction scheme is applied to the function f in the *i*-th node: $\tilde{f}_i = f_i + \kappa\left(\delta f_{i+1/2} - \delta f_{i-1/2}\right)$. The subscript indicates the mesh node number. The notation is used here: $\delta f_{i-1/2} = f_i - f_{i-1}$, $\delta f_{i+1/2} = f_{i+1} - f_i$, $\delta f_{i+3/2} = f_{i+2} - f_{i+1}$ Otherwise: $\tilde{f}_i = f_i$, $\tilde{f}_i$ is the value of the function at the i-th node after the transition to the (n + 1) -th time layer according to the McCormack scheme, κ is the correction factor. The numerical solution of the equations of the mathematical model described here is implemented as a computer program in the Fortran programming language.

4. Calculation results

The channel length was L=1 m. The specific charge of the disperse component q_0=0.001 C/kg, φ_1=15 kV φ_2=0. Computational domain nodes N=200, Courant number C_C=0.5, correction factor κ=0.15. Figure 1 schematically shows the computational domain, positively charged dispersed particles move under the action of gravity and an electric field: since a positive potential is applied to the upper surface of the container, and the lower surface of the container has a potential equal to zero. Physical density of particle material ρ_{10}=2700 kg/m^3, volume content of the dispersed component α=0.0004, the size of particles d=200 μm. The carrier medium has the physical properties of air.

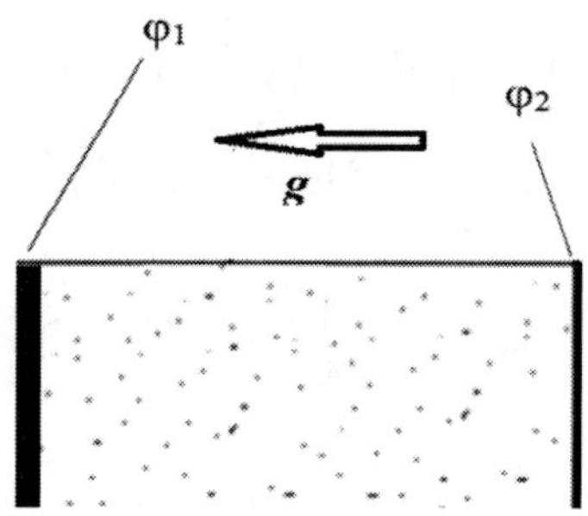

Figure 1. Schematic representation of the simulated process.

Figure 2 shows the distribution of the potential of the electric field, it can be observed that between the two areas of potential reduction there is an area with a uniform distribution of the electric potential.

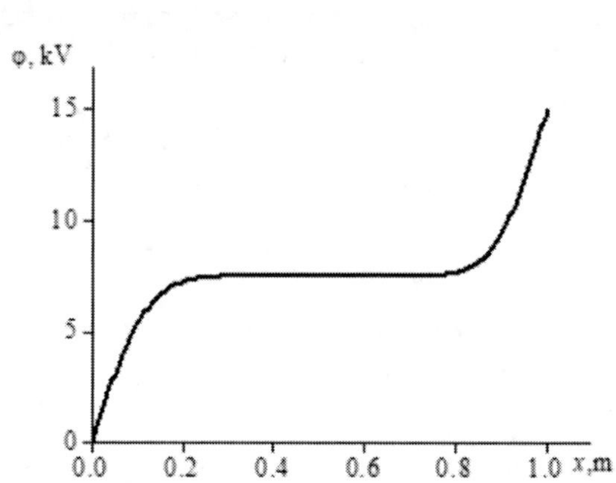

Figure 2. Spatial distribution of the electric field potential. The specific charge of the disperse component q_0=0.001 C/kg, φ_1=15 kV φ_2=0.

The distribution of the potential of the electric field is consistent with the Coulomb force, which reaches the highest value in the areas of the maximum gradient of the potential of the electric field - Figure 3.

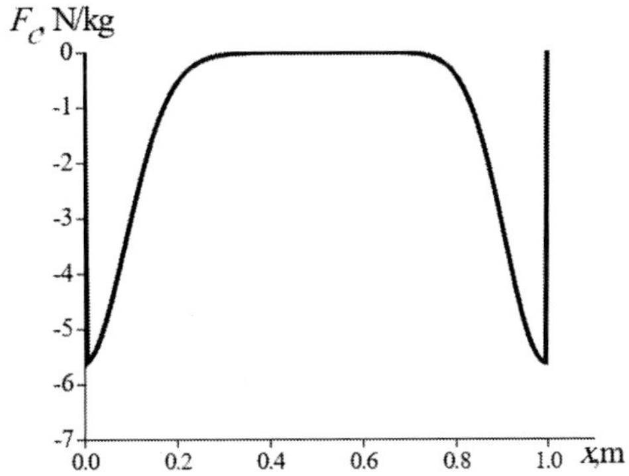

Figure 3. Spatial distribution of the Coulomb force. The specific charge of the disperse component q_0=0.001 C/kg, φ_1=15 kV φ_2=0.

The velocity of the dispersed component reaches the highest value in the areas of the highest value of the Coulomb force – Figure 4.

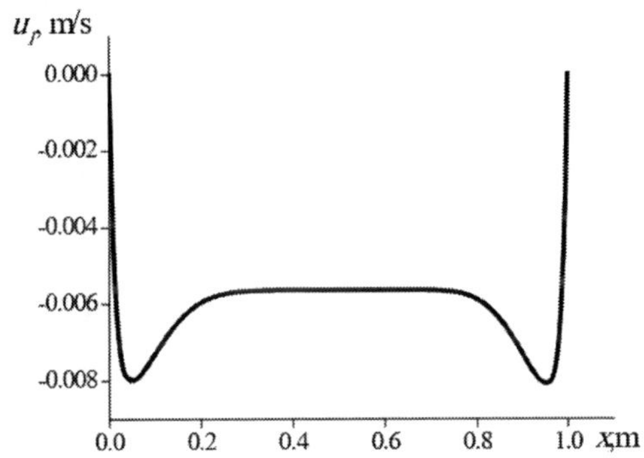

Figure 4. Spatial distribution of particle velocity. Physical density of particle material ρ_{10}=2700 kg/m^3, volume content of the dispersed component α=0.0004, size of particles d=200 μm

Due to the intercomponent exchange of momentum, the movement of dispersed particles results in the movement of gas - Figure 5.

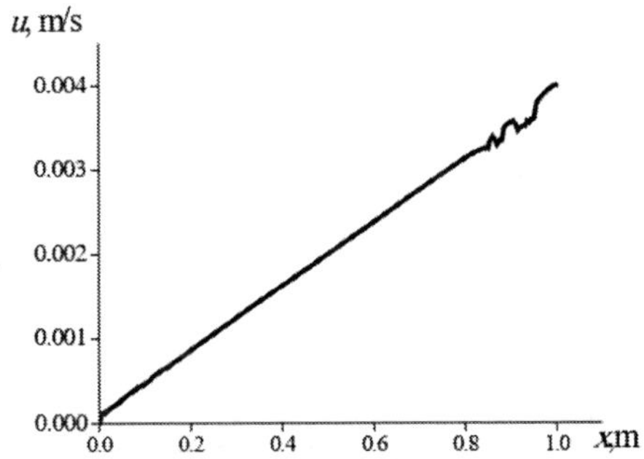

Figure 5. Spatial distribution of gas velocity. Initial gas temperature T_0=293 K. Initial gas density ρ_0=1.21 kg/m^3.

Due to the movement of gas in the channel, an uneven distribution of gas pressure will form Figure6.

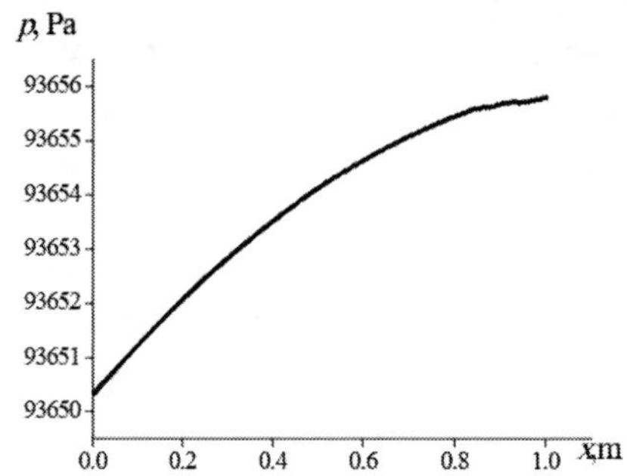

Figure 6. Spatial distribution of gas pressure. Initial gas temperature T_0=293 K. Initial gas density ρ_0=1.21 kg/m^3.

5. Conclusion

The article describes the results of scientific research - the development of a theoretical model based on the theory of fluid, gas and plasma dynamics. It can be more definitely said that the methods of the dynamics of multiphase media and electrodynamics are used. The results of formalization of theoretical calculations are presented in the form of a nonlinear system of partial differential equations of parabolic type. As well as one linear ordinary differential equation implicitly related to the general system of equations for the dynamics of an inhomogeneous medium. The entire mathematical model has the form of a boundary value problem for an ordinary differential equation and a Cauchy boundary value problem for the rest of the model. To obtain the solution, methods of computational mathematics were used. The numerical algorithm is implemented as a computer program. Thus, the results obtained can be used in modeling the dynamics of dusty plasma and in calculating flows of electrohydrodynamics of inhomogeneous media. In the above calculations, a relationship was found between the settling rate of dispersed particles and the magnitude of the gradient of the electric field potential. Also, the use of the continuum modeling technique made it possible to reveal the reciprocal effects of the interaction of the mixture components.

References

[1] Nigmatullin R I 1978 *Fundamentals of the Mechanics of Heterogeneous Media* (Moscow: Nauka)
[2] Kutushev A G 2003 *Mathematical simulation of wave processes in aero dispersed and powdered media* (St. Petersburg: Nedra)
[3] Grishin Y M, Miao L 2020 *High Temperature* 39 p 1.
[4] Tukmakov A L 2014 *Journal of Engineering Physics and Thermophysics* 87 p 38.
[5] Sinkevich O A 2016 *High Temperature* 54 p 775.
[6] Ilgamov M A, Utyashev I M 2022 *Russian Applied mathematics and mechanics* 86 p 66.
[7] Tukmakov A L, Tonkonog V G, Tukmakova N A 2020 *Thermophysics and Aeromechanics* 2 p 269.
[8] Terekhov V I, Kalinina S V, Mshvidobadze Y M 1997 *Journal of Enhanced Heat Transfer* 4 p 131.
[9] Tukmakov D A Ahunov A A 2020 *Russian Chebyshevskii collection* 21, p 257.
[10] Tukmakov D A 2020 *Russian Mathematical Notes of NEFU* 27 p 99.
[11] Urmancheev S F 2020 *Russian Multiphase systems* 15 p 133.
[12] Salyanov F A 1997 *Fundamentals of low-temperature plasma physics, plasma devices and technologies* (Moscow: Nauka)
[13] Fletcher C A 1988 *Computation Techniques for Fluid Dynamics* (Berlin: Springer-Verlang)
[14] Muzafarov I F, Utyuzhnikov S V 1993 *Russian J. of Computational Mechanics* 5 p 74
[15] Krylov V I, Bobkov V V, Monastyrnyj P I 1977 *Computational Methods* 2 (Moscow: Nauka)

STUDY OF THE EFFECT OF LOW-TEMPERATURE PLASMA BETWEEN ELECTROLYTIC ELECTRODES ON POLYMER SURFACES

R R Kayumov[1], A A Nizameev[1], R R Mardanov[1], M F Akhatov[1]

[1]Kazan National Research Technical University named after A.N. Tupolev -KAI, Kazan, Russia

rushan_250189033@mail.ru

Abstract. In the modern world, polymers are used in almost all areas. Modification of polymer surfaces is an urgent task. Increasing adhesion and wettability is one of the important parameters. There are many different methods for increasing adhesion and wettability. The electrolytic plasma method is one of them. After processing the surface of the polymer by this method, the wettability of the surface of modified polymers with water was studied by measuring the size of a drop of technical water on the surface of the polymer. As a result of experimental studies, the modification of polyethylene terephthalate decreased the wetting angle by 11-24%.

1. Introduction

The modification of polymers to solve various problems is becoming an increasingly popular process: an increase quantity of composites in all types of industry, promotional products, etc. requires appropriate materials [1]. Increasing the adhesion and wettability of plastics is one example of modifying the surface of polymers.

2. Main part

At the moment, there are many methods for changing the surface of polymers, for example chemical etching with compounds, treatment with corona discharge, barrier discharge [2] and glow discharge [3]. A method of modification with electrolytic electrodes stands out among the methods of processing with low-temperature plasma [4-6]. The essence of the method is that an electrolytic jet-cathode is poured into an electrolytic bath-anode, at the interface of which a discharge burns. The polymer on the surface of the bath is exposed to low-temperature plasma. This changes the characteristics of its upper layer. This method is good because it has no toxic waste and works at atmospheric pressure. After processing the surface of the polymer with this type of discharge (treatment time 5 s), the wettability of the surface of modified polymers with water was studied by measuring the size of a drop of technical water on the surface of the polymer. The cosine of the wettability angle was calculated from the results of drop size measurements. The data are given in Table 1.

Content from this work may be used under the terms of the Creative Commons Attribution 3.0 licence. Any further distribution of this work must maintain attribution to the author(s) and the title of the work, journal citation and DOI.
Published under licence by IOP Publishing Ltd

Table 1. Wetting angles of samples

Polymer	Wetting angle of the original sample	Wettability angle after processing in the first mode of the power supply	Wettability angle after processing in the second mode of the power supply
Polyethylene	80,8	72,1	68,1
Polyethylene terephthalate	76,8	68,7	58,6

Based on the results obtained, it can be assumed that the modified polyethylene has a lower wettability angle by 11-15%. The difference between 1 and 2 modes of the power supply is not significant. Modification of polyethylene terephthalate reduced the contact angle by 11-24%. Here the second mode is much more effective. The applied drops of water adhere much better to the modified surface.

The modified samples were photographed with a microscope at 1000x magnification. Photos are shown below:

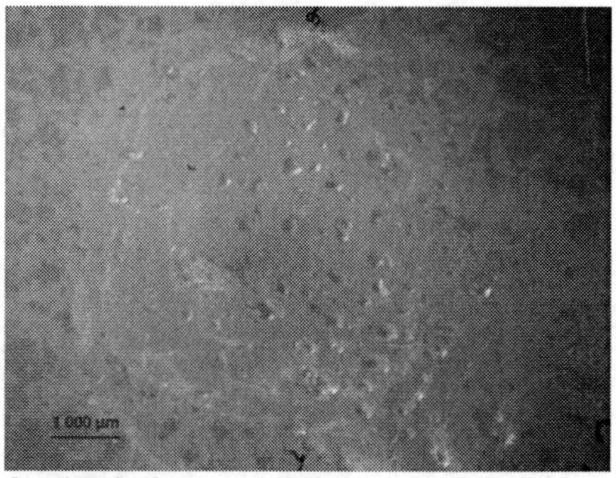

Figure 1 – photo of polyethylene modified in the second mode of power supply

The photograph shows that the main part of the polyethylene has become more homogeneous, but at the same time, areas of individual bulges and lines of irregularities have appeared. This is due to the unevenness of the plasma and electrolyte jet.

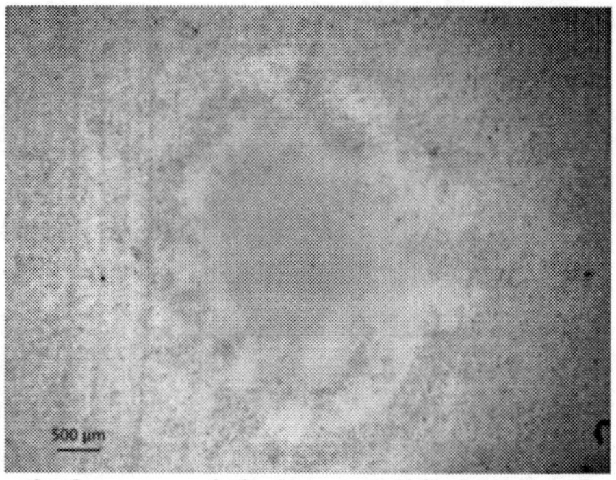

Figure 2 - photo of polyethylene terephthalate modified in the first mode of power supply

It can also be seen here that medium-sized depressions disappear, but the surface becomes covered with microcracks and becomes matte. As a result, the transparency of the material deteriorates.

In the second mode, the effect of plasma on the PET surface is more pronounced: large irregularities appear due to the melting of the polymer by hot electrolyte areas, and the number of small degradations (cracks) increases. The transparency of the polymer is greatly reduced. This is especially pronounced in polyethylene terephthalate: areas of plastic melting and microcracks appear, the surface becomes matte.

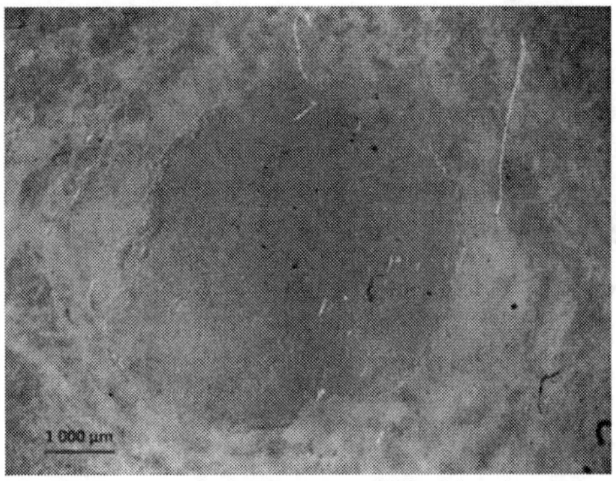

Figure 3 - photo of polyethylene terephthalate modified in the second mode of power supply

Low-temperature plasma can be used for the synthesis of various nanostructures and is well suited for the modification of various surfaces. This is shown in many works [7-9].

The study was supported by a grant from the Russian Science Foundation (project No. 22-29-00021)

References

[1] Production of plastics worldwide from 1950 to 2019 (in million metric tons). – URL: https://www.statista.com/statistics/282732/global-production-of-plastics-since-1950/

[2] Mardanov R.R., Kayumov R.R., Akhatov M.F., Gaisin A.F. A review of use of barrier discharge for modification of surface of polymers // Journal of Physics: Conference Series. – 2020. v. 1588., 012029.

[3] Šira M., Trunec D., Stahel P., Buršíková V., Navrátil Z., Buršík J. Surface modification of polyethylene and polypropylene in atmospheric pressure glow discharge // Journal of Physics D: Applied Physics. – 2005. – 38. – P. 621-627.

[4] Gaisin Al.F., Zheltukhin V.S., Yu Chebakova V.//Ring shaped plasma structures in radio-frequency discharge between liquid jet electrodes//Journal of Physics: Conference Series. Scientific Technical Conference on Low Temperature Plasma During the Deposition of Functional Coatings. 2019. C. 012034

[5] Gaisin A.F., Kuputdinova A.I., Mukhametov R.A., Kashapov N.F.//Discharge between liquid jet and metallic electrodes/ Technical Physics. 2018. T. 63. № 5. C. 695-699

[6] Mirkhanov D.N., Gaisina A.I., Korneev A.V., Gaisin Al.F., Kashapov N.F.// Jet effects of homogeneous and multiphase media under conditions of combustion of gas-steam discharges Journal of Physics: Conference Series. Scientific Technical Conference on Low Temperature Plasma During the Deposition of Functional Coatings. 2019. C. 012018

[7] Saifutdinov, A. I., & Sofronitskii, A. O. (2021). Numerical Study of Breakdown and Formation Dynamics of Arc Discharge Plasma Parameters at Ultrahigh Pressures. High Energy Chemistry, 55(3), 228-232.DOI: 10.1134/S0018143921030115

[8] Shemakhin, A. Y., Zheltukhin, V. S., Shemakhin, E. Y., Terentev, T. N., & Sofronitsky, A. O. (2020, July). Experimental installation to study the RF plasma flow at low pressures with experiment data synchronization. In Journal of Physics: Conference Series (Vol. 1588, No. 1, p. 012018). IOP Publishing. DOI: 10.1088/1742-6596/1588/1/012018

[9] Fairushin, I. I., Saifutdinov, A. I., & Sofronitskiy, A. O. (2020). Numerical and Experimental Studies of the Synthesis of Copper Nanoparticles in a High-Pressure Discharge. High Energy Chemistry, 54(2), 150-153. DOI: 10.1134/S0018143920020071

PLASMOCHEMICAL METHOD OF POLYANILINE FILM OBTAINED

M P Danilaev[1], V A Kuklin[1, 2], I N Sidorov[1]

[1]Kazan National Research Technical University named after A. N. Tupolev - KAI, Kazan, Russia
[2]Kazan Federal University, Kazan, Russia

danilaev@mail.ru

Abstract. The results of the polymer film investigation are considered in that paper. These films are obtained in atmospheric pressure barrier discharge from aniline with simultaneous generation of carbon particles in the films. It is demonstrated that the structure of these films corresponds to one of the amorphous polymers. Carbon particles produce agglomerates with a concentration that slightly depends on the discharge energy density. The authors show that the conductivity of aniline films with carbon particles mainly depends on water saturation.

1. Introduction

A variety of properties that conductive polyaniline (PANI) films demonstrate, from electric conductivity and magnetization to photoelectricity and gas permeability, define a wide range of their practical application [1, 2]. PANI films synthesis technique based on electrochemical polymerization of aniline with doping is the most widespread one, as it is the method that ensures the minimum impurities in the final material [1 - 3]. At the same time, there are two issues related to the reproducibility of PANI properties and treatment of powder being synthesized as a result of oxidative polymerization of aniline. In particular, functionalization of the initial monomer solves these two problems [2, 4]. There are other ways to obtain conductive PANI film for practical application that are based either on the galvanic deposition of aniline monomer directly on electrodes, or vacuum deposition methods using a Knudsen effusion cell [4]. Plasma enhanced chemical vapor deposition (PECVD) at pressures up to 100 mTorr significantly simplifies conductive PANI film synthesis [2]. The transition to atmospheric-pressure plasma, using a barrier discharge, makes it possible to obtain a polymer film, generating carbon particles simultaneously [5]. It increases the power conversion efficiency of photovoltaic cells and conductivity [1, 6]. The application of conductive PANI demonstrates how the environment affects the durability of properties associated with a change in conductivity [3, 4]. Taking into account studies [7], it is possible to assume that the change in conductivity is associated with the doping of PANI molecules with hydroxyl groups from the environment.

This paper presents the results of the study on how the conditions of PANI film synthesis in atmospheric-pressure plasma of a barrier corona discharge affect its morphology.

Content from this work may be used under the terms of the Creative Commons Attribution 3.0 licence. Any further distribution of this work must maintain attribution to the author(s) and the title of the work, journal citation and DOI.
Published under licence by IOP Publishing Ltd

2. Methods

PANI film was synthesized from aniline vapor over argon with simultaneous carbon particles generation in AC barrier corona discharge using a method described in [3,5,8] on a slide (SP-7102 glass, 1mm thick). It was then polymerized in a vacuum (the pressure up to 10^{-1} Torr) at 20-25°C after preliminary 4-hour exposure at 95±5°C. There was no nitrogen, oxygen, or moisture contained in the atmosphere, so there were no redox reactions during PANI film polymerization. Four types of PANI film samples were obtained in the course of experiments at the following densities of the energy released in the streamer: sample no. 1 – 10±2 mJ/m; sample no. 2 – 20±2 mJ/m; sample no. 3 – 30±2 mJ/m; sample no. 4 – 40±2 mJ/m.

The surface morphology of the film samples after polymerization was studied by scanning electron microscopy using a Zeiss Libra 120 electron microscope with a built-in OMEGA filter. Carbon particles generated in PANI film were identified by transmission electron microscopy using the same Zeiss Libra 120 electron microscope with a built-in OMEGA filter at the following conditions: accelerating voltage of 120 kV, LaB6 electron source. For that purpose, a polymer film was dissolved in acetone (99,9%) to obtain a suspension. After that, the solution was exposed to an ultrasonic disperser for up to 10 minutes at the frequency of 44 kHz and ultrasonic power of ~10 W. Then, 2µl of the solution was applied onto the 3mm copper mesh with Formvar/Carbon coating (Ted Pella, Inc.).

The color of the films helps to distinguish PANI molecules structures (leukoemeraldine, emeraldine, pernigraniline) that ensure different conductivity [3]. That is why conductivity absorption spectra were identified using a StellarNet EPP2000 spectrometer with a resolution of 0.5nm when studying the dynamics of PANI film.

3. Results and discussion

The surface morphology of PANI film corresponds to one of the amorphous polymers (Fig. 1). Carbon agglomerates of ~1µm are evenly distributed along the film area. It should be noted that carbon agglomerates concentration slightly increases with an increase in the energy input of the discharge and is estimated at $(4 \div 7) \cdot 10^4$ 1/cm^2. The growth rate is $\sim 2 \cdot 10^3$ particles/cm^2·mJ.

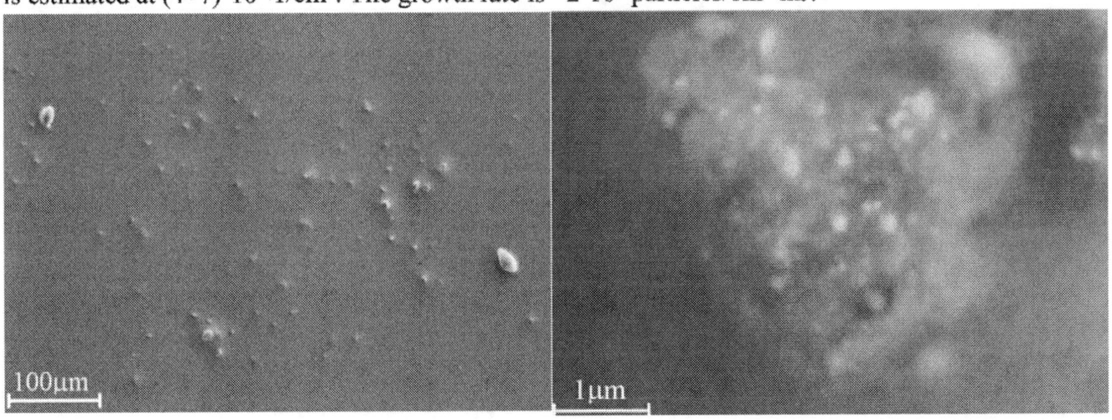

Figure 1. PANI film surface morphology – sample no. 3.

The type of the particles is identified by transmission electron microscopy based on the analysis of diffraction patterns (Fig. 2 [5,9,10].

Journal of Physics: Conference Series **2270** (2022) 012018 doi:10.1088/1742-6596/2270/1/012018

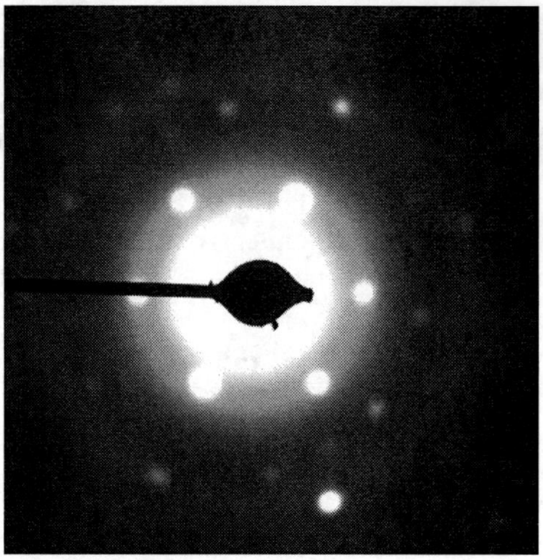

Figure 2. Typical diffraction pattern on carbon particles.

The halo (Fig. 2) indicate that agglomerates contain carbon particles with an amorphous structure. Moreover, the agglomerates contain the carbon particles with crystalline fractions, that indicated by the reflections correspond to the hexagonal structure: $d_{100}/d_{110} \approx 1.73$ [11]. It should be noted that the percentage of crystalline particles increases with an increase in the energy input into the discharge at a rate close to the rate of growth in the number of agglomerates themselves. The increase in the percentage of crystalline carbon particles is due to the temperature increase in corona hood and streamers [5, 12].

The absorption spectrum of the PANI film after polymerization corresponds to the leukoemeraldine form of polyaniline (Fig. 3): sample no. 1 – (1), sample no. 2 – (2), sample no. 3 – (3), sample no. 4 – (4) [8].

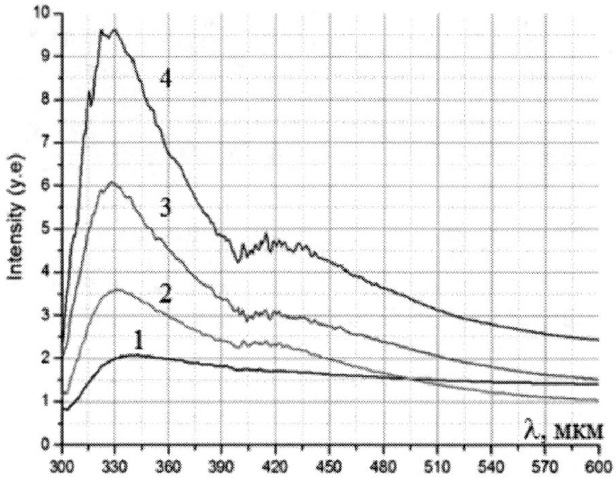

Figure 3. Film samples absorption spectra.

This is characterized by the absence of an absorption band in the long-wavelength part of the spectrum, which is determined by the oxidized block in the polyaniline unit [13]. Atmospheric impact on PANI film leads to an increase of absorption at the wavelength of 500 nm and up, which indicates the oxidation of a part of the reduced blocks and the transition of leukoemeraldine into emeraldine [7]. The reason is the interaction of PANI molecules with hydroxyl groups and the water saturation of the film with atmospheric moisture. Oxidation of leukoemeraldine into emeraldine is apparently carried out according to the scheme illustrated in Fig. 4 [7]: leukoemeraldine-conductive emeraldine oxidation scheme in the form of two polarons – conductive emeraldine in the form of a polaron – non-conductive emeraldine in the form of a base.

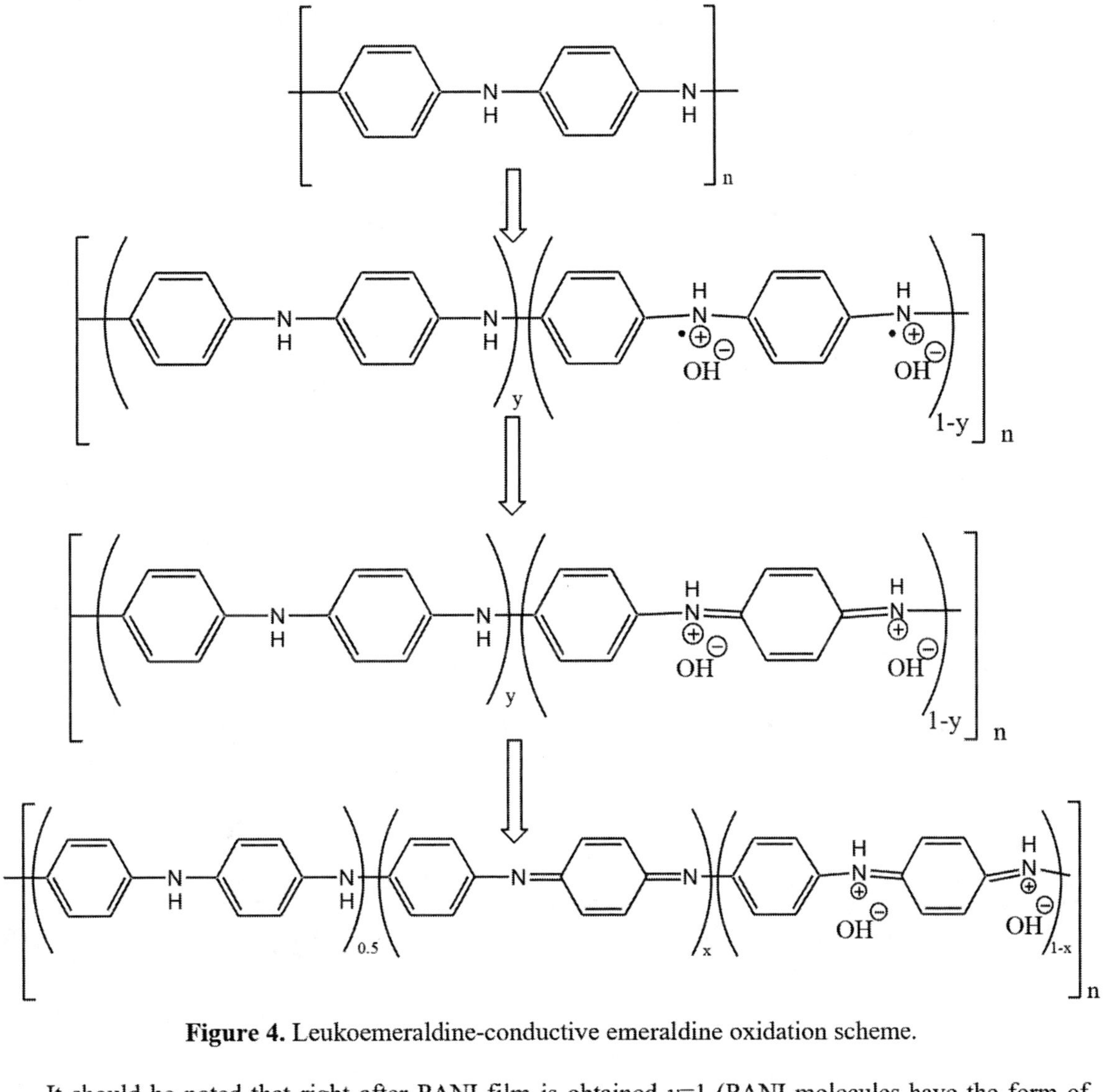

Figure 4. Leukoemeraldine-conductive emeraldine oxidation scheme.

It should be noted that right after PANI film is obtained *y*=1 (PANI molecules have the form of leukoemeraldine) because the film was polymerized at the pressure of up to 10^{-1} Torr. Oxidation is initiated after the film has been withdrawn from the vacuum chamber and saturated with atmospheric hydroxyl groups [14]. During oxidation, first, the *y* decreases, and the *x* increases [7]. The absolute value of the conductivity is determined, on our opinion, by the number of doped blocks of PANI polymer molecules, which depends on the humidity of the atmosphere, as well as the permeability of the film itself. It should also be noted that a separate absorption band in the region of 980 nm (Fig. 3) can be associated with an increase in the number of OH-groups absorbed from the atmosphere [14].

4. Conclusion

Thus, plasma-enhanced chemical deposition from aniline vapor provides a possibility to obtain homogenous films without granular fractions. A percentage of allotropic crystalline carbon particles increases with an increase in the density of discharge energy which is associated with temperature increase in the corona streamers and hood. It is established that the surface structure of the films corresponds to the one specific to amorphous polymers. Moreover, the coating continuity is achieved at energy densities above ~35 J/m, and the average film growth rate changes from ~1.2 ± 0.2 μm/min (at ~25 mJ/m) to ~3 ± 0.5 μm/min (at ~45 J/m).

References

[1] Mezhuev Ya O et al 2011 *Plastics* **3** 25-30 (in Russian)
[2] Ndiaye A A et al 2018 *Plasma Chem. Plasma Process.* **38(4)** 887-902
[3] Boeva Zh A and Sergeev V G 2014 *Polym Sci Ser C* **56(1)** 153-164
[4] Salikhov R B, Biglova Yu N, Yumaguzin Yu M, Salikhov T R, Miftakhov M S, and Mustafin A G 2013 *Tech Phys Let* **39(19)** 25-31
[5] Danilaev M P et al. 2020 *Plasma Process Polym* **17(4)** 1900204
[6] Wang R X, Huang L F, and Tian X Y 2012 *J Phys Chem C* **116(24)** 13120-13126
[7] Wang J 2002 *Synth Met* **132** 49-52
[8] Danilaev M P et al. 2022. *Physics and chemistry of material treatment* **1** (in press) (in Russian)
[9] Hirahara K, Bandow S, Suenaga K, Kato H, Okazaki T, Shinohara H, and Iijima S 2001 *Phys rev B* **64** 115420-1 – 115420-5
[10] Colomer J-F, Henrard L, Launois P, Tendeloo G Van, Lucas A A, and Lambin Ph 2004 *Phys rev B* **70** 075408-1 – 075408-5
[11] Müller U 2013 *Symmetry Relationships between Crystal Structures. Applications of Crystallographic Group Theory in Crystal Chemistry* (Oxford University Press) p 332
[12] Bogoslov E A, Danilaev M P, Drobyshev S V, and Kuklin V A 2020 *J Eng Phys and Thermophys* **93(6)** 1591-1597.
[13] Kompan M E, Sapurina I Y, Babayan V, and Kazantseva N E 2012 *Phys Solid State* **54(12)** 2400-2406
[14] Zuev V E, Krekov G M 1986 *Optical Models of the Atmosphere* (Leningrad: Gidrometeoizdat) p 256

Acknowledgments

The research is carried out under the financial support of the Ministry of Science and Higher Education of Russia Reg. No.AAAA-A20-120102190039-6. The authors thank the Applied Nanotechnology Multiple Access Center at KNRTU-KAI for the images of PANI films surfaces and electron diffraction on agglomerates of carbon particles (the work is carried out under the Priorities-2030 program) and M.S. Pudovkina personally for the measurement of absorption spectra of PANI films.

Creation of plasma flow from a liquid electrolyte cathode with an arbitrary orientation in space

G K Tazmeev[1], K K Tazmeev[1] and B K Tazmeev[2]

[1]Kazan Federal University, Naberezhnye Chelny Institute, Naberezhnye Chelny, Russian Federation
[2]Kuban State Agrarian University named after I.T. Trubilin, Krasnodar, Russian Federation

E-mail: GHTazmeev@kpfu.ru

Abstract. The gas discharge between a metal electrode and a liquid electrolyte supplied through a porous refractory material has been experimentally studied. The studies were carried out in the current range of 5–15 A. Aqueous solutions of sodium chloride were used as a liquid electrolyte. The specific electrical conductivity of the solutions was in the range of 6–18 mS/cm. The conditions necessary for the formation of a plasma flow are revealed. Plasma flows oriented in different directions in space have been created. The energy and spectral properties of the gas discharge are investigated.

1. Introduction

Gas discharges with a liquid electrolyte cathode can be used to create intense plasma flows. As experiments show, in the current range of 10–20 A, flows with a mass velocity in the range of 1–2 g/s can be obtained. In terms of energy parameters, such plasma flows are comparable to electric arc plasma flows [1]. They are quite acceptable for energy-intensive technologies, in particular, for the gasification of waste polymer materials. From the open surface of the liquid electrolyte cathode, the plasma flow goes vertically upward. This is a significant drawback for the plasma source. In practice, plasma flows directed not only vertically but also in other directions are required. One of the options for solving the problem is to change the spatial orientation of the "liquid electrolyte cathode - solid-state anode" system. In this version, experimental studies were carried out in [2]. The authors have developed a cathode assembly using porous refractory material. The electrolyte was fed into the internal cavity of the cathode assembly and partially entered through the porous wall into the discharge gap. Such a cathode assembly was called a "porous electrolyte cathode", abbreviated as PEC. With the help of PEC, discharges with a horizontal orientation of the plasma column were obtained. The experiments were carried out in the current range of 1–4 A. The power deposited in the discharge did not exceed 2 kW. Naturally, with such relatively small energy parameters, it is impossible to form intense plasma flows. This requires more powerful discharges. In this regard, the purpose of this work was to continue research at increased currents and powers.

2. Experiment

Figure 1 shows a diagram of the experimental setup. The gas discharge was ignited between metal electrode 1 and liquid electrolyte, which was supplied through porous insert 2 of electrode assembly 3.

Content from this work may be used under the terms of the Creative Commons Attribution 3.0 licence. Any further distribution of this work must maintain attribution to the author(s) and the title of the work, journal citation and DOI.
Published under licence by IOP Publishing Ltd

A detailed description of electrode assembly 2 is given in [3]. Structural elements 2 and 3, assembled together, formed a porous electrolyte cathode (PEC). Electrode 1 was made of a copper rod with a diameter of 25 mm and served as an anode. It was cooled with water. The PEC and the anode were located at different positions in space. In the figure, both are shown in a horizontal position. The interelectrode distance between electrodes *l* varied between 3-10 cm. Aqueous solutions of sodium chloride served as a liquid electrolyte. The specific electrical conductivity of the solutions σ was in the range of 6–18 mS/cm.

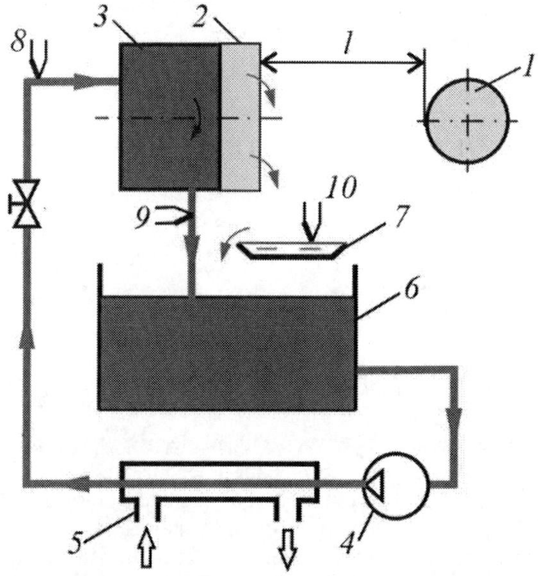

Figure 1. Scheme of the experimental setup.

The electrolyte was supplied by hydraulic pump 4 through heat exchanger 5. From the cathode assembly, the electrolyte flowed into reservoir 6. A part of the electrolyte leaving through porous insert 2 was used to form a plasma flow. The excess electrolyte flowed into ditch 7 and then also into tank 6. The electrolyte temperature was measured using thermocouples 8, 9 and 10, which were placed as shown in figure 1. Heat expended to heat the electrolyte was determined by calorimetry. The electrolyte loss rate in the hydraulic system was assumed to be numerically equal to the plasma flow mass velocity *G*. A three-phase full-wave rectifier with an output voltage of 1700 V served as a power source. Voltage ripples were smoothed out by a capacitive-inductive filter. Current was regulated using a ballast resistor. The current and voltage were measured with pointer instruments of accuracy class 0.2 and recorded with an AKIP-15/1 two-channel oscilloscope.

The photography was carried out with a high-speed video camera VIDIOSCAN. The emission spectra were recorded with an AvaSpec-3648 high-speed fiber-optic spectrometer with an optical resolution of 0.15 nm.

3. Experimental results

Experiments have shown that plasma flows can be created under certain conditions. It was found that the discharge current *I* should be at least 6–7 A, and the electrolyte flow rate *m* through the porous insert should be within 4–10 g/s. When these conditions were met, plasma flows were created in almost all possible orientations in space. Figure 2 shows instantaneous photographs showing plasma flows oriented horizontally and at an angle downward.

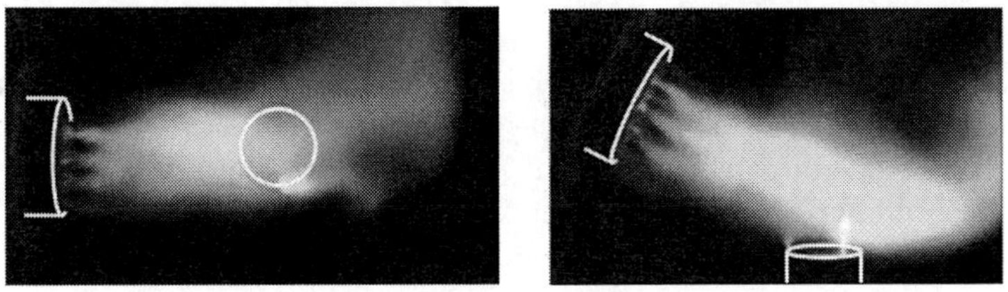

Figure 2. Instant photos of discharges. Exposure 0.2 ms. I = 8 A. l = 6 cm. The white lines show the contours of the ends of the PEC and metal anode.

Table 1 presents some results of experimental studies of gas discharge and plasma flow.

Table 1. Energy characteristics of gas discharge and plasma flow.

	I, A	m, g/s	G, g/s	N, kW	Q_c/N
1	7.8	4.0	0.9	11.9	0.09
2	8.5	6.8	1.0	13.3	0.08
3	13.0	8.8	1.2	20.8	0.12

Here, N is the electrical power supplied to the discharge device; Q_c – is the power of heat losses through the PEC, i.e., the power spent on heating the electrolyte. Note that the electrolyte heating losses are relatively small.

Figure 3 shows the oscillograms of the voltage and current of the discharge. As can be seen, ripples occur at elevated currents. The most probable cause of such current pulsations is the uneven flow of the electrolyte substance into the gas discharge.

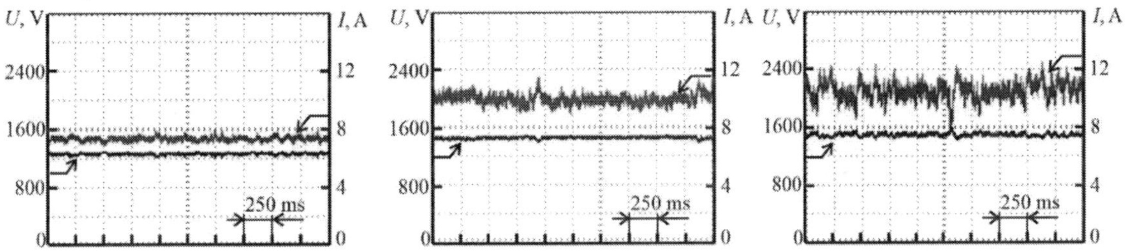

Figure 3. Current and voltage. l = 5 cm. m = 5.8 g/s. σ = 10 mS/cm.

In the experiments, the radiation spectrum of various sections of the plasma column was studied. Figure 4 shows the emission spectrum of the near-cathode zone. Spectral lines of hydrogen were registered in this zone. They were not found in other areas.

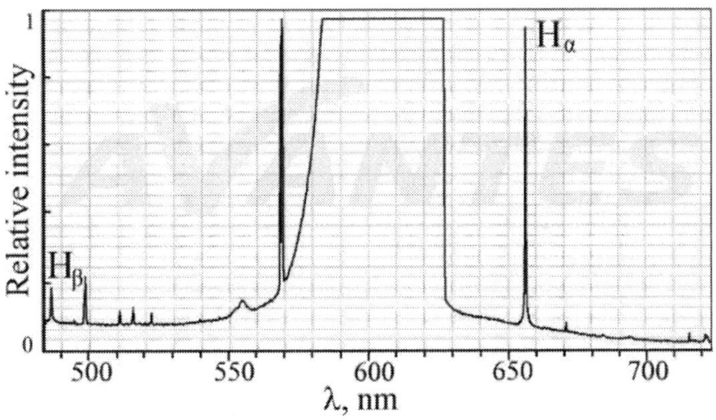

Figure 4. The emission spectrum of the plasma near the cathode. I = 8.8 A. σ = 6 mS/cm.

The presence of spectral lines of hydrogen in the radiation near the cathode zone is a characteristic feature of a gas discharge with a liquid electrolyte cathode. It manifests itself at both low and high currents. The thickness of the near-cathode plasma layer, in which hydrogen atoms emit, is ~ 2 mm.

Table 2 shows the results of calculations of the temperature T_e and density n_e of the electron gas, performed using spectral lines of hydrogen. Details of the calculation procedure are considered in [4]. The results of the calculation agree with the data from the work [4-5].

Table 2. Temperature and concentration of electron gas.

	T_e (K)	$\Delta_{\lambda/2}(H_\alpha)$ (нм)	$\Delta_{\lambda/2}(H_\beta)$ (нм)	n_e (10^{15} см$^{-3}$)
1	5180	0.394	0.380	2.62
2	5100	0.365	0.379	2.61
3	4900	0.341	0.364	1.82
4	4700	0.335	0.363	1.82

4. Conclusions

Plasma flows with different orientations in space have been created. It was found that the most important parameter influencing the formation of the plasma flow is the electrolyte flow through the porous insert on the cathode assembly. It has been established that the porous insert practically does not affect the temperature and concentration of the electron gas in the near-cathode zone.

Acknowledgments

This work is supported by the Russian Federation President's grant for the state support of young Russian scientists - MK-111.2022.1.2.

References

[1] Dautov G Y, Timoshevskii A N, Urjukov B A et al. 2004 *Generacija nizkotemperaturnoj plazmy i plazmennye tehnologii: Problemy i perspektivy* (Novosibirsk: Nauka) p 345
[2] Tazmeev G K, Tazmeeva R N and Tazmeev B K 2020 *J. Phys.: Conf. Ser.* **1588** 012050
[3] Tazmeev G K, Timerkaev B A and Tazmeev K K 2019 *J. Phys.: Conf. Ser.* **1328** 012075
[4] Bruggeman P et al. 2008 *Plasma Sources Science and Technology* **17**(2) 025012
[5] Petrov A Y, Kalbenin D A, Titov V A et al. 2011 *Goreniye i plazmokhimiya* **9**(3) pp 160-168

INVESTIGATION OF VOLT-FARAD CHARACTERISTICS OF THIN MULTIFERRROIC FILMS

Q B Saidzoda[1], A A Semenov[2], M M Safarov[3]

[1]Tajik State Pedagogical University named after S. Aini, Dushanbe, Tajikistan
[2]St.Petersburg State Electrotechnical Univer-sity "LETI" named after Ulyanova (Lenina), St.Petersburg, Russia
[3]Tajik technical university after named by academic M.C.Osimi, Dushanbe, Tajikistan

mahmad1@list.ru

Abstract. In this work, the influence of the concentration of manganese ions in BST films on the capacitance–voltage characteristics (CV characteristics, I–V characteristics, tgδ) of the Cu-Cr/BST/α-Al2O3 and Cu-Cr/BST/GGG structures was studied. To measure the electrophysical characteristics, multiferroic thin-film the samples were made in the form of planar capacitors. Measurements of the electrical properties of the formed planar structures were carried out at a frequency of 1 MHz.

1. Experimental section

Measurements of conduction currents, current-voltage characteristics (CVC) were carried out using a V7-30 electrometer in the range of 10-13-10-5A. The limits of permissible values of the main current measurement error did not exceed 10%. It has been found that the lowest stresses occur in BSTO films grown on polycor substrates and on YIG films (Figs. 1–7, Table 1).

Another method of formation of a medium with multiferroic properties investigated in this work is associated with the introduction of magnetic elements into materials with ferroelectric properties. In this work, alloying of a solid solution of barium titanate, strontium, and manganese in high concentrations was studied [1–6]. The results of the study of capacitance-voltage characteristics are shown in Figures 1 - 7.

Content from this work may be used under the terms of the Creative Commons Attribution 3.0 licence. Any further distribution of this work must maintain attribution to the author(s) and the title of the work, journal citation and DOI.
Published under licence by IOP Publishing Ltd

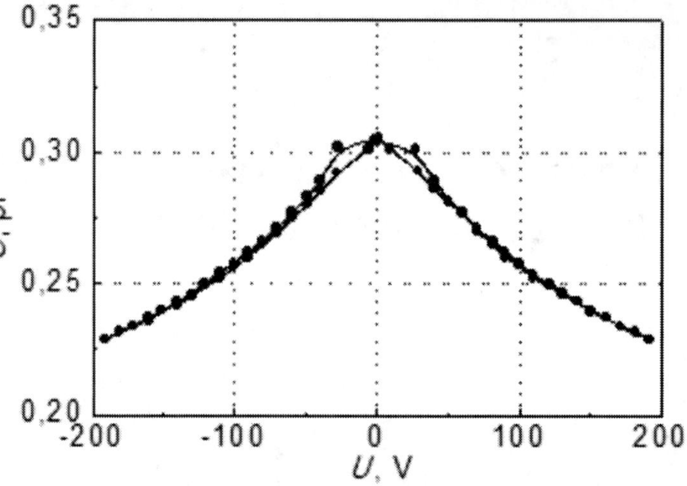

Figure 1. Capacitance-voltage characteristic of the Cu-Cr/BST/α-Al_2O_3 structure, manganese content 10%

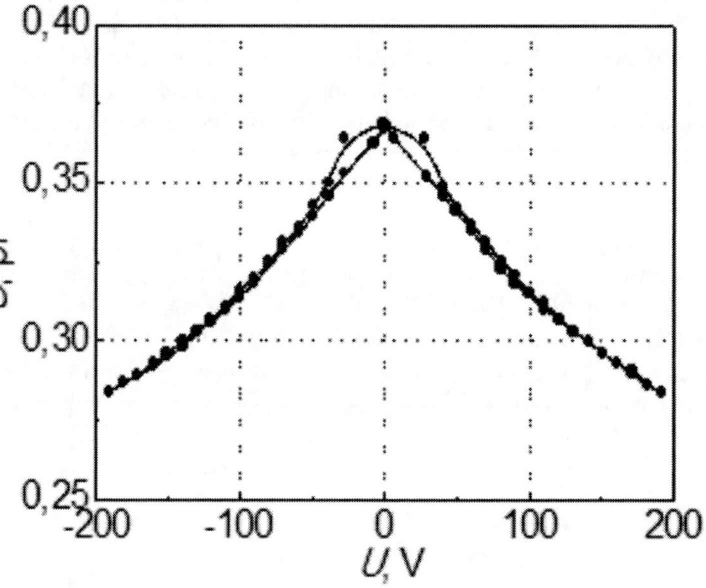

Figure 2. Capacitance-voltage characteristic of the Cu-Cr/BST/GGG structure, manganese content 10%

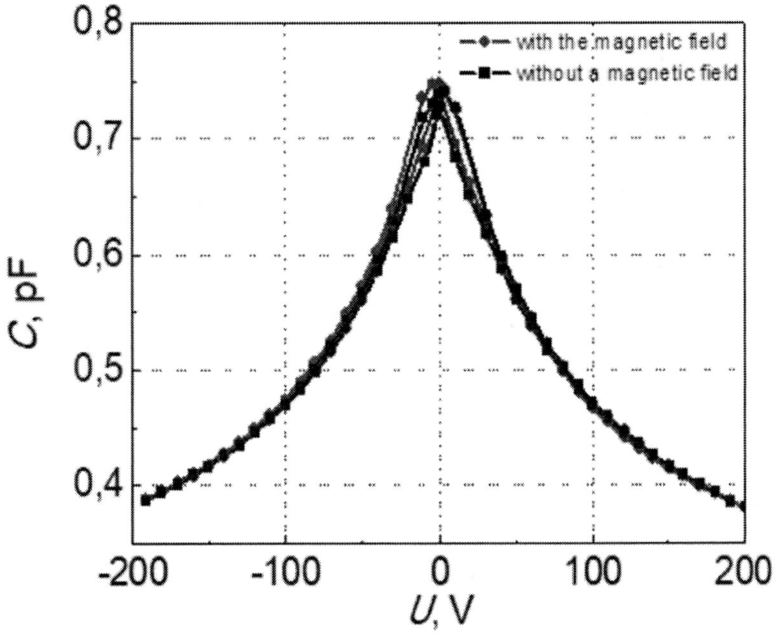

Figure 3. Capacitance-voltage characteristic of the Cu-Cr/BST/GGG structure, manganese content 15%

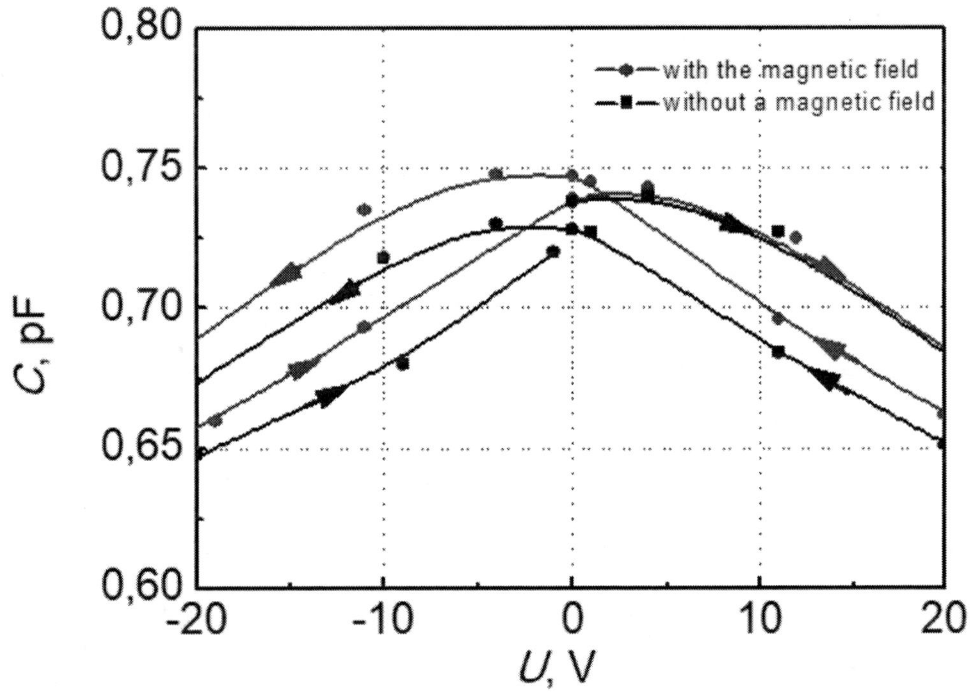

Figure 4. Volt-farad characteristic of Cu-Cr/BST/GGG structure, manganese content 15%.

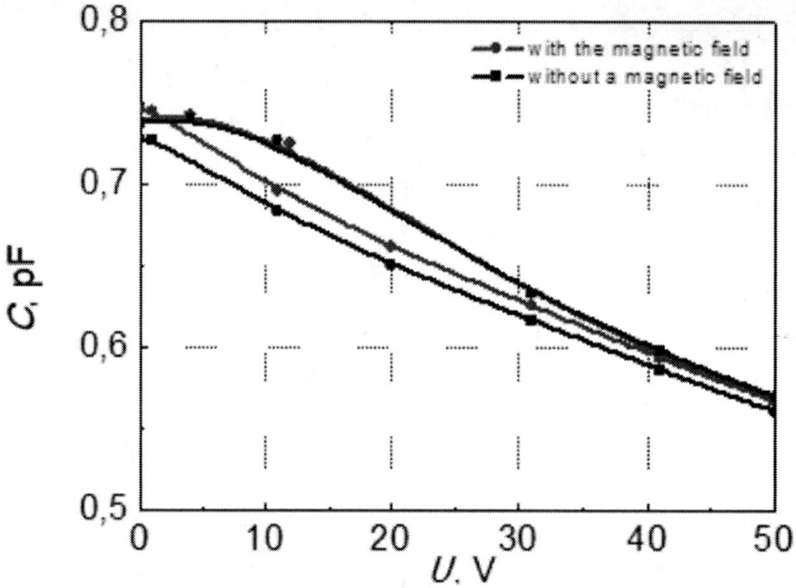

Figure 5. Volt-farad characteristic of Cu-Cr/BST/α-Al_2O_3 structure, manganese content 15%

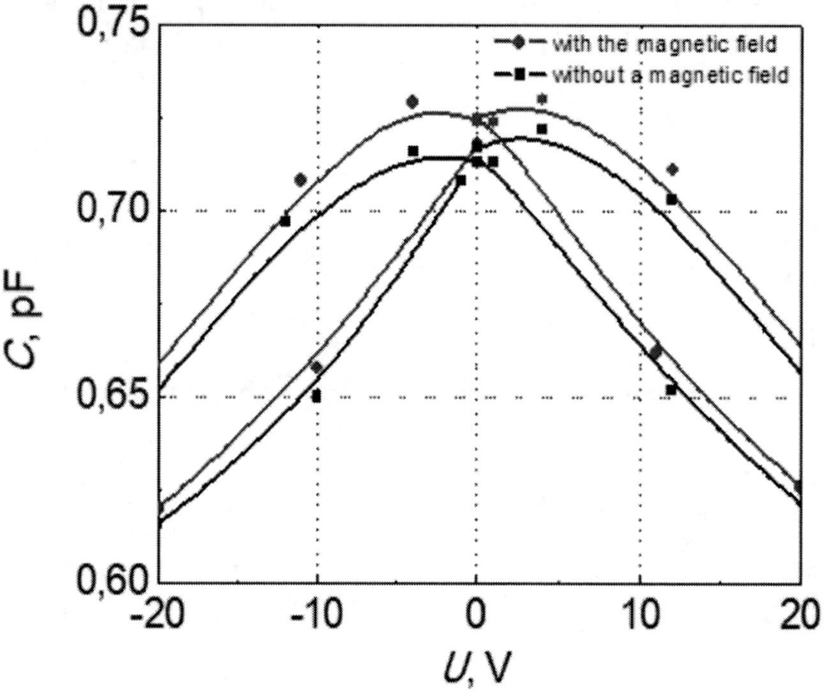

Figure 6. Volt-farad characteristic of Cu-Cr/BST/GGG structure, manganese content 15%

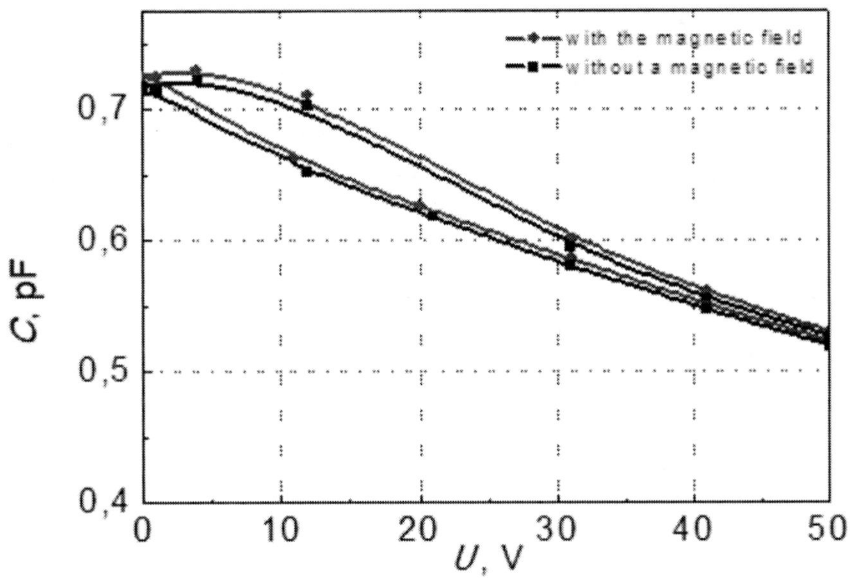

Figure 7. Volt-farad characteristic of Cu-Cr/BST/α-Al_2O_3 structure, manganese content 15%

2. Conclusion

The discussion of the results. When BST is doped with 10% manganese, the capacitance-voltage characteristics have the form of a Gaussian curve (Fig. 1 - 3). Hysteresis manifests itself at low electric field strengths, at high intensities the capacitance-voltage characteristics become unambiguous. The temperature depen-dences show a transition to the paraelectric state, which corresponds to the Curie-Weiess theory (Fig. 5-7). The magnetic field did not affect the capacitance–voltage characteristics of such samples.

Low-temperature plasma can be used for the synthesis of various nanostructures and is well suited for the modification of various surfaces. This is shown in many works [7-19].

References

[1] Belov K.P. Physics of new ferromagnetic substances /K.P. Belov // Bulletin of the Academy of Sciences of the USSR. –1961. -WITH. 5-33.

[2] Vendik, O. G. Ferroelectrics in microwave technology / O. G. Vendik [et al.]; resp. ed. O. G. Vendik; -M.: Soviet Radio, 1979. -272 p.

[3] Subramanyam, G. Microwave Dielectric Properties of Mn:BSTO and PST Thin-Films/ G. Subramanyam, C. Chen, S. Dey// Integrated Ferroelectrics. – Vol.77. - 2005. - P. 189-197.

[4] O. G. Vendik, L. T. Ter-Martirosyan, "Ubungakanani besiphumo kwi-ferroelec-tric layer," i-Zh. - 1996. - T. 66, N 4. - P. 92-98.

[5] A. I. Dedyk, A. I. Dedyk, S. F. Karmanenko, M. N. Malyshev, kunye noL. T. Ter-Martirosyan, "Uphando lwe-dielectric hysteresis kwizakhiwo ezininzi ezis-ekelwe kwi-strontium titanate," iPhys. - 1995. - T. 37. - N 11. - P. 3470-3476.

[6] I-Pronin, I. P. I-self-polarization kunye nepolarization efudukayo kwiifilimu ezincinci ze-lead zirconate-titanate / I. P. Pronin, E. Yu. A. V. Pankrashkin // FTT. - 2002. - T. 44. - V. 4. - P. 739-744.

[7] Saifutdinov, A. I., & Sofronitskii, A. O. (2021). Numerical Study of Breakdown and Formation Dynamics of Arc Discharge Plasma Parameters at Ultrahigh Pressures. High Energy Chemistry, 55(3), 228-232.DOI: 10.1134/S0018143921030115

[8] Shemakhin, A. Y., Zheltukhin, V. S., Shemakhin, E. Y., Terentev, T. N., & Sofronitsky, A. O. (2020, July). Experimental installation to study the RF plasma flow at low pressures with experiment data synchronization. In Journal of Physics: Conference Series (Vol. 1588, No. 1, p. 012018). IOP Publishing. DOI: 10.1088/1742-6596/1588/1/012018

[9] Saifutdinova, A. A., Sofronitskiy, A. O., Timerkaev, B. A., & Saifutdinov, A. I. (2020). Plasma-Chemical Decomposition of Hydrocarbons on the Basis of the Micro-Arc Discharge with Disc Electrodes Rotating in the Bulk of Raw Materials. Russian Physics Journal, 62(11), 2132-2136.DOI: 10.1007/s11182-020-01957-0

[10] Fairushin, I. I., Saifutdinov, A. I., & Sofronitskiy, A. O. (2020). Numerical and Experimental Studies of the Synthesis of Copper Nanoparticles in a High-Pressure Discharge. High Energy Chemistry, 54(2), 150-153. DOI: 10.1134/S0018143920020071

[11] Asadullin, T. Y., Galeev, I. G., Sofronitskiy, A. O., & Gizeev, M. M. (2019, November). Acoustic impact on electric discharge parameters during sterilization of freeze-dried products. In Journal of Physics: Conference Series (Vol. 1370, No. 1, p. 012016). IOP Publishing. DOI: 10.1088/1742-6596/1370/1/012016

[12] Fairushin, I. I., Saifutdinov, A. I., Sofronitskiy, A. O., Timerkaev, B. A., & Dautov, G. Y. (2019, October). Development of plasma reactor design for synthesis of copper nanoparticles using multi-scale simulation. In Journal of Physics: Conference Series (Vol. 1328, No. 1, p. 012088). IOP Publishing. DOI: 10.1088/1742-6596/1328/1/012088

[13] Asadullin, T. Y., Galeev, I. G., & Sofronitskiy, A. O. (2019, October). The application of pulsed discharge for sterilization of freeze-dried product. In Journal of Physics: Conference Series (Vol. 1328, No. 1, p. 012071). IOP Publishing. DOI: 10.1088/1742-6596/1328/1/012071

[14] Timerkaev, B. A., Ganieva, G. R., Kaleeva, A. A., Israfilov, Z. K., & Sofronitskii, A. O. (2019). Growing of carbon nanotubes from hydrocarbons in an arc plasma. Journal of Engineering Physics and Thermophysics, 92(5), 1248-1252. DOI: 10.1007/s10891-019-02040-3

[15] Dautov, G. Y., Kashapov, N. F., Dautov, I. G., & Sofronitskiy, A. O. (2018, July). Research of the influence of the geometry of the discharge chamber on the characteristics of the arc plasmatron. In Journal of Physics: Conference Series (Vol. 1058, No. 1). IOP Publishing. DOI: 10.1088/1742-6596/1058/1/012035

[16] Timerkaev, B. A., Andreeva, A. A., & Sofronitskiy, A. O. (2017, November). Discharge creeping along the surface in the process for producing nanomaterials. In Journal of Physics: Conference Series (Vol. 927, No. 1, p. 012068). IOP Publishing. DOI: 10.1088/1742-6596/927/1/012068

[17] Sadikov, K. G., Sofronitskiy, A. O., & Larionov, V. M. (2017, November). The effect of electrically conductive additives on the plasma pyrolysis of heavy hydrocarbons. In Journal of Physics: Conference Series (Vol. 927, No. 1, p. 012046). IOP Publishing. DOI: 10.1088/1742-6596/927/1/012046

[18] Timerkaev, B. A., Andreeva, A. A., & Sofronitskiy, A. O. (2017). Discharge creeping along the surface in the process of cleaning and strengthening of the materials surface. In Journal of Physics: Conference Series (Vol. 789, No. 1). IOP Publishing. DOI: 10.1088/1742-6596/789/1/012063

[19] Sadikov, K. G., Sofronitskiy, A. O., & Dautov, I. G. (2017). Functional plasma sprayed coatings on magnesium ceramic substrates. In Journal of Physics: Conference Series (Vol. 789, No. 1, p. 012043). IOP Publishing. DOI: 10.1088/1742-6596/789/1/012043

Features of a burning electric arc superimposed on gas discharge with liquid electrolyte cathode

K K Tazmeev and A K Tazmeev

Kazan Federal University, Naberezhnye Chelny Institute, Naberezhnye Chelny, Russian Federation

tazmeevh@mail.ru

Abstract. The conditions for burning an electric arc in a discharge plasma with a solution cathode have been studied. Both discharges were connected to the same metal anode. The electric arc burned in the current range of 5–10 A. The discharge current with a solution cathode varied within 8-15 A. In the experiments, aqueous solutions of common salt were used. The specific electrical conductivity of the solutions was 10–15 mS/cm. Spectral investigations of arc plasma radiation and discharge with a solution cathode have been carried out. Calculations of the parameters of the electron gas in the plasma column have been performed.

1. Introduction

As is known, when different types of gas discharges are superimposed on each other, combinations with their properties are formed. As a result, the possibilities for practical application of gas discharges are expanding. For example, the authors of [1] created an electric discharge for practical use. It has the properties of a microwave discharge. It has the properties of a DC gas discharge, which is a non-self-sustained discharge. The effect of such an electric discharge on the surface of specimens made of mild and hard steel has been studied. It is shown that the use of an electric discharge can significantly improve the performance properties of products made from such steels. There are examples of a combination of other types of gas discharges [2]. The purpose of this work was to study the process of burning an electric arc when it is applied to a discharge with a solution cathode.

2. Experiment

Metal electrodes 1 and 3 were used for the electric arc (figure 1). They were made from copper rods. These electrodes were cooled with water. A discharge with a solution cathode was ignited between the electrolyte flowing from electrode assembly 2 and metal electrode 3. The discharges were powered according to the scheme shown in figure 1. A C-L-C filter was connected to the rectifier output. The currents of the electric arc I_1 and the discharge with a solution cathode I_2 were regulated using resistors 4 and 5. The resistor resistances changed during the burning of the discharges. Thus, both the electric arc and the discharge with a solution cathode received electrical power from the same power source. The output voltage varied within 1200–1700 V. Metal electrode 3 served as an anode for both electrical discharges. In the experiments, the electric arc current I_1 varied within 5-10 A and the discharge current with a solution cathode I_2 was set in the range of 8–15 A.

The studies were carried out at various interelectrode distances. Metal electrodes 1 and 3 were located above the electrode assembly 2. The distance l between electrodes 1 and 3 was established within 5–30 mm. They were placed horizontally. The end face of electrode 3 was displaced from the axis of electrode

Content from this work may be used under the terms of the Creative Commons Attribution 3.0 licence. Any further distribution of this work must maintain attribution to the author(s) and the title of the work, journal citation and DOI.
Published under licence by IOP Publishing Ltd

assembly 2 by a distance of 5 mm. The height h of the location of this electrode-anode 3 above the electrode assembly 2 varied within 2–7 cm.

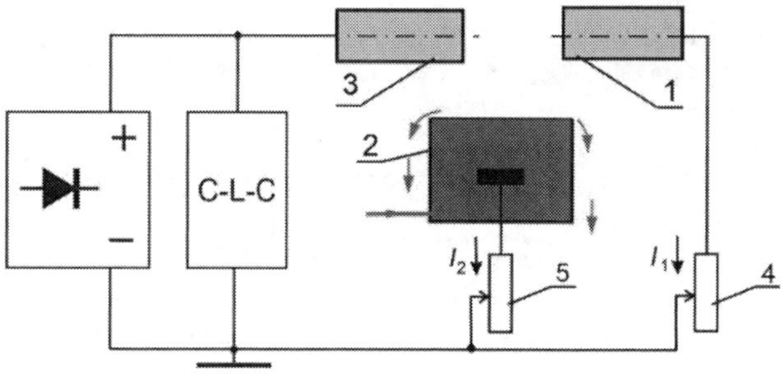

Figure 1. Electric circuit of the experimental setup.

Aqueous solutions of table salt were used as a liquid electrode. They were prepared with the same electrical conductivity σ equal to 10 mS/cm. Measurement of σ was carried out using an ANION 4150 conductometer before and after the experiments. The measurement results showed that there is an increase of σ. The aqueous solutions were reused if the electrical conductivity σ did not exceed 15 mS/cm. Currents and voltages were measured with pointer instruments of accuracy class 0.2 and also recorded with an AKIP-15/1 two-channel oscilloscope. Emission spectra were obtained with an AvaSpec-3648 spectrometer. The optical resolution was 0.15 nm. Visualization was carried out using a VIDEOSKAN-415 high-speed camera. Video frames were obtained with an exposure of 1 μs to 0.2 ms. The luminous flux was controlled by a diaphragm. Thus, photographs of the arc channel against the background of the plasma column of the discharge with a solution cathode were obtained.

3. Results of experiments

Figure 2 shows photographs of discharges in different versions. The oscillograms of currents are also presented here during the joint burning of an electric arc and a discharge with a solution cathode.

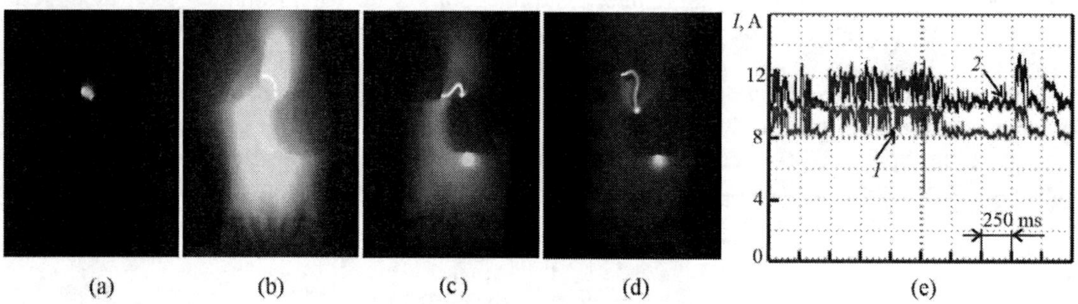

Figure 2. Instant photographs (a, b, c, d) and current oscillograms (e). l = 5 mm, h = 5.5 cm. σ = 15 mS/cm. Exposure 0.2 ms.

The first frame on the left (figure 2a) presents a variant of autonomous burning of an electric arc. As can be seen, in this embodiment, the electric arc burns in the diffusion mode. There is no contracted channel. In the second frame (figure 2b), autonomous discharge burning with a solution cathode is recorded. In this variant, the discharge also burns in the diffusion mode and forms a volumetric plasma column. The discharge plasma fills the interelectrode space between the electrodes. The third and fourth frames show options for the joint burning of an electric arc and a discharge with a solution cathode. In these frames, clear images of the contracted channel of the electric arc were recorded. Thus, the discharge plasma with a solution cathode contributes to the formation of a contracted channel in the electric arc. Under the influence of the plasma flow, this channel is blown away and lengthened. A shunting phenomenon occurs, due to which ripples of the electric arc current I_1 are formed. Such pulsations were recorded in oscillograms (figure 2e). Since the power source was the same, simultaneously with the pulsations of the electric discharge current I_1, there were pulsations of the gas discharge with a solution cathode I_2. As can be seen from the oscillograms presented in figure 2e, the large-scale ripple components of the currents I_1 and I_2 are identical to each other.

The emission spectrum of the plasma was studied in the visible region. During the joint burning of an electric arc and a discharge with a solution cathode, the spectral lines of sodium and hydrogen were recorded (figure 3). As can be seen, in the visible region of radiation, the *D* line of sodium and the Balmer lines of hydrogen H_α and H_β are the most intense. With an increase in the discharge current I_2 with a solution cathode, the intensity of the sodium *D* line increased. Therefore, the pixels of the CCD detector were saturated (figure 3).

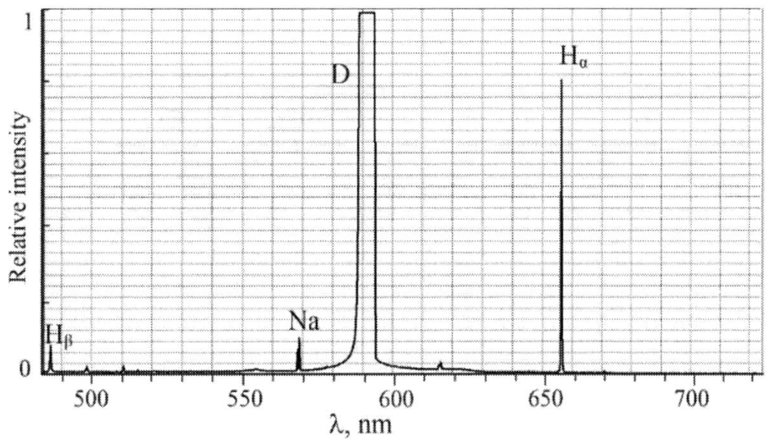

Figure 3. Spectrum of radiation. l = 5 mm, h = 5,5 cm. σ = 15 mS/cm. I_1 = 10 A, I_2 = 9 A.

The spectral parameters of the H_α and H_β lines were used to study the properties of the electron gas. The numerical values of the electron gas temperature T_e and its density n_e were obtained. The temperature of T_e was calculated using the relative intensity method. The statistical parameters required for the calculations were taken from [3]. To calculate the density of the electron gas, formula (1) was used, which is used in electric arc studies [4].

$$\Delta_{\lambda/2} = C \cdot n_e^{2/3} \tag{1}$$

Here C – is a coefficient depending on the electron gas parameters. In the calculations, this coefficient was taken equal to $2.0 \cdot 10^{-11}$. The results of the calculation are presented in table 1.

Table 1. Spectral line and electron gas parameters.

	T_e (K)	$\Delta_{\lambda/2}(H_\alpha)$ (nm)	$\Delta_{\lambda/2}(H_\beta)$ (nm)	n_e (10^{15} sm^{-3})
1	4590	0.261	0.408	2.99
2	4580	0.309	0.495	3.99
3	4600	0.287	0.417	3.09
4	4610	0.292	0.421	3.05

It can be seen that the parameters of the contracted arc channel are similar to those parameters of discharge with the metal cathode located in a water flow [5]. In both cases, intense emission of hydrogen is observed. The temperature T_e and the concentration n_e of the electron gas in both discharges have numerical values close to each other.

4. Conclusions

When an electric arc is applied to a discharge with a solution cathode, its spatial structure changes. Combustion regimes arise in which a narrow arc channel is formed. An electric arc can be used to increase the temperature in the plasma discharge stream with a solution cathode.

References

1. Brzhozovskii B et al. 2021 *Heliyon* **7** 5 e07006
2. Yoshida K 2019 *IEEE Transactions on Industry Applications* **55** 5 5261
3. Raizer Y P 1991 *Gas Discharge Physics* (Berlin: Springer) p 460
4. Ochkin V N 2006 *Spectroscopy of low-temperature plasma* (Moscow: Fizmatlit) p 472
5. Tazmeev G K, Timerkaev B A and Tazmeev K K 2017 *Plasma Physics Reports* **43**(7) 771

Studies of a monatomic Lennard-Jones system at slow cooling by molecular dynamics simulations and regression analysis

A R Farkhutdinov[1]

[1]Kazan Federal University, Kazan, Russia

farhutdinovalbert@yandex.ru

Abstract. The procedure of radial distribution function calculation at low cooling rates without a substantial increase in the computational cost is proposed in this paper. Results of its application to monatomic Lennard-Jones system at cooling rates $\gamma = 4\times10^{-4}...2\times10^{-2}\varepsilon/k_{B}\tau$, density $\rho \approx 0.75\sigma^{-3}$ and temperature range $T = 0.3...1.0\varepsilon/k_{B}$ using various regression models are given. It is shown that k-nearest neighbors regression yields the minimum deviation from the results of molecular dynamics simulations in comparison with other models.

1. Introduction

Molecular dynamics simulations allow obtaining comprehensive information about condensed matters. These methods complement the results of traditional experiments in spectroscopy, diffraction, microscopy, etc. On the other hand, they provide a reliable basis for the development of theoretical concepts [1–8].

However, some phenomena computation is challenging due to a huge gap in time scales and cooling rates between experiments and conventional molecular dynamics simulations. Thus, the development of methods allowing simulating a many-body system at realistic conditions is relevant.

This work is concerned with predicting Lennard-Jones fluid structural properties, in particular, radial distribution function (RDF). First attempts to model RDF of the Lennard-Jones fluid have been based on fitting empiric expressions to simulation data. Goldsman equation [9] accurately predicts RDF in certain limited conditions however its usage is mathematically complicated. The expression by Morsali et al. [10] involves fewer parameters however its accuracy descends with density decrease. Bamdad et al. [11] introduced two different expressions for the first peak of RDF and the rest. Lack of continuity is also a problem in the expressions proposed by Matteoli and Mansoori [12].

Another approach to predict the RDF of simple fluids is to use integral equation methods [13–16]. However, they have not produced acceptable results for the Lennard-Jones fluid. This is because the functional form of the integral equation must be modified specifically to treat the Lennard-Jones system.

Well-Tempered metadynamics (WTmetaD) approach to compute nucleation rates in the Lennard-Jones system was proposed by Salvalaglio et al [17]. WTmetaD is an enhanced sampling method conventionally used to compute free energy surfaces in a variety of contexts. According to this work, the timescale limitations that plague the simulation of nucleation phenomena can be overcome using WTmetaD simulations. In the case of argon condensation, this implies being able to reach timescales of the order of hours with ordinary computational resources.

Content from this work may be used under the terms of the Creative Commons Attribution 3.0 licence. Any further distribution of this work must maintain attribution to the author(s) and the title of the work, journal citation and DOI.
Published under licence by IOP Publishing Ltd

Machine learning (ML) approaches have been applied to examine structural correlations in a Lennard-Jones fluid. Moradzadeh and Aluru [18] have trained a neural network to predict what values of the Lennard-Jones parameters will give a specific form of the RDF. They have also trained an autoencoder network to determine the RDF of simple fluids from a limited number of atomistic configurations. Craven et al [19] have previously developed a ML method based on segmented linear regression and multivariate function decomposition that can predict the RDF of simple fluids with significantly increased accuracy in comparison to traditional theoretical approaches.

In this work, we propose the RDF calculation algorithm for the monatomic Lennard-Jones system using various regression models. It is shown below that *k*-nearest neighbors regression yields the minimum deviation from the results of molecular dynamics simulations in comparison with other models.

The paper is structured as follows; at first, the details of the studied system and simulation parameters are reported. Then regression procedure is described. Finally, results are discussed.

2. Model and simulation

A monatomic system of $N = 1372$ atoms in three dimensions is considered. Particles interact with each other isotropically via the truncated and shifted Lennard-Jones potential [20]:

$$U(r_{ij}) = \begin{cases} U_{LJ}(r_{ij}) - U_{LJ}(r_C), & r_{ij} < r_C \\ 0, & r_{ij} \geq r_C \end{cases} \tag{1}$$

$$U_{LJ}(r_{ij}) = 4\varepsilon\left[\left(\frac{\sigma}{r_{ij}}\right)^{12} - \left(\frac{\sigma}{r_{ij}}\right)^{6}\right], \tag{2}$$

where r_{ij} is the distance between *i*-th and *j*-th particle, $r_c = 2.5\sigma$, ε is the depth of the potential well, σ is the effective radius of an atom.

Henceforth the physical properties are expressed in the normalized units. Thus, energy is measured in ε, length in σ, temperature in ε/k_B with k_B being the Boltzmann constant, pressure in ε/σ^3 and time in $\tau = \sigma\sqrt{m/\varepsilon}$ with m being the mass of a particle.

The Newtonian equations of motion were integrated using the velocity form of the Verlet algorithm [21] under periodic boundary conditions and with a time step of 0.005τ.

The simulation technique used in the present work is molecular dynamics in the *NVT* ensemble, using the velocity scaling thermostat [22], where $V = L^3 \approx 1838\sigma^3$, $L = 12.25\sigma$ is a linear dimension of the system. Thus, the system is characterized by constant density $\rho = N/V \approx 0.75\sigma^{-3}$.

The initial state of the system is prepared an equilibrium liquid at $T_i = 1.3\varepsilon/k_B$ above the melting temperature $T_m \approx 0.75\varepsilon/k_B$ [23]. Then, the system was cooled to temperature $T_f = 10^{-4}\varepsilon/k_B$ at cooling rates $\gamma_1 = 2\times10^{-2}\varepsilon/k_B\tau$ and $\gamma_2 = 4\times10^{-4}\varepsilon/k_B\tau$. The number of steps for this procedure was $M_1 = 4\times10^4$ and $M_2 = 2\times10^6$, respectively. The temperature change was stopped every $M_{1,2}/20$ steps. Then, the system was equilibrated in 15×10^3 steps and radial distribution function (RDF) [24] was calculated in $n = 5\times10^3$ steps:

$$g(\mathbf{r}) = \frac{2V}{N^2}\langle\textstyle\sum_{i=1}^{N-1}\sum_{j>i}^{N}\delta(\mathbf{r} - \mathbf{r}_{ij})\rangle, \tag{3}$$

where $\mathbf{r}_{ij} = \mathbf{r}_i - \mathbf{r}_j$ is a distance between *i*-th and *j*-th particles and angle brackets denote n ensembles averaging. If the system is also isotropic the function can be averaged over angles without loss of information, so $g(\mathbf{r}) \equiv g(r)$. It should be noted, that r_{ij} in (3) is implicitly dependent on the temperature of the system, therefore $g(r) \equiv g(r, T)$. Then, cooling was resumed after RDF calculation.

3. Regression procedure

The key idea of the problem solution is shown in figure 1. Let fix a value of interparticle distance $r = r_0$. Then temperature dependence of RDF $g(T, r = r_0)$ can be obtained. The next step is to choose an appropriate regression model that is fitted to data from molecular dynamics simulations. In terms of supervised machine learning, the system temperature is a feature and RDF is a target. This model is used to predict RDF value at the temperature of interest. After repeating this procedure for r from 0 to $L/2$, RDF at desired temperature can be obtained. This RDF needs to be compared with an analogous one calculated with direct molecular dynamics simulations to test the learned model. The obtained RDF, if it is valid, can be used in the further determination of the system configuration.

In this work following regression models are used: linear model [25], *k*-nearest neighbors (*k*-NN) algorithm [26], decision tree [27]. *k*-NN regression is presented in two forms. The first one is based on a fixed number of neighbors. The second one is based on all particles within a fixed radius. Moreover, these algorithms can be uniform with an equal contribution for each neighbor, and weighted by the inverse of their distance.

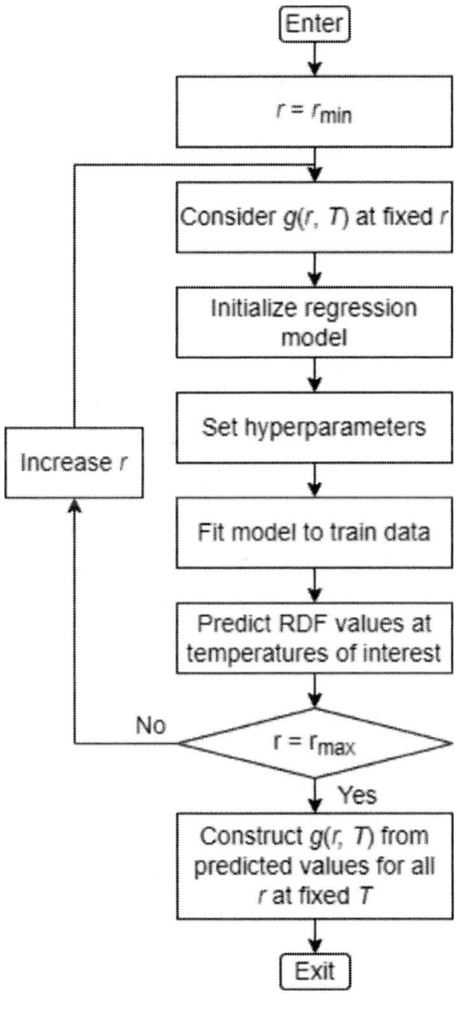

Figure 1. Flowchart of the algorithm for radial distribution function prediction.

4. Results and discussions

An example of the described algorithm implementation for the case of *k*-NN regression is shown in figure 2. Radial distribution functions at various temperatures and cooling rate γ_2, obtained by classical molecular dynamics simulations as well as regression methods are shown.

Mean-square error (MSE) is used as metrics of models quality:

$$MSE(r,T) = \frac{1}{n}\sum_{i=1}^{n}[g_i(r,T) - \hat{g}_i(r,T)]^2, \quad (4)$$

where $g_i(r, T)$ are observed values, $\hat{g}_i(r,T)$ are predicted values, n is the number of samples.

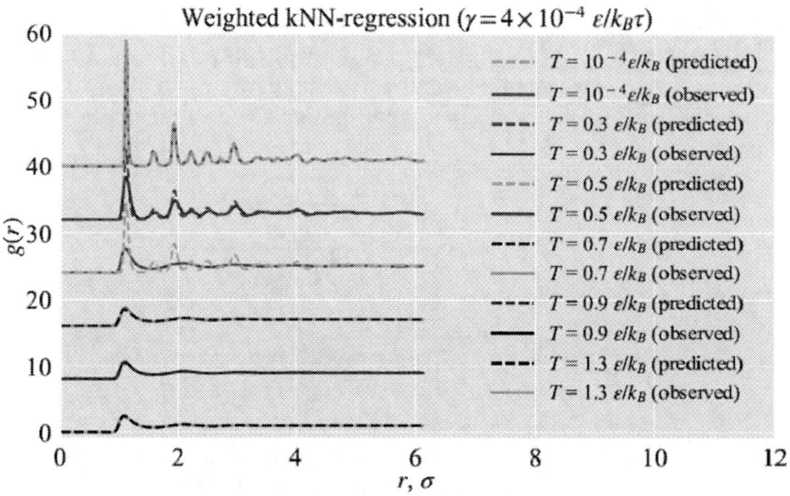

Figure 2. Example of the described algorithm implementation for the case of *k*-NN regression weighted by the inverse of distances.

A comparison of the mentioned above regression models is shown in figure 3. Their names are displayed on the horizontal axis and maximum MSE values for each of them are displayed on the vertical axis. Regression models that should be noted are *k*-NN regression based on a fixed number of neighbors and a decision tree algorithm. These algorithms have minimum MSE almost in all temperature ranges and interparticle distance ones.

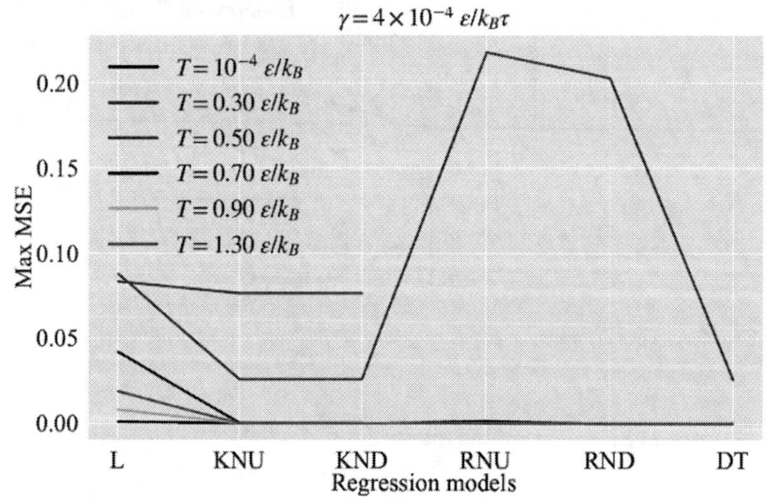

Figure 3. A comparison of the considered regression models. *L* is linear, *KN* is *k*-NN regression based on a fixed number of neighbors, *RN* is variation based on a fixed radius. *U* means uniform regression and *D* means one weighted by the inverse of their distance. *DT* is a decision tree.

5. Acknowledgements
This study is supported by the Kazan Federal University Strategic Academic Leadership Program.

References
[1] Khusnutdinov R M, Mokshin A V and Yulmetev R M 2009 *J. Exp. Theor. Phys.* **108** 417
[2] Khusnutdinoff R M and Mokshin A V 2010 *Bull. Russ. Acad. Sci.: Phys.* **74** 640
[3] Mokshin A V, Chvanova A V and Khusnutdinoff R M 2012 *Theor. Math. Phys.* **171** 541
[4] Mokshin A V and Galimzyanov B N 2014 *J. Chem. Phys.* **140** 024104
[5] Mokshin A V *et al.* 2015 *J. Exp. Theor. Phys.* **121** 828
[6] Mokshin A V 2015 *Theor. Math. Phys.* **183** 449
[7] Mokshin A V and Galimzyanov B N 2018 *J. Phys.: Condens. Matter* **30** 085102
[8] Khusnutdinoff R M, Galimzyanov B N and Mokshin A V 2018 *J. Exp. Theor. Phys.* **126** 83
[9] Goldman S 1979 *J. Phys. Chem.* **83** 3033
[10] Morsali A, Goharshadi E K, Mansoori G A and Abbaspour M 2005 *Chem. Phys.* **310** 11
[11] Bamdad M, Alavi S, Najafi B and Keshavarzi E 2006 *Chem. Phys.* **325** 554
[12] Matteoli E and Mansoori G A 1995 *J. Chem. Phys.* **103** 4672
[13] Broyles A A, Chung S U and Sahlin H L 1962 *J. Chem. Phys.* **37** 2462
[14] Duh D M and Haymet A D J 1995 *J. Chem. Phys.* **103** 2625
[15] Duh D M and Henderson D 1996 *J. Chem. Phys.* **104** 6742
[16] Choudhury N and Ghosh S K 2002 *J. Chem. Phys.* **116** 8517
[17] Salvalaglio M *et al.* 2016 *J. Chem. Phys.* **145** 211925
[18] Moradzadeh A and Aluru N R 2019 *J. Phys. Chem. Lett.* **10** 7568
[19] Craven G T, Lubbers N, Barros K and Tretiak S 2020 *J. Chem. Phys.* **153** 104502
[20] Jones J E 1924 *Proc. R. Soc. Lond. A* **106** 441
[21] Swope W C, Andersen H C, Berens P H and Wilson K R 1982 *J. Chem. Phys.* **76** 637
[22] Woodcock L V 1971 *Chem. Phys. Lett.* **10** 257
[23] Smit B 1992 *J. Chem. Phys.* **96** 8639
[24] Rapaport D C 1988 *The Art of Molecular Dynamics Simulation* (Cambridge: Cambridge University Press)
[25] Yan X and Su X G 2009 *Linear regression analysis: theory and computing* (Singapore: World Scientific Publishing Co. Pte. Ltd)
[26] Altman N S 1992 *Am. Stat.* **46** 175
[27] Breiman L, Friedman J, Olshen R and Stone C 1984 *Classification and Regression Trees* (Belmont: Wadsworth)

A STABLE NUMERICAL ALGORITHM FOR INVESTIGATING THE PROPERTIES OF TWO-DIMENSIONAL LASER WAVEGUIDES IN MULTILAYER QUANTUM-DIMENSIONAL SEMICONDUCTOR HETEROSTRUCTURES

H.Sh. Juraev[1], N.O. Mamatkulova[1], B.I. Makhsudov[1]

[1]Tajik National University, Dushanbe, Tajikistan

hayrullo_58@mail.ru

Abstract. The transfer of heat through nanostructures differs significantly from the corresponding processes inside macroscopic bodies. Various research methods, both theoretical and experimental, are applied to such objects. This approach to nanostructures gives great advantages. The hyperbolic equations we have obtained describe the wave process of the thermal signal flow and the properties of heat transfer in nanostructures across a thin layer, in which the simplest task is to determine the thermal conductivity.

1. Introduction

Multilayer nanostructures are artificially formed layered structures consisting of a large number (up to tens of thousands) of alternating nanolayers of various materials. The list of materials used to create multilayer nanostructures is quite wide – it can be pure metals, alloys, semiconductor materials, materials with dielectric properties in various combinations. The thickness of the layers of multilayer nanostructures, as a rule, is in the range from 1 to 100 nm. It is worth noting that the greatest interest is the study of multilayer nanostructures with layers of thickness from units to several tens of nanometers. This is due to the fact that it is in this range that the degree of influence of the dimensional factors of nanolayers on the structure and properties of materials acquires the greatest importance. Thus, the ultra-small thickness of the layers in combination with a large number of interfaces between the layers and the alternation of layers of different materials leads to the appearance of multilayer nanostructures with unique structural features that are not present in the materials that make up the nanostructures in massive samples [1]. The manifestation of such features is associated with the detection of such physical properties in multilayer nanostructures as extremely high hardness [2], low transverse thermal conductivity [3], the effect of giant magnetoresistance (GMR) [4], divergence of the radiative characteristics of heterolasers with increasing temperature [5-8], etc.

Heteronanolasers have high radiative characteristics and different variants, for example, strained quantum-dimensional heterostructures, which are currently being actively investigated [5-10].

Today's technologies for the production of semiconductor lasers make it possible to grow quantum-sized heterostructures with dozens of layers, which, in any case, requires a more natural description of

Content from this work may be used under the terms of the Creative Commons Attribution 3.0 licence. Any further distribution of this work must maintain attribution to the author(s) and the title of the work, journal citation and DOI.
Published under licence by IOP Publishing Ltd

the parameters of their structures in terms of the temperature dependence of the radiative characteristics. Previously developed models do not always provide the required accuracy.

In [11-13], on the example of multilayer planar optical waveguides, the waveguide properties of which are determined not by the total internal reflection from the core-shell boundary, as in conventional optical waveguides, but by the antiresonance reflection from the multilayer shell, ARROW waveguides, the effectiveness of a new numerical method for calculating the spectrum and attenuation coefficient of the resulting electromagnetic modes is demonstrated. The method proposed in [11-13] is equally applicable both to the calculation of electromagnetic modes in dielectric waveguides and to the calculation of quantum states of electrons in multi-barrier semiconductor heterostructures. The results of comparing the characteristics of multilayer waveguides calculated on the basis of the proposed method with the known ones obtained by solving a complex dispersion equation within the framework of the transfer matrix method are presented. As an example, the results of calculating the spectral dependence of the radiation attenuation of the first flowing TE mode of a planar optical waveguide with 52 pairs of layers are given.

The authors of [5, 6] for the first time experimentally detected a decrease in the divergence of InGaAs-heterolaser radiation with an increase in temperature. Based on the model of a waveguide with varying parameters, a calculation is made that allows us to explain the temperature decrease in the width of the radiation pattern and the increase in the threshold current. The results of the work are determined in favor of a waveguide mechanism for abnormal temperature growth of the threshold current of long-wave InGaAs-heterolasers.

The paper [13] presents a simple matrix technique for calculating the radiation characteristics of various modes, including losses for an arbitrary stepped planar waveguide structure with complex refractive indices.

In this regard, this article proposes a new approach to the development of a more convenient and optimal method for numerical calculation of the properties of the waveguide of heteronostructures, which allows optimizing the temperature dependence of the intensity distribution in the far field of injection lasers at different parameters of nanostructures.

2. A model study of the characteristics of two-dimensional laser waveguides

It is known [10] that the electromagnetic field in the medium is determined by Maxwell's equations:

$$rot\vec{E} = -\mu\frac{\partial\vec{H}}{\partial t}, \quad div\vec{H} = 0, \tag{1}$$

$$rot\vec{H} = \varepsilon\frac{\partial\vec{E}}{\partial t}, \quad div\vec{E} = 0, \tag{2}$$

where $\vec{E}, \vec{H}$ are the electric and magnetic field strengths, respectively; μ is the magnetic permeability, and ε is the dielectric constant of the medium.

As is known [10] (see page 155), the electromagnetic field propagating in a vacuum is called an electromagnetic wave. Vector $\vec{E}$ and $\vec{H}$ satisfy the wave equation:

$$\Delta\vec{E} - \mu\varepsilon\frac{\partial^2\vec{E}}{\partial t^2} = 0,$$
$$\Delta\vec{H} - \mu\varepsilon\frac{\partial^2\vec{H}}{\partial t^2} = 0, \tag{3}$$

where $v = \frac{1}{\sqrt{\varepsilon\mu}}$ is the velocity of propagation of an electromagnetic wave in a material medium. The propagation velocity of an electromagnetic wave in a medium is related to the propagation region in a vacuum by the expression: $\vartheta = \frac{c}{\varepsilon'}$ In this expression, ε' is the relative permittivity. The relative permittivity is related to the refractive index by the expression: $n = \sqrt{\varepsilon'}$.

The task is to study the influence of the field propagating along the axis on the nature of the change in the components E_x and E_y of the field. Suppose that an electromagnetic wave in a semiconductor laser propagates along the axis Oz. Then, for example, equation (3) for the electric field strength can be divided into two equations:

$$\frac{\partial^2 E_z(x,y,t)}{dx^2} - \mu\varepsilon \frac{\partial^2 \vec{E}_z(x,y,t)}{\partial t^2} = 0, \quad (4)$$

$$\frac{\partial^2 E_z(x,y,t)}{dy^2} - \mu\varepsilon \frac{\partial^2 \vec{E}_z(x,y,t)}{\partial t^2} = 0, \quad (5)$$

We will look for the general solution of equations (4) and (5) in the following form:

$$E_z(x,y,t) = E_z(x,y)exp(i(\omega t + \beta z)), \quad (6)$$

where ω is the frequency of the electromagnetic field oscillation.

In multilayer semiconductor laser structures, only the polarization TE modes are considered. This is due to the fact that TM modes have significantly greater attenuation than TE modes. In the TE modes of polarization the only spatial component $E_z(x,y)$ is different from zero.

Thus, according to (4), (5) and (6), the wave function can be searched for as a multiplication of two functions:

$$E_z(x,y) = E_z(x)E_z(y),$$

each of which satisfies the one-dimensional Helmholtz equation:

$$\frac{d^2 E_z(x)}{dx^2} + q_1^2 E_z(x) = 0, \quad q_1 = \sqrt{\beta_x^2 - \frac{1}{2}n^2 k_0^2}, \quad (7)$$

$$\frac{d^2 E_z(y)}{dy^2} + q_2^2 E_z(y) = 0, \quad q_2 = \sqrt{\beta_y^2 - \frac{1}{2}n^2 k_0^2}, \quad (8)$$

where (x, y) is the coordinate along the transverse axis to the layers, $k_0 = \frac{\omega}{c}$ is the wavenumber in vacuum, ω is the frequency of optical radiation, $\varepsilon = n^2$ is the complex permittivity, n is the complex refractive index, $\beta_z = \beta_x + \beta_y$ is the longitudinal propagation constant, $E_z(x)$, $E_z(y)$ are the amplitude profiles of modes – own functions (OF). It is enough to solve one of these equations to write a solution for the others by analogy.

The solution of equation (7) ((8)) makes it possible to determine a singular point in the phase plane. Depending on the value of β_x (β_y) - the complex wave propagation constant, it is possible to have an infinite number of equilibrium states, that is, for equation (7) ((8)) there are an infinite number of solutions that can be in the far field of multilayer injection lasers or near.

The constant factor at $q_1 = \sqrt{\beta_x^2 - \frac{1}{2}n^2 k_0^2}$ is equal to one here, which is possible if we choose the scale along the x axis accordingly. Let the potential be a given piecewise continuous real periodic function with a period T:

$$E_z(x+T) = E_z(x).$$

We are looking for the eigenvalue of β_z - the longitudinal constant of wave propagation, for which the differential equation has an identically non-zero solution of $E_z(x)$ and which remains bounded for all real x.

We use matrix calculus for the sake of clarity. Let $E_z^1(x)$, $E_z^2(x)$ be a system of fundamental functions of equation (7). Then, due to the periodicity of the differential equation $E_z^1(x+T)$, $E_z^2(x+T)$ are also solutions: they can therefore be represented as linear combinations of $E_z^1(x)$, $E_z^2(x)$:

$$\begin{cases} E_z^1(x+T)=C_{11}E_z^1(x)+C_{12}E_z^2(x), \\ E_z^2(x+T)=C_{21}E_z^1(x)+C_{22}E_z^2(x). \end{cases} \tag{9}$$

The Wronskian of this system has a value independent of x and nonzero:

$$\begin{vmatrix} E_z^1(x+T) & E_z^2(x+T) \\ E_z^{1'}(x+T) & E_z^{2'}(x+T) \end{vmatrix} = \begin{vmatrix} E_z^1(x) & E_z^2(x) \\ E_z^{1'}(x) & E_z^{2'}(x) \end{vmatrix} \bullet \begin{vmatrix} C_{11} & C_{12} \\ C_{21} & C_{22} \end{vmatrix};$$

Therefore,

$$C_{11} \bullet C_{22} - C_{12} \bullet C_{21} = 1. \tag{10}$$

If you enter a column vector

$$\delta_k = \begin{pmatrix} E_z^1(x+kT) \\ E_z^2(x+kT) \end{pmatrix}$$

and a matrix of coefficients

$$D = \begin{pmatrix} C_{11} & C_{12} \\ C_{21} & C_{22} \end{pmatrix},$$

then (8) can be written as:

$$\delta_k = D \cdot \delta_{k-1} = D^2 \cdot \delta_{k-2} = \dots = D^k \cdot \delta_0. \tag{11}$$

Here D is a degenerate matrix, (11) valid for all integer (also negative) values of *k*.

Let η_1, η_2 be the characteristic numbers of the matrix D, that is, by virtue of (10) the root of the equation:

$$q^2 - q(C_{11}+C_{22})+1=0. \tag{12}$$

Then we can deduce from the matrix calculus theorem:

a) If $|C_{11}+C_{22}|>2$, then the roots q_1, q_2 are real, their multiplication is 1, that is, in particular, if $|q_1|>1$, then $q_2<1$; $D^k \cdot \delta_0$ for $k \to \infty$ or $k \to -\infty$ is unlimited; the corresponding value β_1 is not an eigenvalue in this case.

b) If $|C_{11}+C_{22}|<2$, then the roots q_1, q_2 are complex conjugate, $|q_1|=|q_2|=1$; $D^k \cdot \delta_0$ for any δ_0 remains bounded; each solution of equation (7) is an eigenfunction; the corresponding quantities β_1 are doubly degenerate eigenvalues.

c) If $|C_{11}+C_{22}|=2$, that is $C_{11}+C_{22}=\pm 2$; there is at least one vector δ_0 satisfying equality $D \cdot \delta_0 = \pm\delta_0$, that is, a solution $E_z(x)$ of equation (7) having the property $E_z(x+T)=\pm E_z(x)$. Then there is equality $E_z(x+2T)=E_z(x)$.

The general solution of the first equation in the inner domain has the form:

$$E_z(x) = C_1 \exp(q_1 x) + C_2 \exp(-q_1 x). \tag{13}$$

The first term in (13) describes the free movement of the wave in the positive direction of the x axis, and the second in the opposite direction. The squares of the modules of the coefficients C_1 and C_2 are equal to the probabilities of detecting a wave in the corresponding states.

According to the problem statement, the wave function should vanish at the boundary of these regions. The derivative of the wave function at such a boundary may experience a jump. When solving equation (7), the boundary conditions at infinity must be correctly set. In this case, all wave propagation processes can be divided into two classes:

1) the negligible probability of the wave staying at infinity, in which the wave function can be normalized by one condition:

$$\int |E(r)|^2 dV = 1; \tag{14}$$

2) the wave state is finite, the probability of staying in infinity is nonzero. In this case, one can see by example the propagation of a wave in a limited volume (quantum well) of the simplest geometric shape (see Fig. 1 in [11]):

$0 \le x \le d, \quad 0 \le y \le b.$

The specified conditions for the propagation of an electromagnetic wave will be met at the following potential:

$$E_z(x,x) = \begin{cases} 0 & inside\ the\ rectangle, \\ \infty & outside\ the\ rectangle. \end{cases}$$

Boundary conditions 2) on the walls of the parallelepiped $E_z(0) = 0, \quad E_z(d) = 0$ lead to the following relations:

$$\begin{cases} C_1 + C_2 = 0, \\ C_1 \exp(q_1 d) + C_2 \exp(-q_1 d) = 0. \end{cases}$$

According to the third condition c) of matrix calculus, the wave eigenfunction has the form:

$E_z(x) = C_1(\cos(q_1 x) - \sin(q_1 x)) + 2\sin(q_1 x),$

according to the second one, it is possible to find the eigenvalue of the wave vector and the energy of one-dimensional motion:

$$q_{1,1} = \frac{2}{3+\sqrt{2}} \quad \left(q_{1,2} = \frac{2}{3-\sqrt{2}}\right), \quad \beta_x = \sqrt{q_1^2 - \frac{1}{2} n^2 k_0^2},$$

where n is any integer greater than or equal to zero.

Conditions (2) can be satisfied if we normalize each one-dimensional function by one. From here we have:

$$1 = \int_0^a |E_z(x)|^2 dx = C_1^2 \left(a + \frac{1}{q_1} \sin^2(q_1 a) \right) +$$

$$+ 2C_1 \left(\frac{1}{q_1} \sin(q_1 a)(\cos(q_1 a) - \sin(q_1 a)) - a \right) + 2a - \frac{2}{q_1} \sin(q_1 a)\cos(q_1 a).$$

It follows that

$$C_1^2 \left(q_1 a + \sin^2(q_1 a)\right) + 2C_1 \left(\sin(q_1 a)(\cos(q_1 a) - \sin(q_1 a)) - q_1 a\right) +$$

$$+ 2(aq_1 - \sin(q_1 a)\cos(q_1 a)) - q_1 = 0.$$

where

$$C_1^1 = -\frac{2a(1+\cos(q_1 a))-1}{a(\cos(aq_1)-aq_1-1)+\sqrt{a^2(\cos^2(aq_1)-1+a(1-aq_1)}},$$

$$C_1^2 = -\frac{2a(1+\cos(q_1 a))-1}{a(\cos(aq_1)-aq_1-1)-\sqrt{a^2(\cos^2(aq_1)-1+a(1-aq_1)}}.$$

Let us consider only the case of the finite propagation of a traveling wave in the field of multilayer injection lasers, which must correspond to the positive (negative) states of the propagation constant β_x in the system, depending on the thickness a. Let's divide the entire area of charge motion into three intervals. In the first region, the wave function is $E_{z1}=0$ (or $E_{z2}=0$, or $E_{z3}=0$). In the second region (the inner part of the pit), the Helmholtz equation and its general solution have the form:

$$\frac{d^2E_{z2}}{dx^2}+q_{x2}^2E_{z2}(x)=0, \quad q_{x2}=\sqrt{\beta_{x2}^2-\frac{1}{2}n^2k_0^2},$$

$$E_{z2}(x)=C_1\exp(iq_{x2}x)+C_2\exp(-iq_{x2}x).$$

In the third area:

$$\frac{d^2E_{z3}}{dx^2}+q_{x3}^2E_{z3}(x)=0, \quad q_{x3}=\sqrt{\beta_{x3}^2-\frac{1}{2}n^2k_0^2},$$

$$E_{z3}(x)=C_3\exp(iq_{x3}x)+C_4\exp(-iq_{x3}x).$$

To satisfy the normalization condition, the condition of the wave function decreasing at $x\to\infty$ is necessary so that $C_4=0$.

The boundary value at x=0 gives the ratio:

$$C_1+C_2=0. \tag{15}$$

For $x=a$, the wave function and its derivative must be continuous, that is, they must be crosslinked:

$$E_{z2}(a)=E_{z3}(a), \quad \left.\frac{dE_{z2}(x)}{dx}\right|_{x=a}=\left.\frac{dE_{z3}(x)}{dx}\right|_{x=a}. \tag{16}$$

The conditions of "crosslinking" (15), considering (14), give the following relations:

$$2iC_1\sin(q_{x2}a)=C_3\exp(q_{x3}a),$$
$$2iC_1q_{x2}\cos(q_{x2}a)=C_3q_{x3}\exp(q_{x3}a). \tag{17}$$

From these relations and the normalization conditions, the coefficients C_1, C_3 and the modulus of the wave vector q_{x2} can be found. By dividing the ratios (17) one by the other, it is possible to exclude the coefficients C_1, C_3 and determine the following transcendental equation for q_{x2}:

$$tg(q_{x2}a)=\frac{q_{x2}}{q_{x3}}.$$

Thus, for the value of the propagation constant and the wave function of a two-dimensional waveguide, we have:

$$\beta_x=\beta_{x1}+\beta_{x2}=\sqrt{q_{x1}^2-\frac{1}{2}n^2k_0^2}+\sqrt{q_{x2}^2-\frac{1}{2}n^2k_0^2},$$

$$E_z(x,y)=E_z(x)E_z(y),$$

where $n=0,1,2,\cdots$.

The algorithm for calculating the task with the help of computer modeling allows optimizing the parameters of multilayer semiconductor injection lasers based on heteronostructures from the point of view of stabilizing the temperature dependence of the radiative characteristics of these types of lasers. Calculation of the intensity distribution of laser radiation in the near and far field, considering the temperature regime, allows to increase the efficiency of radiation input into optical cables of special laser models.

3. Conclusion

The new approach allows us to consider the influence of the temperature mode of operation of semiconductor quantum-dimensional injection lasers on their radiative characteristics and quantum semiconductor nanostructures. It is based on two ideas: the replacement of the traditionally used Transfer Matrix Method (TMM) with the exact analytical method proposed by us and the use of a new method for calculating the attenuation coefficient of the resulting modes, approximate, but asymptotically tending to be accurate with decreasing attenuation.

The possibility of accurate calculation of multilayer ARROW waveguides is extremely important, in particular, for their prospective application as elements of optical integrated devices.Other paragraphs are indented (BodytextIndented style).

Low-temperature plasma can be used for the synthesis of various nanostructures and is well suited for the modification of various surfaces. This is shown in many works [16-35].

References

[1] Bazaleeva, K.O. Structural changes in multilayer nanofilms Ti/Al / K.O. Bazaleeva, V.S. Kraposhin, P.A. Tsygankov et al. // Materials Science. - 2008. - No. 8. - pp. 35-39.
[2] Yashar P. C. Nanometer scale multilayered hard coatings / P.C. Yashar, W.D. Sproul // Vacuum. – 1999. – Vol. 55. – P. 179–190.
[3] Teixeira, V. Nanostructured ceramic coatings produced by magnetron sputtering / Y.G. Gogotsi // Nanostructured materials and coatings in biomedical and sensor applications. – Boston: Kluwer academic pub., 2003. – P. 131–149.
[4] Hartmann, U. Magnetic multilayers and giant magnetoresistance: fundamentals and industrial applications. / U. Hartmann // – New York: Springer verlag, 2000. – 336 p. [3] More references
[5] Bogatov, A. P. The influence of temperature on the radiation pattern of InGaAsP heterolasers / A. P. Bogatov, P. G. Eliseev, B. I. Makhsudov // Quantum Electronics. - 1988. - volume 15, No. 2. - pp. 253-258.
[6] Schweikin, V.I. Radiation radiation pattern of InGaAs/GaAs quantum-dimensional lasers operating on the "flowing" mode / V.I. Schweikin, A.P. Bogatov, A.E. Drakina, Yu.V. Kuriyavko // Quantum Electronics. -1999. - volume 26, No. 1. - pp.33-36.
[7] Schweikin, V.I. New diode lasers with leaking radiation in an optical resonator. / Shveykin V.I., Gelovani V.A. // Quantum electronics. - 2002. - volume 32, No. 8. - pp.683-688.
[8] Bogatov, A.P. Optimization of waveguide parameters of InGaAs/AlGaAs/GaAs laser heterostructures in order to maximize the beam width in the resonator and obtain maximum laser power / A.P. Bogatov, T.I. Guschik, A.E. Drakin, A.P. Nekrasov, V.V. Popovich // Quantum Electronics. - 2008. - No. 10. - pp.935-939.
[9] Vinokurov, D.A. Influence of the thickness of the active region on the characteristics of semiconductor lasers based on asymmetric AlGaAs/GaAs/InGaAsc heterostructures with an expanded waveguide / D.A. Vinokurov, V.V. Vasilyeva [et al.] // Physics and Technology of Semiconductors. - 2010. - Vol.44. - Vol.2. - p.246-250.
[10] Juraev, H.Sh. On a study of a mathematical model of radiation intensity propagation in the far field of multilayer injection lasers / H.Sh. Juraev, B.I. Makhsudov, N.O. Mamatkulova // Bulletin of the Tajik National University. - 2019. - №3/1(134) - pp.152-159.
[11] Golant, E.I. A new method for calculating the spectrum and radiation losses of the resulting modes of multilayer optical waveguides / E.I. Golant, K.M. Golant // Journal of Technical Physics. - 2006. - volume 76. - issue 8. - pp.99-106.

[12] Golant, E.I. A new approach to the calculation of the resulting modes of multilayer waveguide structures based on the exact method of finite differences / E.I. Golant // Letters of VZHTF. - 2005. - volume 31. - issue 24. - pp.81-87.

[13] Juraev, H.S. Model study of the properties of volumetric waveguides in multilayer quantum-dimensional semiconductor heterostructures / H.S. Juraev, N.O. Mamatkulova, B.I. Makhsudov, N. Narzulloev // Materials of the international scientific and practical conference "Problems and prospects of physics, engineering and technology" dedicated to "The twentieth anniversary of the study and development of natural, exact and mathematical sciences in the field of science and education" and the 70th anniversary and 45th anniversary of the activities of Ph.D., Professor A.L. Kadyrov. - Khujand: KhSU im. Academician B.Gafurov. - 2021. - p.17-23.

[14] Landau, L.D., Lifshits E.M. Theoretical physics: Textbook: For universities. In 10 vols. Vol. II. Field theory. 8th ed., stereotype. / L.D. Landau, E.M. Lifshits // -Moscow: FIZMATLIT. - 2003. -536 p.-ISBN 5-9221-0056-4 (Vol. II)

[15] Bogatov A.P. Efficiency and intensity distribution in a semiconductor laser operating on the "flowing" mode. /A.P. Bogatov, A.E. Drakin, V.I. Shveykin // Quantum Electronics. - 1999. - volume 26, No. 1. - pp. 28-32.

[16] Saifutdinov, A. I., & Sofronitskii, A. O. (2021). Numerical Study of Breakdown and Formation Dynamics of Arc Discharge Plasma Parameters at Ultrahigh Pressures. High Energy Chemistry, 55(3), 228-232.DOI: 10.1134/S0018143921030115

[17] Shemakhin, A. Y., Zheltukhin, V. S., Shemakhin, E. Y., Terentev, T. N., & Sofronitsky, A. O. (2020, July). Experimental installation to study the RF plasma flow at low pressures with experiment data synchronization. In Journal of Physics: Conference Series (Vol. 1588, No. 1, p. 012018). IOP Publishing. DOI: 10.1088/1742-6596/1588/1/012018

[18] Saifutdinova, A. A., Sofronitskiy, A. O., Timerkaev, B. A., & Saifutdinov, A. I. (2020). Plasma-Chemical Decomposition of Hydrocarbons on the Basis of the Micro-Arc Discharge with Disc Electrodes Rotating in the Bulk of Raw Materials. Russian Physics Journal, 62(11), 2132-2136.DOI: 10.1007/s11182-020-01957-0

[19] Fairushin, I. I., Saifutdinov, A. I., & Sofronitskiy, A. O. (2020). Numerical and Experimental Studies of the Synthesis of Copper Nanoparticles in a High-Pressure Discharge. High Energy Chemistry, 54(2), 150-153. DOI: 10.1134/S0018143920020071

[20] Asadullin, T. Y., Galeev, I. G., Sofronitskiy, A. O., & Gizeev, M. M. (2019, November). Acoustic impact on electric discharge parameters during sterilization of freeze-dried products. In Journal of Physics: Conference Series (Vol. 1370, No. 1, p. 012016). IOP Publishing. DOI: 10.1088/1742-6596/1370/1/012016

[21] Fairushin, I. I., Saifutdinov, A. I., Sofronitskiy, A. O., Timerkaev, B. A., & Dautov, G. Y. (2019, October). Development of plasma reactor design for synthesis of copper nanoparticles using multi-scale simulation. In Journal of Physics: Conference Series (Vol. 1328, No. 1, p. 012088). IOP Publishing. DOI: 10.1088/1742-6596/1328/1/012088

[22] Asadullin, T. Y., Galeev, I. G., & Sofronitskiy, A. O. (2019, October). The application of pulsed discharge for sterilization of freeze-dried product. In Journal of Physics: Conference Series (Vol. 1328, No. 1, p. 012071). IOP Publishing. DOI: 10.1088/1742-6596/1328/1/012071

[23] Timerkaev, B. A., Ganieva, G. R., Kaleeva, A. A., Israfilov, Z. K., & Sofronitskii, A. O. (2019). Growing of carbon nanotubes from hydrocarbons in an arc plasma. Journal of Engineering Physics and Thermophysics, 92(5), 1248-1252. DOI: 10.1007/s10891-019-02040-3

[24] Dautov, G. Y., Kashapov, N. F., Dautov, I. G., & Sofronitskiy, A. O. (2018, July). Research of the influence of the geometry of the discharge chamber on the characteristics of the arc plasmatron. In Journal of Physics: Conference Series (Vol. 1058, No. 1). IOP Publishing. DOI: 10.1088/1742-6596/1058/1/012035

[25] Timerkaev, B. A., Andreeva, A. A., & Sofronitskiy, A. O. (2017, November). Discharge creeping along the surface in the process for producing nanomaterials. In Journal of Physics: Conference Series (Vol. 927, No. 1, p. 012068). IOP Publishing. DOI: 10.1088/1742-6596/927/1/012068

[26] Sadikov, K. G., Sofronitskiy, A. O., & Larionov, V. M. (2017, November). The effect of electrically conductive additives on the plasma pyrolysis of heavy hydrocarbons. In Journal of Physics: Conference Series (Vol. 927, No. 1, p. 012046). IOP Publishing. DOI: 10.1088/1742-6596/927/1/012046

[27] Timerkaev, B. A., Andreeva, A. A., & Sofronitskiy, A. O. (2017). Discharge creeping along the surface in the process of cleaning and strengthening of the materials surface. In Journal of Physics: Conference Series (Vol. 789, No. 1). IOP Publishing. DOI: 10.1088/1742-6596/789/1/012063

[28] Sadikov, K. G., Sofronitskiy, A. O., & Dautov, I. G. (2017). Functional plasma sprayed coatings on magnesium ceramic substrates. In Journal of Physics: Conference Series (Vol. 789, No. 1, p. 012043). IOP Publishing. DOI: 10.1088/1742-6596/789/1/012043

[29] Timerkaev, B. A., Sofronitskiy, A. O., & Andreeva, A. A. (2016). Carbon nanotubes formation in the decomposition of heavy hydrocarbons creeping along the surface of the glow discharge. In Journal of Physics: Conference Series (Vol. 669, No. 1, p. 012062). IOP Publishing. DOI: 10.1088/1742-6596/669/1/012062

[30] Timerkaev, B. A., & Shamsutdinov, R. S. (2020, July). The Influence of Supersonic Gas Stream on Spatial Structure of Glow Discharge. In Journal of Physics: Conference Series (Vol. 1588, No. 1, p. 012061). IOP Publishing. DOI 10.1088/1742-6596/1588/1/012061

[31] Galeev I G, Asadullin T Y 2017 Improving of stability of the volumetric glow discharge in the gas flow J. of Phys. Conf. Ser. 789(1) 012012

[32] Asadullin T Y, Galeev I G 2017 Combined method of stabilization of the glow discharge in gas flow J. of Phys. Conf. Ser. 927(1) 012003

[33] Galeev I G, Asadullin T Ya 2016 Obtaining fullerene-containing soot during combustion of gaseous hydrocarbons in an external electric field J. of Phys. Conf. Ser. 669 012016

[34] Galeev I G, Asadullin T Y 2016 Enhancing of the glow discharge stability in chamber with cathode sections coated by a discontinuous dielectric coating J. of Phys. Conf. Ser. 669(1) 012015

[35] Shakirov A S, Galeev I G and Asadullin T Ya 2020 Features of formation of fullerene-containing soot during combustion gaseous and liquid hydrocarbons J. of Phys. Conf. Ser. 1588(1) 012013

Synthesis of nickel coatings with high surface area on the ceramic powders

M V Morozov[1], K Y Nagulin[1], R N Mansurov[2], A K Gilmutdinov[1]

[1]Kazan National Research Technical University named after A.N.Tupolev-KAI, Kazan, Russia
[2]FRC Kazan Scientific Center, Russian Academy of Sciences, Kazan, Russia

misha617@mail.ru

Abstract. The study of the dynamic of the synthesis processes of nickel nanostructured coatings on the ceramic powders is presented. It has used the approach with the minimal set of reagents for the deposition of nickel coatings with high surface area. The Al_2O_3 is selected as a ceramic powder. As a result of the study, it has found the optimal conditions for the synthesis of Ni/Al_2O_3 composite powders with high surface area. It has studied the dynamic of the synthesis process of nickel coating by the scanning electron microscopy that allowed us to clarify the mechanism of the synthesis of nickel nanostructured coatings. The surface area of the synthesized composite powders was studied by nitrogen adsorption. The obtained results may have wide applications in catalysis and electrochemistry.

1. Introduction

A creation of new architectures of functional materials is important for combination of their unique properties, including heat conductivity and electroconductivity, strength and plasticity, stability and reactivity. This makes it necessary to synthesize new composite materials, as well as research of the conditions for the manufacture of new products based on them [1-2]. Nickel combined with alumina extremely interesting for use in many applications, such as catalysis, soldering, corrosion-resistant coating, functionally graded metal-ceramic composites, fuel cells and others [3-7]. Powder composites are interesting as building blocks for creation of new electrodes of chemical current sources based on them [8-10]. The presence of such a wide range of applications of nickel-alumina materials requires the development of existing methods of their synthesis [11-17].

In this work, the dynamic of the process of synthesis of nickel nanostructured coatings on ceramic powders has studied. The scientific novelty of this work is that the synthesis of coatings was carried out according to a procedure with a minimum set of starting reagents to avoid occurrence of residual impurities apart from known approaches [18-20]. Also, the nickel-coated powders with high surface area were synthesized.

Content from this work may be used under the terms of the Creative Commons Attribution 3.0 licence. Any further distribution of this work must maintain attribution to the author(s) and the title of the work, journal citation and DOI.
Published under licence by IOP Publishing Ltd

2. Experimental section

The initial powders of the Al_2O_3 composition were spheroidized by inductively coupled plasma (ICP). Plasma system used spheroidization of powder materials consists of lamp high-frequency generator with frequency 5.2 MHz and power up to 60 kW with control unit, two-circuit water cooling system, gas supply system and plasma-chemical reactor.

For the synthesis of nickel-coated powders chemicals from Acros Organics were used: nickel chloride $NiCl_2 \cdot 6H_2O$, hydrazine hydrate $N_2H_4 \cdot H_2O$; purity of reactants is 99+; no additional purification of reactants was used. Reaction was realized in distilled water solution. 18 g $NiCl_2$*$6H_2O$, 4 g of Al_2O_3 powder and 200 ml distilled water were putted at room temperature into a 500 ml round-bottom flask under stirring with magnetic stirrer. A green color of the solution was observed. After that mixture of 25 ml of $N_2H_4 \cdot H_2O$ in 200 ml distilled water was dropped into the reaction zone under magnetic stirring, and the color changed to blue. The reaction temperature was 80 °C during 2 h. The reaction resulted in formation of nickel coating on the alumina powder. After the reaction the product was washed with distilled water and isopropyl alcohol. Also, the synthesis of nickel-coated powders with the concentration of nickel 6 times less has realized.

Optical microscopy, scanning electron microscopy (SEM) and X-ray diffraction (XRD) were used to characterize the structure and morphology of the powders. SEM has done on Auriga Crossbeam workstation (Zeiss). Optical microscopy has done on Axio Z2 optical microscope from Zeiss. XRD was studied on XRD-7000 diffractometer (SHIMADZU). Also, the specific surface area of the powders was determined by nitrogen adsorption on Autosorb-iQ MP surface area analyzer (Quantachrome). Brunauer-Emmet-Teller (BET) method was used to calculate the specific surface area of the powders.

3. Results and discussion

In figure 1 SEM-images of spheroidized powders of Al_2O_3 before and after nickel coating are shown. In figure 2 photos of coated powders at intermediate and final stages of synthesis are shown.

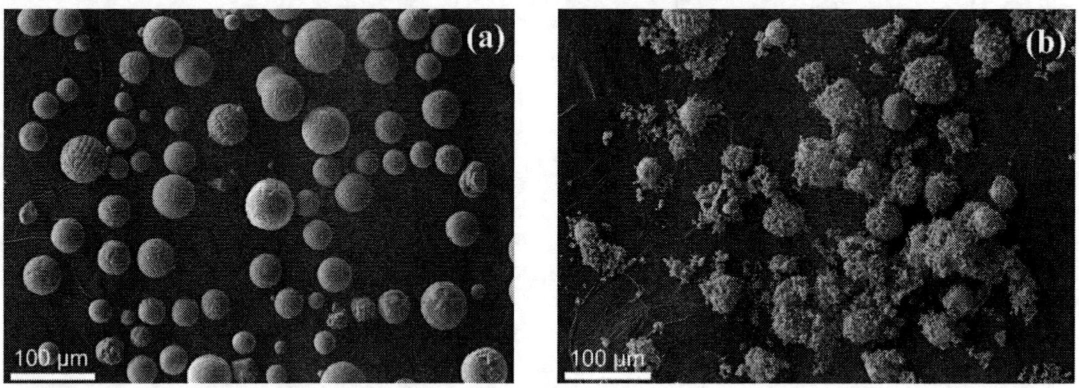

Figure 1. SEM-images of spheroidized powders of Al_2O_3 before (a) and after (b) nickel coating.

During the synthesis we can see the changing of the color of the synthesized powder from blue to black. In figure 3 SEM-images of coated powders at intermediate (25 min after heating) and final stages of synthesis are shown. The presence of loose uncrystallized phase around the Al_2O_3 particles could be explained as nickel complexes formed at intermediate stages of the synthesis [20-22]. This confirmed by XRD measurements that shown pure nickel for the final nickel product. It could be proposed the formation of $[Ni(NH_3)_6]Cl_2$ and $[Ni(N_2H_4)_3]Cl_2$ complexes for the intermediate product. The detailed study of intermediate substances will be done in future.

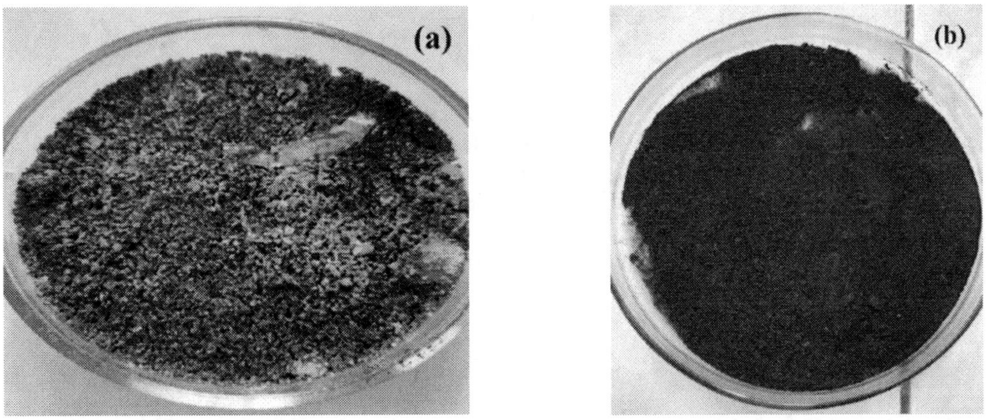

Figure 2. Al_2O_3 powders coated at intermediate (a) and final (b) stages of synthesis.

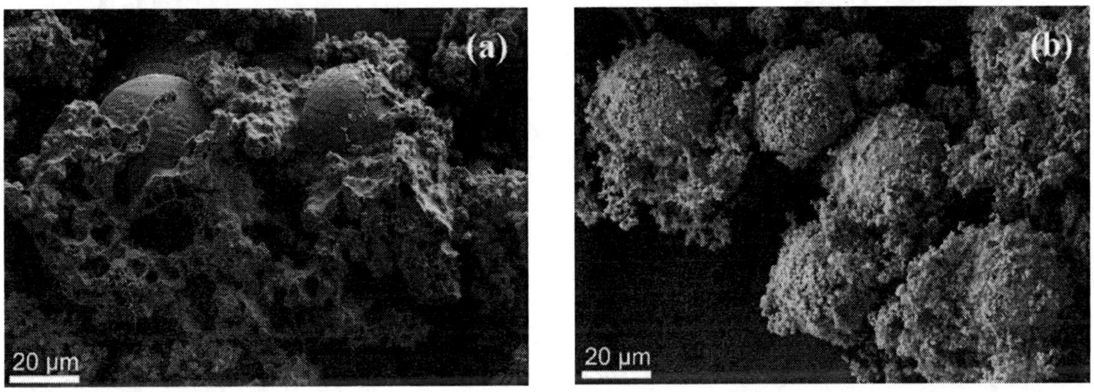

Figure 3. SEM-images of Al_2O_3 powders coated at intermediate (a) and final (b) stages of synthesis.

According to literature, blue solid product should be complex $[Ni(NH_3)_6]Cl_2$ [22-23]. It is known that reduction potential of Ni^{2+}/Ni is higher than reduction potential $[Ni(NH_3)_6]^{2+}$/Ni, so formation of nickel from this complex in solution is thermodynamically more favorable in comparison with nickel chloride [24]. The known approach is the decomposition of the complex in alkaline solution (pH~13) that resulted to the formation of $Ni(OH)_2$ [25-26], that could be reduced by N_2H_4 by the next reaction [27-30]:

$$2Ni^{2+} + N_2H_4 + 4OH^- \rightarrow 2Ni + N_2 + 4H_2O \quad (1).$$

We have proposed the next stoichiometric reaction for our approach with minimal set of reagents:

$$5N_2H_4 * H_2O + NiCl_2 * 6H_2O \rightarrow Ni + 2N_2 + 4NH_3 + 2NH_4Cl + 8H_2O \quad (2).$$

As discussed above, the growth of nickel included intermediate reactions with formation of nickel complexes also.

In figure 4 SEM-images of Al_2O_3 powders coated at optimal (a) and reduced (b) concentration of oxidizer $NiCl_2 \cdot 6H_2O$ are shown. It has found that at reduced concentration of nickel salt the nickel coating is not dense. Also, we have studied the specific surface area of the received products.

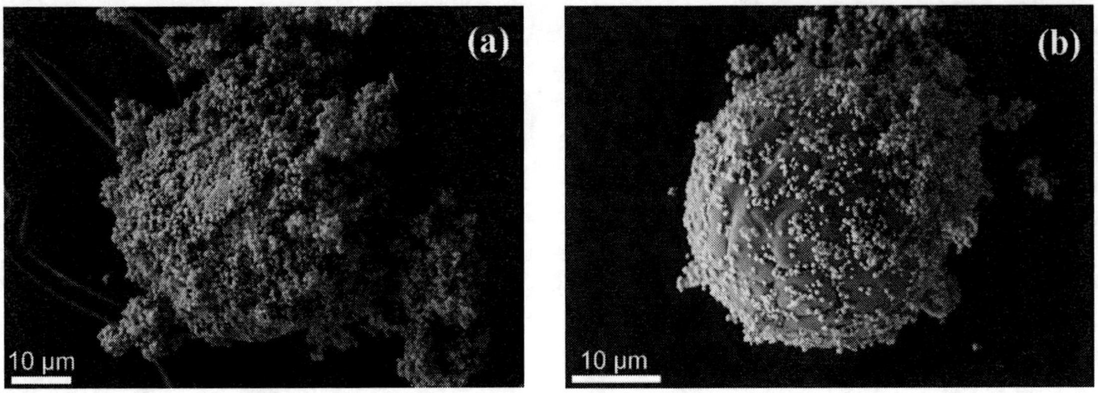

Figure 4. SEM-images of Al_2O_3 powders coated at optimal (a) and reduced (b) concentration of oxidizer $NiCl_2$.

It has measured the specific surface area of Al_2O_3 powder with and without nickel nanostructured coatings synthesized at different concentrations of nickel salt (table 1). The Al_2O_3 ceramic powder has a specific surface area 0.75 m^2/g, that much smaller than specific surface area of Al_2O_3 powder coated with nickel at optimal concentration of nickel oxidizer – 10.8 m^2/g. The specific surface area of Al_2O_3 powder coated with nickel at reduced concentration of nickel oxidizer is 2.6 m^2/g.

Table 1. Comparison of specific surface area of Al_2O_3 powders before and after nickel coating.

Sample	Specific surface area, m^2/g
Al_2O_3 ceramic powder	0.75
Nickel-coated Al_2O_3 powder at reduced concentration	2.6
Nickel-coated Al_2O_3 powder at optimal concentration	10.8

The increased specific surface area of nickel-coated Al_2O_3 powder could be explained by roughened nanostructured surface of nickel coating. SEM of the surface of nickel coatings synthesized at different concentrations of nickel salt are shown in figure 5. It is visible that the surface morphology of nickel coatings is rough and formed by homogeneous distributing of nanocones on the coating surface. The height of nanocones is up to 600 nm – 750 nm for the coating synthesized at optimal concentration of nickel salt, and about 200 nm for the coating synthesized at reduced concentration of nickel salt. These results are fully correlates with the experimental data about specific surface area of the Al_2O_3 powders before and after nickel coating.

Figure 5. SEM of the surface of Al_2O_3 powders with nickel coatings deposited at optimal (a) and reduced (b) concentration of oxidizer $NiCl_2$.

4. Conclusion

The nickel coatings with high surface area have synthesized on ceramic spheroidized Al_2O_3 powder. The new approach of synthesis of nickel coatings by procedure with a minimum set of starting reagents is presented. It has studied the dynamic of the synthesis process of nickel coating by the scanning electron microscopy (SEM). The surface area of the synthesized composite powders was studied by nitrogen adsorption. It has found that the specific surface area of the nickel coated Al_2O_3 powder could be reach 10.8 m^2/g. That is related with roughened surface of the nickel coating formed by homogeneous distributing of nanocones with the height up to 600 nm – 750 nm on the Al_2O_3 powder surface. The obtained results may have wide applications in catalysis and electrochemistry.

Acknowledgments
The reported study was supported by the Ministry of Science and Higher Education of the Russian Federation, government assignment with the registration number AAAA-A20-120122490071-1. SEM measurements were performed using the device of the Center «Applied Nanotechnologies» of KNRTU-KAI by Sergey Drobyshev.

References

[1] Török T I, Pázmán J, Szabó M and Janó V 2013 Laboratory preparation and characterization of electroless nickel coated powders of industrially produced aluminium and iron (III) oxides *Mater. Sci. Forum* **752** 284-93
[2] Gromov A A, Nalivaiko A Y, Ambaryan G N, Vlaskin M S, Buryakovskaya O A, Kislenko S A, and Arnautov A N 2019 Aluminum-alumina composites: Part I: Obtaining and characterization of powders *Mater.* **12**(19) 3180
[3] Kim P, Kim Y, Kim H, Song I. K and Yi J. 2004 Synthesis and characterization of mesoporous alumina with nickel incorporated for use in the partial oxidation of methane into synthesis gas *Appl. Catal., A* **272**(1-2) 157-66
[4] Latiff M I A, Nuruzzaman D M, Basri S, Ismail N M, Jamaludin S N S and Kamaruzaman F F 2018 Preparation and characterization of 6-layered functionally graded nickel-alumina (Ni-Al_2O_3) composites *IOP Conference Series: Materials Science and Engineering* **342**(1) 012063
[5] Hanyaloglu S C, Aksakal B and McColm I J 2001 Reactive sintering of electroless nickel-plated aluminum powders. *Mater. Charact*. **47**(1) 9-16
[6] Wen G, Guo Z X and Davies C K L 2000 Microstructural characterisation of electroless-nickel coatings on zirconia powder *Scripta mater.* ***43***(4) 307-11
[7] Karayannis V G and Moutsatsou A K 2012 Synthesis and characterization of nickel-alumina composites from recycled nickel powder *Adv. in Mater. Sci. and Eng* **2012** 1-9
[8] Holzer L, Iwanschitz B, Hocker T, Münch B, Prestat M, Wiedenmann D and Graule T 2011 Microstructure degradation of cermet anodes for solid oxide fuel cells: Quantification of nickel grain growth in dry and in humid atmospheres *J. Power Sources* **196**(3) 1279-94
[9] Morozov M, Ivanov S, Kadirov M and Bund A 2021 Facile synthesis of a binder-free 3D Ni/NiO microwire network with a nanostructured fiber surface for a negative electrode in Li-ion battery *J. Appl. Electrochem.* **51**(5) 815-28
[10] Liu C, Li C, Ahmed K, Mutlu Z, Ozkan C S and Ozkan M 2016 Template free and binderless NiO nanowire foam for Li-ion battery anodes with long cycle life and ultrahigh rate capability *Scientific reports* **6**(1) 1-8
[11] Luo L M, Lu Z L, Tan X Y, Ding X Y, Huang L M, Cheng J G and Wu Y C 2013 A specific chemical activation pretreatment for electroless nickel plating on SiC ceramic powders *Powder technol.* **249** 431-5
[12] Luo L, Wu Y, Li J and Zheng Y 2011 Preparation of nickel-coated tungsten carbide powders by room temperature ultrasonic-assisted electroless plating *Surf. Coat. Technol.* **206**(6) 1091-5

[13] Lucchini M A, Testino A, Ludwig C, Kambolis A, El-Kazzi M, Cervellino A and Canepa F 2014 Continuous synthesis of nickel nanopowders: Characterization, process optimization, and catalytic properties *App. Cat. B: Environmental* **156** 404-15

[14] Hirata Y, Hatano H, Kyoda H and Hamasaki K 1995 Synthesis of alumina/nickel composite by electrodeposition of nickel. *J. mat. res.* **10**(11) 2697-9

[15] Haag S, Burgard M and Ernst B 2006 Pure nickel coating on a mesoporous alumina membrane: preparation by electroless plating and characterization *Surf. Coat. Technol.* **201**(6) 2166-73

[16] Lin Y J and Jiang B F 1998 Sintering and Phase Evolution of Electroless-Nickel-Coated Alumina Powder. *J. Am. Chem. Soc.* **81(9)** 2481-4

[17] Uysal M, Karslioğlu R, Alp A and Akbulut H 2013 The preparation of core–shell Al_2O_3/Ni composite powders by electroless plating *Ceram. int.* 39(5) 5485-93

[18] Choi J Y, Lee Y K, Yoon S M, Lee H C, Kim B K, Kim J M and Lee J H 2005 A chemical route to large-scale preparation of spherical and monodisperse Ni powders *J. Am. Chem. Soc.* **88**(11) 3020-3

[19] Liang-Hsing C H O U, Wen-Ching H U N G and Maw-Tien L E E 2016 On the preparation of nickel nanoparticles by chemical reduction method: X-ray absorption spectroscopy. *Mater. Sci.* **22**(2) 305-8

[20] Eluri R and Paul B 2012 Synthesis of nickel nanoparticles by hydrazine reduction: mechanistic study and continuous flow synthesis *J. nanopart. res.* **14**(4) 1-14

[21] Nicholls D and Swindells R 1968 Hydrazine complexes of nickel (II) chloride *J. Inorg. Nucl. Chem.* **30**(8) 2211-7

[22] Guo L, Liu C, Wang R, Xu H, Wu Z and Yang S 2004 Large-scale synthesis of uniform nanotubes of a nickel complex by a solution chemical route *J. Am. Chem. Soc.* **126**(14) 4530-1

[23] Rejitha K S, Ichikawa T and Mathew S 2011 Thermal decomposition studies of $[Ni(NH_3)_6]$ X_2 (X= Cl, Br) in the solid state using TG-MS and TR-XRD J. Therm. Anal. Calorim. **103**(2) 515-23

[24] Ningthoujam R S, Gajbhiye N S and Sharma S 2009 Reduction mechanism of Ni^{2+} into Ni nanoparticles prepared from different precursors: magnetic studies *Pramana* **72**(3) 577-86

[25] Li Y D, Li C W, Wang H R, Li L Q and Qian Y T 1999 Preparation of nickel ultrafine powder and crystalline film by chemical control reduction *Mater. chem. phys.* **59**(1) 88-90

[26] Gao J, Guan F, Zhao Y, Yang W, Ma Y, Lu X and Kang J 2001 Preparation of ultrafine nickel powder and its catalytic dehydrogenation activity *Mater. chem. phys.* **71**(2) 215-9

[27] Chen D H and Wu S H 2000 Synthesis of nickel nanoparticles in water-in-oil microemulsions *Chem. Mater.* **12**(5) 1354-60

[28] Boudjahem A G, Monteverdi S, Mercy M and Bettahar M M 2004 Study of support effects on the reduction of Ni^{2+} ions in aqueous hydrazine *Langmuir* **20**(1) 208-13

[29] Zheng H G, Liang J H, Zeng J H and Qian Y T 2001 Preparation of nickel nanopowders in ethanol-water system (EWS) *Mater. Res. Bull.* **36**(5-6) 947-52

[30] Park J W, Chae E H, Kim S H, Lee J H, Kim J W, Yoon S M and Choi J Y 2006 Preparation of fine Ni powders from nickel hydrazine complex *Mater. chem. phys.* **97**(2-3) 371-8

EFFECT OF TEMPERATURE, PRESSURE, AND FRACTION OF NANOPARTICLES ON CHANGES IN THE THERMAL CONDUCTIVITY OF COLLOID NANOFLUIDS

T R Tilloeva[1], M M Safarov[1], M A Zaripova[1]

[1]Tajik technical university after named by academic M.C.Osimi, Dushanbe, Tajikistan

mahmad1@list.ru

Abstract. The paper presents the results of studies on the thermal diffusivity of colloidal nanofluids depending on the change in one of the state parameters (pressure). Studies were carried out both with a change in the pressure of the experiment, and with a change in the nanometric size of silver particles in a liquid. Based on the results obtained, the corresponding empirical equations were derived and a proper comparative analysis was performed by comparing the experimental and calculated data on thermal diffusivity. The suitability of the derived equations is evidenced by the satisfactory agreement between the obtained data and the calculated ones within 1.79% at α=0.95.

1. Introduction

Colloidal systems refer to dispersed systems in which one substance in the form of various particle sizes is evenly distributed in another substance. The dispersed system consists of a dispersed phase (finely divided substance) and a dispersed medium (homogeneous substance), in which the dispersed phase is distributed. Dispersed systems include ordinary and colloidal solutions, as well as suspensions and emulsions, which, in turn, differ from each other in particle size or degree of dispersion. Disperse systems are classified according to their state of aggregation, according to particle size, according to the nature of the dispersed phase and medium. According to the degree of dispersion, they mainly distinguish between coarse and colloidal systems, the first of which include particles of the dispersed phase with a size of 10^{-7}m, and the second particles of the dispersed phase with a size of 10^{-7}m - 10^{-9}m [6]. In this work, we studied a colloidal aqueous solution of nanosilver with different particle diameters from 2 to 10 nm. The work is devoted to the experimental study of the thermal diffusivity of a given substance depending on the change in the pressure of the experiment.

2. Experimental section

In recent years, to study the process of heat transfer, more and more people resort to the laser pulse method (Parker method [1]), which is based on the absorption of an energy pulse in a thin layer of the surface of the substance under study and fixing the temperature change of its opposite surface over time [2,3].

The values obtained using this method make it possible to calculate such thermophysical characteristics as heat conductivity, specific heat capacity, thermal conductivity of a wide range of materials in the form of solids, powders, liquids, pastelike substances, fibrous materials, films [4].

Content from this work may be used under the terms of the Creative Commons Attribution 3.0 licence. Any further distribution of this work must maintain attribution to the author(s) and the title of the work, journal citation and DOI.
Published under licence by IOP Publishing Ltd

Table 1. Experimental values of thermal diffusivity ($\alpha.10^7$,m^2/s) of a colloidal aqueous solution of nanosilver with a concentration of nanometallic particles of 0.05% depending on the particle diameter and pressure at room temperature [5].

d,nm \ *P, MPa*	0,101	0,108	0,114	0,121	0,128	0,135	0,141
2	1,372	1,449	1,511	1,57	1,661	1,732	1,772
5	1,245	1,282	1,317	1,331	1,443	1,553	1,583
10	0,964	0,994	1,029	1,059	1,056	1,135	1,176

According to the values given in Table 1, it can be seen that with an increase in pressure from 0.101MPa to 0.141MPa, the thermal conductivity of the studied samples increases by~(20-29)%, and as the diameter of silver nanoparticles incre-ases at a pressure of 0.141MPa with a concentration of 0.05% thermal conductivity decreases by~33.6%, This can be explained by the fact that, like thermal con-ductivity, temperature conductivity depends on the intermolecular forces of inte-raction and the distance between these particles, through which the transfer of thermal energy is actually carried out, and since thermal diffusivity characterizes the rate of propagation of thermal energy, these factors have a significant impact on this parameter.

In the sample we are studying, particles of a liquid and particles of a solid body uniformly distributed in it interact. Due to the fact that small particles have a larger specific surface, compared to a solution of the same concentration, but a large size of silver nanoparticles, the contact surface with liquid particles increases accordingly, and the rate of thermal energy transfer also increases [5].

3. Conclusion

After performing the appropriate processing of the experimental data obtained by using the law of the corresponding states and thermodynamic similarity, we obtained an empirical equation in the form.

$$\alpha = \left[0{,}8152\left(\frac{P}{P^*}\right) + 0{,}2307\right] \cdot [-0{,}0545 \cdot 10^{-7}d + 1{,}61 \cdot 10^{-7}],\ \mathrm{m^2/s}$$

Using the proposed equation (at n=0.05% and d=2÷10nm), it is possible to calculate the thermal diffusivity of the investigated colloidal aqueous solutions of nanosilver depending on the change in the experimental pressure in the range (0.101-0.141)MPa at room temperature with an error of 1.79%.

Low-temperature plasma can be used for the synthesis of various nanostructures and is well suited for the modification of various surfaces. This is shown in many works [7-27].

References

[1] Kuznetsov, G.V. On the conditions for the application of impulse methods determination of thermophysical characteristics of structural materials / V.G. Kuznetsov, M.D. Katz // Bulletin of the Tomsk Polytechnic University - 2008. V. 312. No. 4. P.10-13.

[2] Miroshnichenko, V.I. Experimental setup with pulsed laser heating to study the thermal conductivity and heat capacity of solids at high temperatures/V.I. Miroshnichenko, V.V.Makhrov, M.V. Rebrov//Thesis document. 9th Thermophysical Conference of the CIS. Makhachkala, 1992. P. 22-23.

[3] Shashkov, A.G. Methods for determining thermal conductivity and temperature conductivity./A.G.Shashkov,G.M.Volokhov, T.N. Abramenko//M.: Energy, 1973. -335p.

[4] Klyuev, V.V. Unbrakable control. / V.V.Klyuev // Handbook M: Mashinostro-enie 2004.-679 p.

[5] Tilloeva, T.R. Thermophysical and thermodynamic properties of a colloidal aqueous solution of nanosilver depending on temperature and pressure. /Tahmina Rustamovna Tilloeva// Diss. can. technical sciences - Kazan, 2016, 164p.

[6] Zimon, A.D. Colloidal chemistry./A.D. Zimon //M.: Agar, 2003. - 320 p.
[7] Saifutdinov, A. I., & Sofronitskii, A. O. (2021). Numerical Study of Breakdown and Formation Dynamics of Arc Discharge Plasma Parameters at Ultrahigh Pressures. High Energy Chemistry, 55(3), 228-232.DOI: 10.1134/S0018143921030115
[8] Shemakhin, A. Y., Zheltukhin, V. S., Shemakhin, E. Y., Terentev, T. N., & Sofronitsky, A. O. (2020, July). Experimental installation to study the RF plasma flow at low pressures with experiment data synchronization. In Journal of Physics: Conference Series (Vol. 1588, No. 1, p. 012018). IOP Publishing. DOI: 10.1088/1742-6596/1588/1/012018
[9] Saifutdinova, A. A., Sofronitskiy, A. O., Timerkaev, B. A., & Saifutdinov, A. I. (2020). Plasma-Chemical Decomposition of Hydrocarbons on the Basis of the Micro-Arc Discharge with Disc Electrodes Rotating in the Bulk of Raw Materials. Russian Physics Journal, 62(11), 2132-2136.DOI: 10.1007/s11182-020-01957-0
[10] Fairushin, I. I., Saifutdinov, A. I., & Sofronitskiy, A. O. (2020). Numerical and Experimental Studies of the Synthesis of Copper Nanoparticles in a High-Pressure Discharge. High Energy Chemistry, 54(2), 150-153. DOI: 10.1134/S0018143920020071
[11] Asadullin, T. Y., Galeev, I. G., Sofronitskiy, A. O., & Gizeev, M. M. (2019, November). Acoustic impact on electric discharge parameters during sterilization of freeze-dried products. In Journal of Physics: Conference Series (Vol. 1370, No. 1, p. 012016). IOP Publishing. DOI: 10.1088/1742-6596/1370/1/012016
[12] Fairushin, I. I., Saifutdinov, A. I., Sofronitskiy, A. O., Timerkaev, B. A., & Dautov, G. Y. (2019, October). Development of plasma reactor design for synthesis of copper nanoparticles using multi-scale simulation. In Journal of Physics: Conference Series (Vol. 1328, No. 1, p. 012088). IOP Publishing. DOI: 10.1088/1742-6596/1328/1/012088
[13] Asadullin, T. Y., Galeev, I. G., & Sofronitskiy, A. O. (2019, October). The application of pulsed discharge for sterilization of freeze-dried product. In Journal of Physics: Conference Series (Vol. 1328, No. 1, p. 012071). IOP Publishing. DOI: 10.1088/1742-6596/1328/1/012071
[14] Timerkaev, B. A., Ganieva, G. R., Kaleeva, A. A., Israfilov, Z. K., & Sofronitskii, A. O. (2019). Growing of carbon nanotubes from hydrocarbons in an arc plasma. Journal of Engineering Physics and Thermophysics, 92(5), 1248-1252. DOI: 10.1007/s10891-019-02040-3
[15] Dautov, G. Y., Kashapov, N. F., Dautov, I. G., & Sofronitskiy, A. O. (2018, July). Research of the influence of the geometry of the discharge chamber on the characteristics of the arc plasmatron. In Journal of Physics: Conference Series (Vol. 1058, No. 1). IOP Publishing. DOI: 10.1088/1742-6596/1058/1/012035
[16] Timerkaev, B. A., Andreeva, A. A., & Sofronitskiy, A. O. (2017, November). Discharge creeping along the surface in the process for producing nanomaterials. In Journal of Physics: Conference Series (Vol. 927, No. 1, p. 012068). IOP Publishing. DOI: 10.1088/1742-6596/927/1/012068
[17] Sadikov, K. G., Sofronitskiy, A. O., & Larionov, V. M. (2017, November). The effect of electrically conductive additives on the plasma pyrolysis of heavy hydrocarbons. In Journal of Physics: Conference Series (Vol. 927, No. 1, p. 012046). IOP Publishing. DOI: 10.1088/1742-6596/927/1/012046
[18] Timerkaev, B. A., Andreeva, A. A., & Sofronitskiy, A. O. (2017). Discharge creeping along the surface in the process of cleaning and strengthening of the materials surface. In Journal of Physics: Conference Series (Vol. 789, No. 1). IOP Publishing. DOI: 10.1088/1742-6596/789/1/012063
[19] Sadikov, K. G., Sofronitskiy, A. O., & Dautov, I. G. (2017). Functional plasma sprayed coatings on magnesium ceramic substrates. In Journal of Physics: Conference Series (Vol. 789, No. 1, p. 012043). IOP Publishing. DOI: 10.1088/1742-6596/789/1/012043

[20] Timerkaev, B. A., Sofronitskiy, A. O., & Andreeva, A. A. (2016). Carbon nanotubes formation in the decomposition of heavy hydrocarbons creeping along the surface of the glow discharge. In Journal of Physics: Conference Series (Vol. 669, No. 1, p. 012062). IOP Publishing. DOI: 10.1088/1742-6596/669/1/012062

[21] Timerkaev, B. A., & Shamsutdinov, R. S. (2020, July). The Influence of Supersonic Gas Stream on Spatial Structure of Glow Discharge. In Journal of Physics: Conference Series (Vol. 1588, No. 1, p. 012061). IOP Publishing. DOI 10.1088/1742-6596/1588/1/012061

[22] Galeev I G, Asadullin T Y 2017 Improving of stability of the volumetric glow discharge in the gas flow J. of Phys. Conf. Ser. 789(1) 012012

[23] Asadullin T Y, Galeev I G 2017 Combined method of stabilization of the glow discharge in gas flow J. of Phys. Conf. Ser. 927(1) 012003

[24] Galeev I G, Asadullin T Ya 2016 Obtaining fullerene-containing soot during combustion of gaseous hydrocarbons in an external electric field J. of Phys. Conf. Ser. 669 012016

[25] Galeev I G, Asadullin T Y 2016 Enhancing of the glow discharge stability in chamber with cathode sections coated by a discontinuous dielectric coating J. of Phys. Conf. Ser. 669(1) 012015

[26] Shakirov A S, Galeev I G and Asadullin T Ya 2020 Features of formation of fullerene-containing soot during combustion gaseous and liquid hydrocarbons J. of Phys. Conf. Ser. 1588(1) 012013

[27] Shamsutdinov, R. S., & Timerkaev, B. A. (2021, April). The influence of a supersonic flow of gas at glow discharge. In Journal of Physics: Conference Series (Vol. 1870, No. 1, p. 012019). IOP Publishing. DOI 10.1088/1742-6596/1870/1/012019.

DENSITY OF TOLUENE-BASED NANOFLUIDS

Q Mirzoeva[1], J Matlabi[1], M M Safarov[2]

[1]Tajik State Pedagogical University named after S. Aini, Dushanbe, Tajikistan
[2]Tajik technical university after named by academic M.C.Osimi, Dushanbe, Tajikistan

mahmad1@list.ru

Abstract. The purpose of this work is to study the density of two-component titanium dioxide systems (nanosized particles of various fractions (30nm, -50nm and 70nm)) (up to 0.5%) + toluene in the temperature range (293) K at atmospheric pressure (0.101) Mpa.

1. Experimental section

To determine the density of nanofluids of a system of liquid toluene with the addition of nanoparticles at atmospheric pressure, a setup was used [1, 2], which consists of a thermostat and an analytical balance. The results of the obtained data on the density of nanofluids are shown in Tables 1-3.

The results of determining the density showed that with increasing temperature, the density of the studied solutions decreases according to the law of the first degree. The density of the solution with the addition of up to 0.1% titanium dioxide 70nm in toluene at a temperature of 293K increases by 1.39%, and at a temperature of 313K this difference reaches 1.06%. When added to pure toluene up to 0.5% titanium dioxide at 293K, the density increases to 2.67%.

Table 1. Density (ρ, kg/m^3) of a system of liquid toluene and nanosized titanium dioxide particles of various fractions (TiO_2, (ϕ=30nm)), depending on temperature at atmospheric pressure (0.101) MPa.

Objects / T,K	№1[3]	№2	№3	№4	№5	№6
293,2	867,5	870,6	872,8	875,3	878,3	881,5
303,6	858,2	861,5	864,4	867,6	870,3	873,6
313,2	848,3	851,4	854,5	857,4	860,4	862,5

Objects of study: №.1- (chemically pure toluene); №.2- (chemically pure toluene + 0.1%. TiO_2); №.3- (chemically pure toluene + 0.2% TiO_2); №.4- (chemically pure toluene + 0.3% TiO_2); №5 - (reagent grade toluene + 0.4% TiO_2); №.6- (chemically pure toluene + 0.5% TiO_2).

Content from this work may be used under the terms of the Creative Commons Attribution 3.0 licence. Any further distribution of this work must maintain attribution to the author(s) and the title of the work, journal citation and DOI.
Published under licence by IOP Publishing Ltd

Table 2. Density (ρ, kg/m^3) of a system of liquid toluene and nanosized particles of titanium dioxide of various fractions (TiO_2, (ϕ=50nm)), depending on tempe-rature at atmospheric pressure (0.101) MPa.

Objects / T,K	№1 [3]	№2	№3	№4	№5	№6
293,2	867,5	873,6	875,8	878,6	881,6	884,5
303,6	858,2	867,3	869,5	872,7	874,5	877,6
313,2	848,3	856,9	857,5	860,6	863,6	865,5

Objects of study: №.1- (chemically pure toluene); №.2- (chemically pure toluene + 0.1%. TiO_2); №.3- (chemically pure toluene + 0.2% TiO_2); №.4- (chemically pure toluene + 0.3% TiO_2); №5 - (reagent grade toluene + 0.4% TiO_2); №.6- (chemically pure toluene + 0.5% TiO_2)..

Table 3. Density (ρ, kg/m^3) of a system of liquid toluene and nanosized particles of titanium dioxide of various fractions (TiO_2, (ϕ=70nm)) depending on tempera-ture at atmospheric pressure (0.101) MPa.

Objects / T,K	№1 [3]	№2	№3	№4	№5	№6
293,2	867,5	879,6	881,8	885,6	888,6	890,5
303,6	858,2	873,4	876,5	880,7	883,5	885,6
313,2	848,3	856,9	860,5	865,6	869,6	871,5

Objects of study: №.1- (chemically pure toluene); №.2- (chemically pure toluene + 0.1%. TiO_2); №.3- (chemically pure toluene + 0.2% TiO_2); №.4- (chemically pure toluene + 0.3% TiO_2); №5 - (reagent grade toluene + 0.4% TiO_2); №.6- (chemically pure toluene + 0.5% TiO_2)

Thus, using the density values of the studied solutions given in Tables 1–3, it is possible to reveal the contribution of the concentration of toluene introduced into chemically pure liquid toluene depending on the temperature and different fractions of the nanofiller.

$$\rho = \left[1{,}075 - 0{,}075\left(\frac{T}{T_1}\right)\right]\left(-45547m^2 + 1064{,}8m + 1000{,}6\right), \ \frac{\text{kg}}{\text{m}^3} \quad (1)$$

2. Conclusion

This calculation allows us to conclude that when 0 to 0.5% TiO_2 is added to chemically pure toluene, its density increases, respectively, i.e. the greater the concentration of nanosized particles, the more influence is observed on the density of the studied solutions. Using equation (1), knowing the concentration of TiO_2 nanoparticles and the size of nanoparticle fractions, it is possible to calculate the thermodynamic characteristics (ρ,) and experimentally unexplored samples of the system (toluene+titanium dioxide with various fractions) depending on tempe-rature (293-313)K, at atmospheric pressure (0.101 MPa) with an error of up to 3.13%.

Low-temperature plasma can be used for the synthesis of various nanostructures and is well suited for the modification of various surfaces. This is shown in many works [4-24].

References

[1] Safarov, M.M. Thermophysical properties of ethers and aqueous solutions of hydrazine depending on temperature and pressure. / Mahmadali Mahmadievich Safarov//Dis. ... Doctor of Technical Sciences. – Dushanbe, 1993. – 495 p.

[2] Safarov, M.M. P-p-T dependences of aqueous solutions of hydrazine./ M.M. Safarov, A.V. Kartavchenko.//Journal of Physics. Chemistry, 1993, M., T. 67, No. 4.-P. 710-712.

[3] Physical and chemical properties of individual hydrogens. Handbook under. ed. Tatevskiy V.M., M.: -1960.-873p.

[4] Saifutdinov, A. I., & Sofronitskii, A. O. (2021). Numerical Study of Breakdown and Formation Dynamics of Arc Discharge Plasma Parameters at Ultrahigh Pressures. High Energy Chemistry, 55(3), 228-232.DOI: 10.1134/S0018143921030115

[5] Shemakhin, A. Y., Zheltukhin, V. S., Shemakhin, E. Y., Terentev, T. N., & Sofronitsky, A. O. (2020, July). Experimental installation to study the RF plasma flow at low pressures with experiment data synchronization. In Journal of Physics: Conference Series (Vol. 1588, No. 1, p. 012018). IOP Publishing. DOI: 10.1088/1742-6596/1588/1/012018

[6] Saifutdinova, A. A., Sofronitskiy, A. O., Timerkaev, B. A., & Saifutdinov, A. I. (2020). Plasma-Chemical Decomposition of Hydrocarbons on the Basis of the Micro-Arc Discharge with Disc Electrodes Rotating in the Bulk of Raw Materials. Russian Physics Journal, 62(11), 2132-2136.DOI: 10.1007/s11182-020-01957-0

[7] Fairushin, I. I., Saifutdinov, A. I., & Sofronitskiy, A. O. (2020). Numerical and Experimental Studies of the Synthesis of Copper Nanoparticles in a High-Pressure Discharge. High Energy Chemistry, 54(2), 150-153. DOI: 10.1134/S0018143920020071

[8] Asadullin, T. Y., Galeev, I. G., Sofronitskiy, A. O., & Gizeev, M. M. (2019, November). Acoustic impact on electric discharge parameters during sterilization of freeze-dried products. In Journal of Physics: Conference Series (Vol. 1370, No. 1, p. 012016). IOP Publishing. DOI: 10.1088/1742-6596/1370/1/012016

[9] Fairushin, I. I., Saifutdinov, A. I., Sofronitskiy, A. O., Timerkaev, B. A., & Dautov, G. Y. (2019, October). Development of plasma reactor design for synthesis of copper nanoparticles using multi-scale simulation. In Journal of Physics: Conference Series (Vol. 1328, No. 1, p. 012088). IOP Publishing. DOI: 10.1088/1742-6596/1328/1/012088

[10] Asadullin, T. Y., Galeev, I. G., & Sofronitskiy, A. O. (2019, October). The application of pulsed discharge for sterilization of freeze-dried product. In Journal of Physics: Conference Series (Vol. 1328, No. 1, p. 012071). IOP Publishing. DOI: 10.1088/1742-6596/1328/1/012071

[11] Timerkaev, B. A., Ganieva, G. R., Kaleeva, A. A., Israfilov, Z. K., & Sofronitskii, A. O. (2019). Growing of carbon nanotubes from hydrocarbons in an arc plasma. Journal of Engineering Physics and Thermophysics, 92(5), 1248-1252. DOI: 10.1007/s10891-019-02040-3

[12] Dautov, G. Y., Kashapov, N. F., Dautov, I. G., & Sofronitskiy, A. O. (2018, July). Research of the influence of the geometry of the discharge chamber on the characteristics of the arc plasmatron. In Journal of Physics: Conference Series (Vol. 1058, No. 1). IOP Publishing. DOI: 10.1088/1742-6596/1058/1/012035

[13] Timerkaev, B. A., Andreeva, A. A., & Sofronitskiy, A. O. (2017, November). Discharge creeping along the surface in the process for producing nanomaterials. In Journal of Physics: Conference Series (Vol. 927, No. 1, p. 012068). IOP Publishing. DOI: 10.1088/1742-6596/927/1/012068

[14] Sadikov, K. G., Sofronitskiy, A. O., & Larionov, V. M. (2017, November). The effect of electrically conductive additives on the plasma pyrolysis of heavy hydrocarbons. In Journal

of Physics: Conference Series (Vol. 927, No. 1, p. 012046). IOP Publishing. DOI: 10.1088/1742-6596/927/1/012046

[15] Timerkaev, B. A., Andreeva, A. A., & Sofronitskiy, A. O. (2017). Discharge creeping along the surface in the process of cleaning and strengthening of the materials surface. In Journal of Physics: Conference Series (Vol. 789, No. 1). IOP Publishing. DOI: 10.1088/1742-6596/789/1/012063

[16] Sadikov, K. G., Sofronitskiy, A. O., & Dautov, I. G. (2017). Functional plasma sprayed coatings on magnesium ceramic substrates. In Journal of Physics: Conference Series (Vol. 789, No. 1, p. 012043). IOP Publishing. DOI: 10.1088/1742-6596/789/1/012043

[17] Timerkaev, B. A., Sofronitskiy, A. O., & Andreeva, A. A. (2016). Carbon nanotubes formation in the decomposition of heavy hydrocarbons creeping along the surface of the glow discharge. In Journal of Physics: Conference Series (Vol. 669, No. 1, p. 012062). IOP Publishing. DOI: 10.1088/1742-6596/669/1/012062

[18] Timerkaev, B. A., & Shamsutdinov, R. S. (2020, July). The Influence of Supersonic Gas Stream on Spatial Structure of Glow Discharge. In Journal of Physics: Conference Series (Vol. 1588, No. 1, p. 012061). IOP Publishing. DOI 10.1088/1742-6596/1588/1/012061

[19] Galeev I G, Asadullin T Y 2017 Improving of stability of the volumetric glow discharge in the gas flow J. of Phys. Conf. Ser. 789(1) 012012

[20] Asadullin T Y, Galeev I G 2017 Combined method of stabilization of the glow discharge in gas flow J. of Phys. Conf. Ser. 927(1) 012003

[21] Galeev I G, Asadullin T Ya 2016 Obtaining fullerene-containing soot during combustion of gaseous hydrocarbons in an external electric field J. of Phys. Conf. Ser. 669 012016

[22] Galeev I G, Asadullin T Y 2016 Enhancing of the glow discharge stability in chamber with cathode sections coated by a discontinuous dielectric coating J. of Phys. Conf. Ser. 669(1) 012015

[23] Shakirov A S, Galeev I G and Asadullin T Ya 2020 Features of formation of fullerene-containing soot during combustion gaseous and liquid hydrocarbons J. of Phys. Conf. Ser. 1588(1) 012013

[24] Shamsutdinov, R. S., & Timerkaev, B. A. (2021, April). The influence of a supersonic flow of gas at glow discharge. In Journal of Physics: Conference Series (Vol. 1870, No. 1, p. 012019). IOP Publishing. DOI 10.1088/1742-6596/1870/1/012019.

Electrophysical parameters of AC plasma system

I I Oshenko[1], S A Smirnov[1]

[1] Department of technology of devices and materials for electronic equipment, Ivanovo State University of Chemistry and Technology, Ivanovo, 153000, Russia

E-mail: sas@isuct.ru

Abstract. Experimental results related to main characteristics of an underwater discharge with a frequency of 50 Hz and various types of electrodes were reported. The data on 1) discharge geometry; 2) current-voltage characteristics; 3) gas temperature and reduced electric field strength; 4) plasma emission spectra were obtained. Based on these data, the plasma modeling provided the information on densities of plasma active species.

1. Introduction

Investigations of interaction processes between non-equilibrium plasmas and a liquid phase are becoming important for a large number of science and technology applications. Several decades ago, research attention in this field was mainly attracted to both electrolysis and breakdown phenomena in dielectric liquids [1, 2, 3]. Recently, the field of plasma-liquid interaction studies has been expanded significantly and already covers analytical chemistry, water purification and disinfection, material processing, chemical synthesis and some others [4, 5, 6].

Similarly to gas-phase plasmas, there are some classifications of plasma-liquid systems according to the generation method or equipment configuration. At the same time, an important issue is also the type of interaction with a liquid which strongly affects the properties of plasma itself. Generally, all the variety of interactions between the gas discharge and a liquid phase can be divided into three main groups. These are gas-phase plasmas which are in the contact with a liquid, multiphase plasma discharges and plasmas directly excited in the liquid phase. Systems from the third group allow one to use various liquids, both conductive and non-conductive (for example, liquid nitrogen, ethanol or deionized water) as electrolytes. In this case, both the electrode material and ions from the solution may be the sources of solid particles generated in a liquid phase. In particular, when the electrode provides the material for nanoparticles, the corresponding process is called as the "plasma sputtering in solution" [7]. It is known also that plasma discharges in a liquid can be excited using pulsed power sources, direct current (DC) sources as well as alternate current (AC) sources [8].

Though some investigations of plasmas excited in the liquid phase have been made, there many unclear points that need additional research efforts. In particular, at the moment it is not clear how breakdowns are formed in a liquid medium. Obviously, details of the breakdown mechanism depend on both characteristics of the excitation wave and properties of the liquid. At the same time, the excitation by direct or alternate currents may lead to the Joule heating and thus, to the formation of a vapor phase through which a breakdown can occur. As such, it is believed that liquid phase discharges generated by microsecond pulses are initiated due to either gas bubbles which originally present in the liquid or bubbles formed after applying the voltage. At the same time, the ignition of a discharge seems to be possible without bubbles [9], since nanosecond high-voltage pulses are too short to form bubbles.

Content from this work may be used under the terms of the Creative Commons Attribution 3.0 licence. Any further distribution of this work must maintain attribution to the author(s) and the title of the work, journal citation and DOI.
Published under licence by IOP Publishing Ltd

The purpose of this work was to determine principal characteristics of an underwater discharge with using the 50 Hz power supply as well as various types of electrodes.

2. Experimental

Experimental setup is shown in Figure 1. The discharge was excited between two metallic electrodes in ceramic cases immersed in the liquid phase in a form of the distilled water. Each electrode has a thickness of 1 mm while tested electrode materials were molybdenum, copper and steel (St3sp). The angle between electrodes was 45^0 with a possibility of adjustment. The discharge current (i) was 50-450 mA, and the interelectrode distance (L) was 0.1-5 mm.

The discharge power supply system included a step-up transformer (maximum output voltage of 10 kV) and an adjustable laboratory autotransformer. The GWinstek GDS-71022 oscilloscope was used to determine the voltage-current characteristics of the discharge. The geometric dimensions of the discharge (bubbles) were determined using photographs made by a digital high-speed camera. Plasma emission spectra were recorded using an AvaSpec-2048L-2-USB2 spectrophotometer. The spectrophotometer is equipped with diffraction gratings of 600 strokes/mm and an operating wavelength range of 200-900 nm. The radiation was collected on a collimating lens of a fiber light guide, through which it fell into the entrance slit (25 microns wide and 100 microns high).

Optical plasma diagnostics methods are commonly used to determine rotational and vibrational temperatures of excited electronic states as well as densities of both excited and ground-state particles [10, 11]. In this work, in order to obtain rotational and vibrational temperatures of OH′ radicals, we used the Cyber Wit Diatomic program. For the $OH(A^2\Sigma, V' \rightarrow X^2\Pi, V'')$ bands the values of the vibrational and rotational temperature of $OH(A^2\Sigma)$ molecules were adjusted to achieve the minimum deviation of the calculated band profiles from the experimentally measured ones. The arc discharge temperature was determined by analyzing the slope of the continuum on the recorded spectra. As it follows from Planck's formula, the angle tangent of inclination of the radiation spectrum constructed in certain coordinates characterizes the temperature of the emitting object.

The spectrum of thermal radiation of a real object with an arbitrary emissivity $\varepsilon(\lambda,T) \leq 1$ is described by the formula: $I(\lambda,T) = \varepsilon \cdot C_1 \cdot \lambda^{-5}[\exp(C_2/\lambda T)^{-1}]^{-1}$, where C_1 and C_2 are the first and second pyrometric constants ($C_1 = 2\pi hc^2 = 37418$ W·μm^4·cm^{-2}; $C_2 = h \cdot c/k = 14388$ μm·K), h is Planck's constant, k is Boltzmann's constant, c – the speed of light in a vacuum, λ is the wavelength. For the Wine region, the formula takes the form $\ln(\varepsilon C_1 - - C_2/(\lambda T) = \ln(\lambda^5 I)$. For a gray body (ε = const), this expression is the equation of a straight line in coordinates $x=C_2/\lambda$ and $y=\ln(\lambda^5 I)$ (T<3000K) or $x=\ln(1+C_1/\lambda^5 I)$ (T>4500K), while the slope of the straight line is determined by temperature [12].

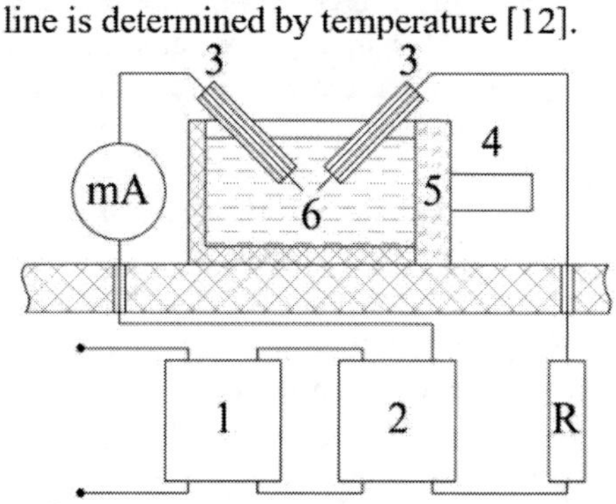

Figure 1. Experimental setup diagram (not to scale): 1 - adjustable laboratory autotransformer, 2 - step-up transformer, 3 - ceramic casings and electrodes, 4 – spectrophotometer lens, 5 - quartz cuvette with a volume of 50 ml, 6 - plasma formation zone, R - ballast resistance.

3. Results and Discussion

The approximate power density of heat sources was determined by analyzing the heating area $(dT/dt)_{Heating}$ and cooling area $(dT/dt)_{Cooling}$ of the liquid phase temperature from time to time (Figure 2).

The calculation of the power density of thermal effluents and sources was carried out according to the formula cm(dT/dt)= ∑Pi – ∑Pj. In this case, ∑Pi represents the total power of all heat sources heating the system, and ∑Pj the power of thermal effluents in the system. The power put into the system by a discharge at a current strength of 100 mA and an interelectrode distance of 5 mm ∑Pi ≈ 141 W, and ∑Pj ≈ 18 W.

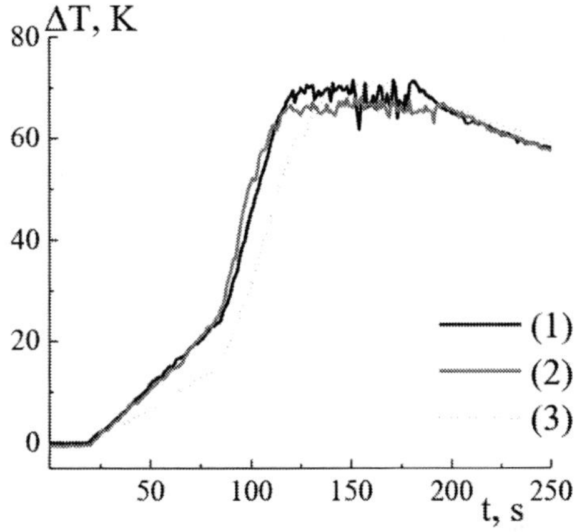

Figure 2. Change in the temperature of the liquid phase at the discharge between two electrodes made of copper (1), steel (2) and molybdenum (3); current 100 mA; interelectrode distance 5 mm.

Bands of hydroxyl radicals and hydrogen molecules, emission lines of iron, copper, molybdenum, hydrogen and oxygen atoms were detected in the discharge emission spectra (Figure 3).
The arc temperature did not depend on the types of electrodes and the interelectrode distance, amounting to 5800 ± 500 K. The effective rotational (Tr) and vibrational (Tv) temperatures of $OH(A^2\Sigma)$ are weakly dependent on external conditions Tr ≈ 3300±100 K, and Tv ≈ 7800±200 K.

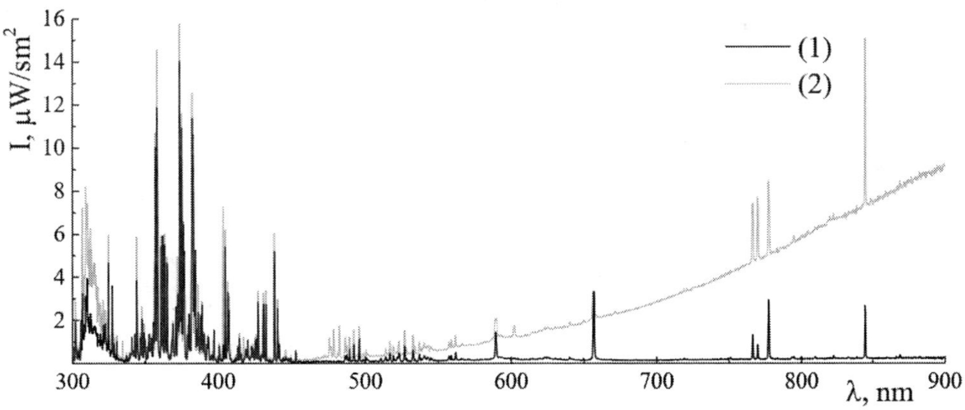

Figure 3. The emission spectrum of discharge in the form of a spark (1) and arc (2).

Based on the registered volt-ampere characteristics of the discharge (Figure 4), the duration of a single discharge was determined, which was 1.22 milliseconds. This value does not depend on the electrode material and the interelectrode distance. The voltage and current of the micro-discharge are determined mainly by the interelectrode distance. The frequency of initiation of single discharges in the vast majority of cases coincides with the frequency of the current. Under the same conditions, individual micro-discharges are almost completely identical, but are accompanied by stochastic current fluctuations. In the interval of several seconds, we can talk about the existence of a stable path along which the bulk of the discharges pass. The initially formed discharge passage area shifts to the surface of the liquid phase, which leads to the elongation of the discharge channel and subsequently its rupture.

Journal of Physics: Conference Series **2270** (2022) 012027 doi:10.1088/1742-6596/2270/1/012027

In the overwhelming majority of cases (≈ 70%), it can be said that the length of the discharge channel is ~ 10-20% longer than the interelectrode distance for all types of electrodes. Table 1 shows plasma electro physical parameters.

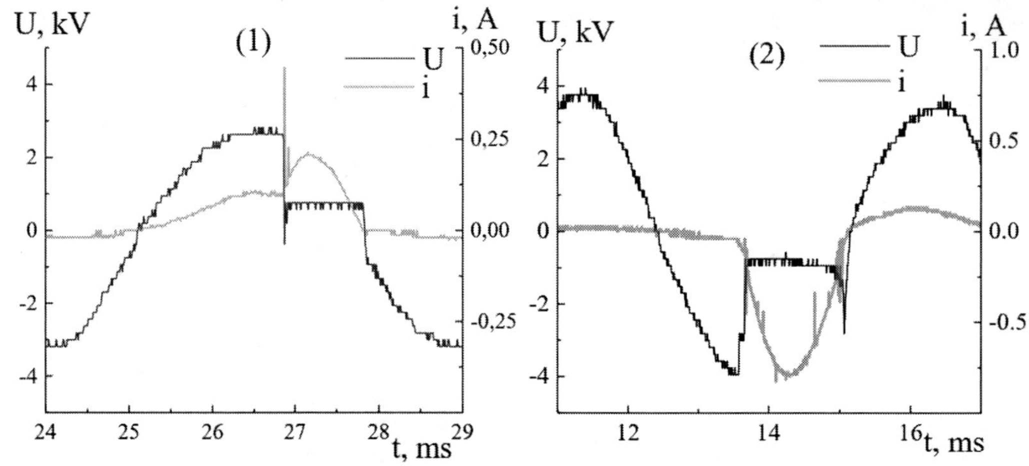

Figure 4. Current and voltage traces at the discharge between two electrodes made of copper (1) and molybdenum (2); interelectrode distance 5 mm.

Table 1. Plasma electro physical parameters [a].

Electrodes	U, V	i, A	E, V/cm	N, cm^{-3}	E/N, $V \cdot cm^2$	D, cm	J, A/cm^2	W, W/cm^3
Mo– Mo	228	0,276	114	$2,1 \cdot 10^{18}$	$0,5 \cdot 10^{-16}$	0,010	3,3	374
Cu– Cu	246	0,257	366	$2,2 \cdot 10^{18}$	$1,7 \cdot 10^{-16}$	0,015	1,4	511
Fe– Fe	311	0,633	239	$2,5 \cdot 10^{18}$	$1,1 \cdot 10^{-16}$	0,014	4,4	1042

[a] *Interelectrode distance 3 mm*

4. Conclusion

The results of the study of the parameters of an electric discharge in the liquid phase when using an alternating power source with a frequency of 50 Hz are presented. The digital photos, radiation spectra and volt-ampere characteristics of the discharge were recorded, the thermal sources of heating of the liquid phase and the geometric parameters of the discharge were evaluated.

References

[1] Kogelschatz U. 2003.*Plasma Chemistry and Plasma Processing* **23** 1
[2] Qi F., Li Y., Zhou R. 2019. *Applied Physics Letters* **115** 194101-1
[3] Hickling A., Ingram M.D. 1964. *Faraday Society* **60** 783
[4] Webb M.R. and Hieftje G.M. 2009. *Analytical Chemistry* **3** 862
[5] Khlyustova A., Sirotkin N., Kraev A. 2021. *Materialia* **16** 1
[6] Khan I., Saeed K., Khan I. 2019. *Arabian Journal of Chemistry* **12** 908
[7] Hu X. / X. Hu, X. Shen, O. Takai, N. Saito 2013. *J. Alloys Compd.* **552** 351
[8] Burakov V. S., Nevar A. A., Nedel'Ko M. I., Tarasenko N. V. 2008. *Tech. Phys. Lett.* **34** 679
[9] Starikovskiy A., Yang Y., Cho Y. 2011. *Plasma Sources Science and Technology* **2** 1
[10] Bruggeman P, Verreycken T, Gonz´alez M A at al. 2010. *Journal of Physics D Applied Physics* **43**(12):124005
[11] Zhang G., Shao X., Chang Z. *High Voltage Engineering* 2013 **9** 2260
[12] Lapshinov B.A., Suvorinov A.V., Timchenko N. *Control and measurements* 2018 **6** 116 (In Russian)

Low-current arc in plasma flow from discharge with liquid electrolyte cathode

K K Tazmeev[1], R N Tazmeeva[1] and B K Tazmeev[2]

[1]Kazan Federal University, Naberezhnye Chelny Institute, Naberezhnye Chelny, Russian Federation

[2]Kuban State Agrarian University named after I.T. Trubilin, Krasnodar, Russian Federation

tazmeevh@mail.ru

Abstract. The effect of plasma flow from a discharge with a solution cathode on the properties of a low-current electric arc has been experimentally studied. Electric power was supplied to the arc from a GORN-type inverter source. A rectifier served as a power source for the discharge with a solution cathode. The output voltage was 1700 V. Aqueous solutions of table salt were used in the experiments. The specific electrical conductivity of the solutions was in the range of 10–15 mS/cm. The electric arc was ignited between horizontally oriented rod electrodes. The interelectrode distance varied within 2–15 mm. The discharge current with a water cathode was established within 8–15 A. The arc burned in the current range of 1–8 A. The conditions under which stable burning of the electric arc is ensured are revealed.

1. Introduction

To solve a number of practical problems, a volumetric plasma of atmospheric pressure is required. In particular, this problem is relevant in the processes of air purification and sterilization in laboratory rooms for the study of nanomaterials. One of the sources of volumetric plasma is gas discharge with a liquid electrolyte cathode [1]. With the help of an electric arc, under certain conditions, it is also possible to obtain a volumetric plasma. Favorable conditions form in the so-called "sliding arc" [2]. The purpose of this work was to study the possibility of forming a volumetric plasma when two sources are used together: discharge with a solution cathode and an electric arc.

2. Experiment

Two variants of electric arc excitation were investigated in experiments. In the first variant, the arc was ignited in the interelectrode space of a gas discharge with a discharge with a solution cathode (figure 1a) and in the second variant, outside this space (figure 1b). Metal electrodes 1 and 2 were used for the electric arc. They were copper rods with diameters of 25 and 16 mm, respectively. Electrodes 1 and 2 were horizontally oriented. They were cooled with water. The distance l between them was set within 2–15 mm. Electric power was supplied to the arc from inverter sources of the GORN type. The maximum output voltage was 3000 V. The discharge was ignited with a solution cathode electrolyte cathode between metal electrode 3 and electrode assembly 4, from which the electrolyte flowed. A detailed description of electrode assembly 4 is given in [3]. Salt solutions were used as a liquid electrode. They were prepared with specific electrical conductivity σ in the range of 10–15 mS/cm. Electrode 3

Content from this work may be used under the terms of the Creative Commons Attribution 3.0 licence. Any further distribution of this work must maintain attribution to the author(s) and the title of the work, journal citation and DOI.
Published under licence by IOP Publishing Ltd

was made of copper in the form of a rod. It was cooled with water. This electrode was placed horizontally and was oriented perpendicular to the rod electrodes 1 and 2. The vertical distance *h* between the cathode of the arc (electrode 1) and the discharge anode with a solution cathode (electrode 3) varied between 2–7 cm. In both variants, the metal electrodes were located at a height of 2 cm above electrode assembly 4 (in the first version – electrodes 1 and 2, and in the second – electrode 3). The end of metal electrode 3 was moved away from the axial line of electrode assembly 4 by 1 cm. With this electrode configuration, the plasma flow from the liquid electrode almost freely rose upward. The gas discharge current with a liquid electrolyte cathode was regulated by the resistor 5. The power source was a rectifier. A C-L-C filter was connected to it. The output voltage was 1700 V.

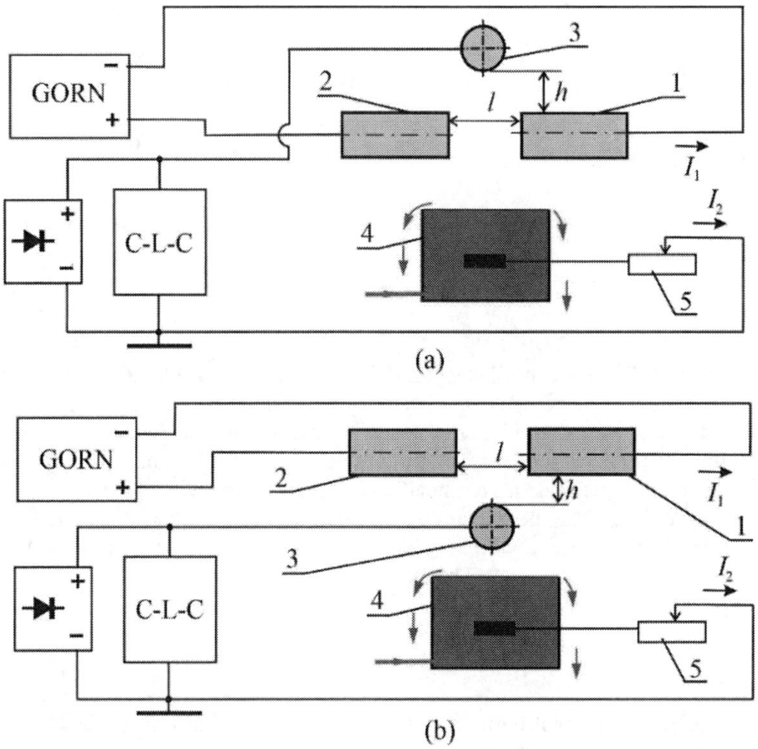

Figure 1. Electrical diagram of the experimental setup. Options for ignition of the arc: (a) – in the combustion zone of a gas discharge with a liquid electrolyte cathode; (b) – in the afterglow region.

Electric arc current I_1 was set within 1–8 A. The discharge current with the solution cathode I_2 varied in the range of 8–15 A. Currents I_1 and I_2 were measured with pointer instruments of precision class 0.2.

The value of σ was performed using an ANION 4150 conductometer before and after the experiments. The measurement results showed that there is an increase in σ. The aqueous solutions were reused if their electrical conductivity σ did not exceed 15 mS/cm.

The photographing was carried out with a video camera VIDEOSKAN-415. It made it possible to obtain frames with an exposure of 1 μs. The luminous flux was controlled by a diaphragm. Thus, photographs of the arc channel against the background of the plasma column of the discharge with a solution cathode were obtained.

3. Experimental results

Figure 2 shows photographs of the discharges obtained with the arrangement of the electrodes according to option (a).

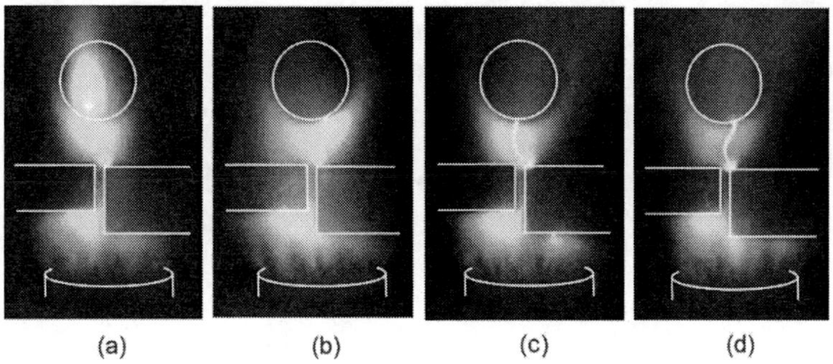

Figure 2. Pictures of discharges. White lines – contours of the electrodes. $l = 5$ mm. $h = 2$ cm. $I_1 = 7$ A, $I_2 = 8.5$ A.

As can be seen from the presented frames, in this variant there are some features of the burning of an electric arc. In the first two frames (figures 2a and b), the arc is not visible against the background of a discharge with a solution cathode. Here, the anode spot on the upper rod electrode stands out noticeably. In the next two frames, a contracted discharge channel was recorded. It is formed between the cathode of an electric arc and the anode of gas discharge with a liquid electrolyte cathode. The appearance of such a channel leads to some decrease in the radiating volume. In general, it can be noted that the ignition of an arc in the interelectrode space of a discharge with a solution cathode does not contribute to an increase in the plasma volume.

Figure 3 shows photographs of the discharges obtained with the arrangement of the electrodes according to option (b).

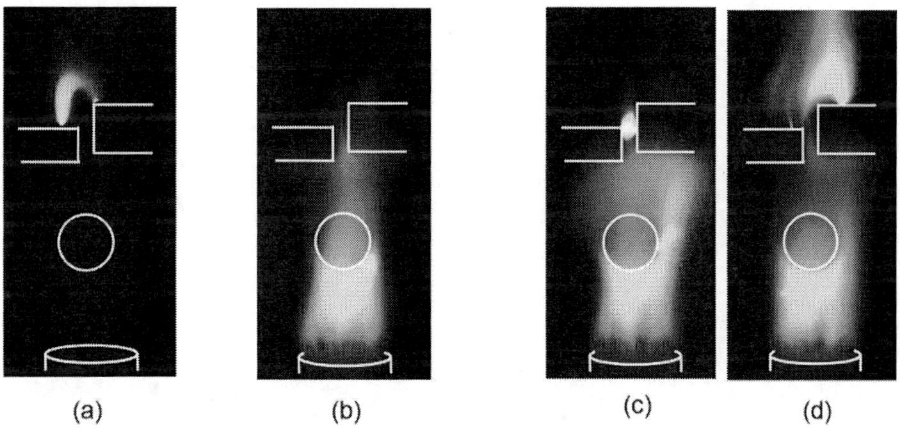

Figure 3. Instant photos of discharges in autonomous and joint combustion modes. (a) – electric arc; (b) – discharge with a solution cathode; (c) and (d) arc in the plasma flow from a discharge with a solution cathode. $l = 5$ mm. $h = 3$ cm. $I_1 = 7$ A, $I_2 = 10$ A.

In the first frame on the left (figure 3a), an electric arc was recorded in the variant of autonomous combustion. In this variant, the geometry of the discharge channel changed continuously. The processes of elongation and shortening of the discharge channel were repeated. Shunting phenomenon was observed. The mechanism of such processes is considered in more detail in [4-6]. In the second frame (figure 3b), autonomous combustion of a discharge with a solution cathode was recorded. It can be seen that the discharge forms a volumetric plasma column. In the third (figure 3c) and fourth (figure 3d) frames, different moments of electric arc burning in the gas discharge plasma flow with a discharge with a solution cathode were recorded: shunting (figure 3c) and upward blowing (figure 3d). The upward blowing of the electric arc was accompanied by the formation of a plasma cloud. At the moments of shunting, the cloud disappeared (figure 3c). Note that blowing out the arc with a plasma flow did not lead to its breakage. With an increase in the current of gas discharge with a discharge with a solution cathode I_2, arc shunting occurred less frequently.

4. Conclusions

Thus, experiments have shown that an electric arc can be used to increase the plasma volume. The greatest effect is achieved when the electric arc burns in the region of the afterglow of the discharge with a solution cathode.

References

[1] Bruggeman P et al. 2008 *Plasma Sources Science and Technology* **17**(2) 025012
[2] Korolev Y D 2013 *Gaodianya Jishu/High Voltage Engineering* **39**(9) 2061-2076
[3] Tazmeev G K, Timerkaev B A and Tazmeev K K 2019 *Journal of Physics: Conference Series* **1328** 012075
[4] Tazmeev K K and Tazmeev B K 2016 *Plasma Physics Reports* **42**(1) 86-90
[5] Yunusov R F and Garipov M M 2020 *Journal of Physics: Conference Series* **1588** 012064
[6] Yunusov R F and Yunusova E R 2020 *Journal of Physics: Conference Series* **1683** 032006

INVESTIGATION OF ELECTRIC DISCHARGE WITH LIQUID ELECTRODES UNDER INFLUENCE ON CARBON FIBER

K I Khazeev[1], R R Kayumov[1], A A Nizameev[1], M F Akhatov[1]

[1]Kazan National Research Technical University named after A.N. Tupolev -KAI, Kazan, Russia

rushan_250189033@mail.ru

Abstract. In this work, experimental reseaches of the effect
of a DC electric discharge with jet electrolytic cathode and liquid anode at
atmospheric pressure on carbon fiber are carried out. Experimental studies were carried out in the voltage range U= 100-1300 V and discharge current I=0.02-1.8 A. A 20% NaCl solution in process water was used as liquid electrodes. The length of the jet varied lc = 10÷50 mm, the diameter of the jet dc = 1÷3 mm. The optimal form of discharge and modes for processing carbon fiber have been established. The optimal mode of electric discharge which is between a jet electrolytic cathode and a liquid anode at atmospheric pressure is revealed. After processing for 30 s, the tensile strength of the modified samples changed by 7% compared to the untreated sample.

1. Introduction

Today, complex materials are widely consumed all over the world, which help to reduce the weight and improve the performance of various parts and assemblies that are used in the space industry, aviation technology, mechanical engineering, energy, shipbuilding, construction, the oil and gas industry, and even among sports goods.

Complex materials which are based on polymer binders strengthened by carbon fibers possess an original set of technically valuable properties, such as high specific effect, low lengthening during destruction, the highest heat resistance and electrical conductivity. One of the most crucial characteristics affecting the strength of carbon fiber is the traction of carbon fiber to the polymer matrix. To increase traction, it is proposed to modify the carbon fiber surface by such methods as thermochemical or plasma fiber recycling surface. These methods increase the endurance of complex materials compared to unprocessed fiber [1].

There are some methods for handling carbon fiber, one of the important spheres is low temperature plasma of an electric discharge using [2-5].

Content from this work may be used under the terms of the Creative Commons Attribution 3.0 licence. Any further distribution of this work must maintain attribution to the author(s) and the title of the work, journal citation and DOI.
Published under licence by IOP Publishing Ltd

Journal of Physics: Conference Series **2270** (2022) 012029 doi:10.1088/1742-6596/2270/1/012029

2. Main part

Experimental studies of carbon fiber modification were carried out with interaction of discharge with jet cathode and a liquid anode at atmospheric pressure in the span of voltage U=100÷1300V and discharge current I=0.02÷1.8A. A 20% NaCl solution in technical water was used as liquid electrodes. The jet length varied lc=10÷50mm, jet diameter dc=1÷3mm.

Experimental researches of the impact of direct current electric discharge with liquid electrodes were carried out on the installation shown in fig. 1. The installation consists of a direct current source, an electrolyte nozzle and a discharge chamber with a coordinate mechanism for adjusting the length of the jet electrolytic cathode, as well as the workpiece processing site. Change of jet length and location is controlled from a personal computer. The electrolyte flow rate is controlled by a motor located in the lower electrolytic bath. The electrolyte circulates between the two baths. For diagnostic purposes, the following devices were used: multimeters for measuring the voltage and current of the discharge; infrared pyrometer «TemPro 300».

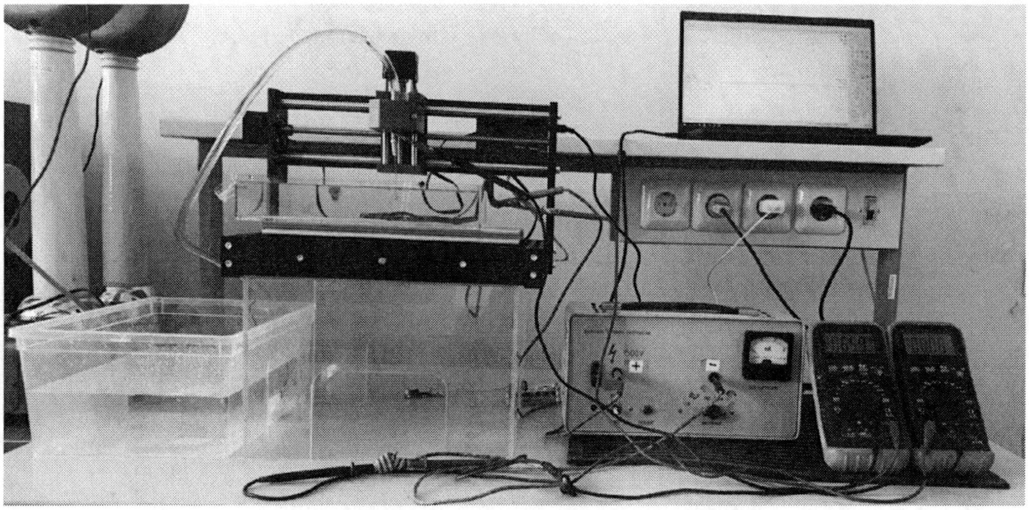

Figure 1. Experimental installation

In the course of experimental studies of electrical discharges between liquid electrodes during the processing of carbon fiber, the optimal form of the discharge was revealed (Fig. 2), the current strength in the range I=0.7÷0.95A; voltage U=450÷580V; jet length lc=20mm, jet diameter dc=2mm.

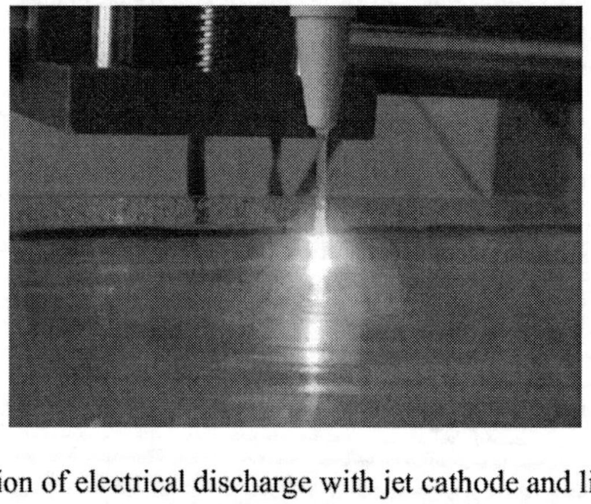

Figure 2. Form of interaction of electrical discharge with jet cathode and liquid anode at atmospheric pressure

In this burning mode of an electric discharge between a jet electrolytic cathode and a liquid anode at atmospheric pressure, carbon fiber tensile strength changed by 7% compared to the untreated sample with a processing time of 30 s.

The research was awarded by a grant from the Russian Science Foundation (project No. 22-29-00021)

References

[1] Strakhov I.S., Gubanov A.A., Ustinova M.S., Krivtsov D.I., Varshavsky V.Ya., Vahramyan T.A., Korshak Yu.V. Electrochemical modification of the surface of carbon fiber based on polyacrylonitrile // "Science and Education" 2013 No. 9. from. 29-42

[2] High frequency electric discharge between jet and solid electrodes Kuputdinova A.I.,Akhmadullina L.I., Zheltukhin V.S., Gaysin F.M., Gaysin Az.F./ Journal of Physics: Conference Series, 2020, 1588(1), 012010

[3] Some Features of Dielectric Barrier Discharge with Liquid and Solid Electrodes Bagautdinova L.N., Sadriev R.S., Gaysin A.F., Leushka M.A., Gaysin A.F. High Temperature, 2019, 57(6), стр. 944–947

[4] Electrolyte-plasma product treatment Akhmadullina L.I., Gaisin Al.F., Gaisin Az.F., Kashapov N.F., Zheltukhin V.S. Journal of Physics: Conference Series, 2020, 1588(1), 012012

[5] Features of a Low-Power, Low-Frequency, AC Arc between Solid and Liquid Electrodes Bagautdinova L.N., Sadriev R.S., Gaysin, A.F., Gaysin F.M., Mastyukov S.C. High Temperature, 2020, 58(3), стр. 441–443

Journal of Physics: Conference Series **2270** (2022) 012030 doi:10.1088/1742-6596/2270/1/012030

SYNTHESIS OF MICRODIAMONDS AND GERMANIUM NANOTUBES IN THE ARGON-GERMANIUM ARC

B A Timerkaev[1], A A Abdullin[1], R M Suleimanov[1], R R Shaikhattarov[1], R K Gevorgyan[1]

[1]KNRTU-KAI, Kazan, Russia

btimerkaev@gmail.com

Abstract. The paper describes a technique for the synthesis of microdiamonds, germanium nanoballs and nanotubes in an argon-germanium arc. The evaporation of germanium atoms was carried out from the surface of a graphite rod, which served as the anode of the arc discharge. The synthesis of microdiamonds was observed on graphite substrates near the anode; germanium nanotubes and germanium nanoballs were also found nearby. It has been established that germanium promotes the synthesis of nano and microdiamonds. Synthesized diamonds have a classic structure with rectangle faces and four hexagons adjacent to the edges of this rectangle. The sizes of microdiamonds reach tens of microns. The diameters of germanium nanotubes are of the order of 50-100 nm, and the length is of the order of 10 μm.

1. Introduction

Germanium is a semiconductor element with a fairly low bandgap and therefore has great application potential. Germanium has a high electronic conductivity and has a large absorption coefficient throughout the visible spectrum. This property has great prospects for the manufacture of photodetectors and solar batteries. In addition, germanium is also attractive for the synthesis of semiconductor nanoparticles, as well as a catalyst for the synthesis of nanodiamonds [1–3]. Nanodiamonds usually include carbon nanostructures ranging in size from 1 to 10 nanometers, the atoms in which are arranged in the same way as in diamond. Externally, nanodiamonds look like a light gray powder, consisting of agglomerates ranging in size from one to hundreds of micrometers. Nanodiamonds today are mainly obtained from natural diamonds by physical methods, by synthesis at ultrahigh pressures and temperatures, by chemical deposition of carbon-containing vapor at high temperatures and pressures, by detonation synthesis, and by irradiating a carbon-containing material with electron beams and argon ions.

In recent years, various approaches to the synthesis of nanostructured Ge have been highlighted, such as laser ablation, chemical reduction, and hydrothermal process. In [1-3], germanium served as a catalyst in the plasma-chemical synthesis of nanodiamonds. In [4], carbon nanostructures were synthesized from liquid hydrocarbons in a recessed arc. The results of plasma-chemical synthesis of carbon nanostructures from hydrocarbons are given in [5–6]. In [7–9], methods for increasing the stability of electric discharges in the processes of synthesis of various nanostructures were studied. In

Content from this work may be used under the terms of the Creative Commons Attribution 3.0 licence. Any further distribution of this work must maintain attribution to the author(s) and the title of the work, journal citation and DOI.
Published under licence by IOP Publishing Ltd

this work, we continued studying the process of synthesis of germanium nanoballs and nanotubes, as well as nano and microdiamonds in electric arc plasma.

2. Materials and methods

The experiment was carried out in a vacuum chamber with water-cooled walls. After evacuation, the chamber was filled with argon to a pressure of 500 Torr. Graphite rods 14 mm in diameter were used as electrodes. The distance between the electrodes could be controlled during the experiment. The graphite cathode was sharpened to precisely localize the current spot on the anode. A hole 5 mm in diameter and 6 mm deep was drilled on the anode. Germanium was crumbled into this hole. The experiment was carried out at a voltage of about 24 V and a current of 50 A. The experiment was carried out according to the following scenario. By rotating the regulator, the graphite cathode was lowered until it came into contact with the anode. After the appearance of the arc, its length was set to about 3 mm. In the first few tens of seconds, evaporation of carbon atoms occurred, and the surface of the graphite anode became hot, while germanium evaporated insignificantly. Subsequently, the evaporation of germanium increased and its intense boiling began. Vapors of germanium and carbon atoms were deposited on the surface of a graphite substrate, on the surface of the cathode, on the surface of the anode, and on the walls of the vacuum chamber. The experiment lasted for 2 minutes. These deposits have been carefully studied with optical and scanning electron microscopes. The evaporation of germanium and the growth of germanium deposits were observed through a special darkened viewing window.

3. Results and discussion

Figures 1-4 show images of the samples synthesized during the experiment. In Figure 1, at a magnification of 5000 times, we see germanium nanotubes, germanium nanoballs and nanodiamonds. Germanium nanotubes and nanoballs are conductors and are therefore black and gray in color, while dielectric nanodiamonds are white. Germanium nanotubes have a length of about 2-4 microns and a diameter of about 50 nm. The main difference between these nanotubes and those previously synthesized is the fact that they grew, as it were, on nanodiamonds. The sizes of nanodiamonds in this picture are about 800-900 nm and they represent a kind of "handle" for these nanotubes. Thus, the presence of such holders, "fasteners" will allow the use of such nanotubes in many applications where it is necessary to deal with individual nanotubes. These nanotubes have high mechanical strength. The germanium nanobeads in this image range in size from 1 μm to 100 nm.

On fig. 2 at the same magnification are germanium nanoballs and nanodiamonds. The sizes of nanoballs here reach up to 4 μm in diameter. The bulk of the nanoballs have dimensions of the order of 100–200 nm. Some nanoballs are connected to each other by nanotubes like dumbbells. The fact that these nanotubes did not break is a testament to their exceptional strength.

On fig. 3 and fig. Figure 4 shows synthesized microscopic diamonds. On Fig. 3 diamonds have a regular geometric shape: 4 faces of four regular hexagons adjoin one face in the form of a square. Such a set of faces forms the upper cap of the crystal. The middle belt consists of four hexagons alternating with four squares. Thus, the middle belt consists of eight faces. On the lower side, the crystal ends with the same cap as the upper cap. There are 18 faces in a diamond crystal, 6 of which are squares and 12 are hexagons. The dimensions of all diamonds shown in the figure are strictly the same. Their cross section is about 1.5 μm. Diamonds are surrounded by other nanostructures. Apparently, carbon, since a deficit of these formations formed near the nanodiamonds.

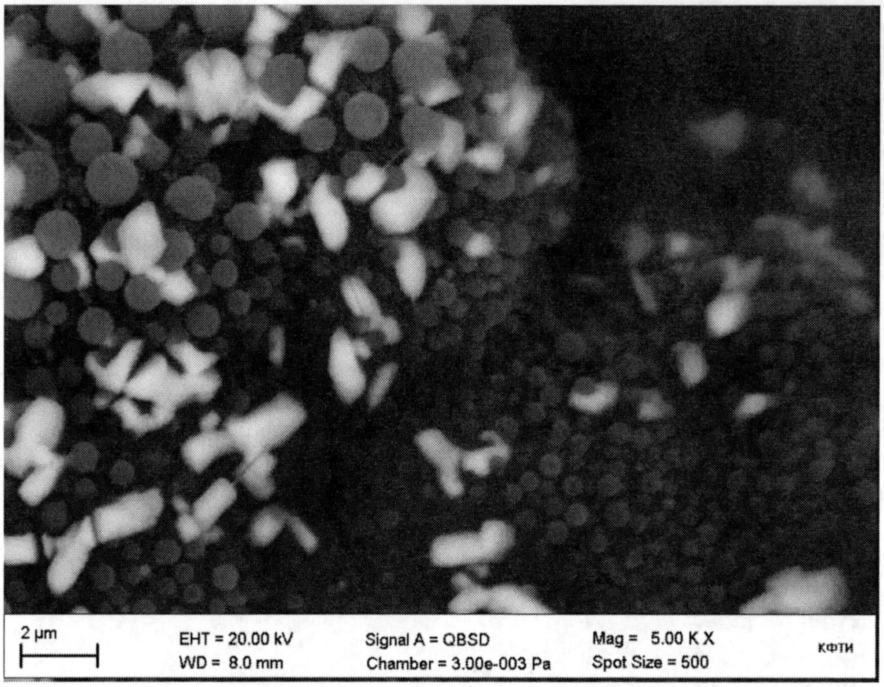

Figure 1. Nanodiamonds, germanium nanotubes and nanoballs. 5000 times magnification

Figure 2. Germanium nanoballs and nanotubes

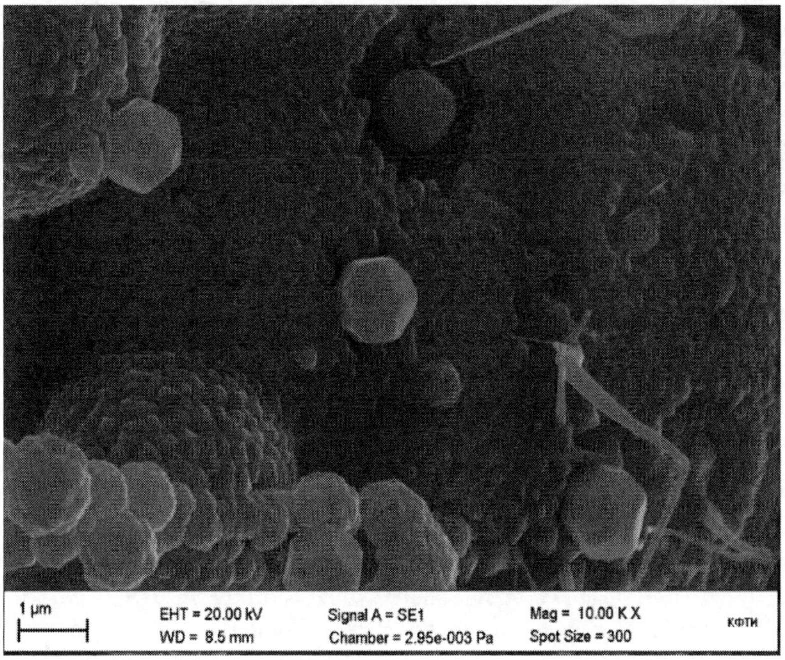

Figure 3. Microdiamonds and nanodiamonds.

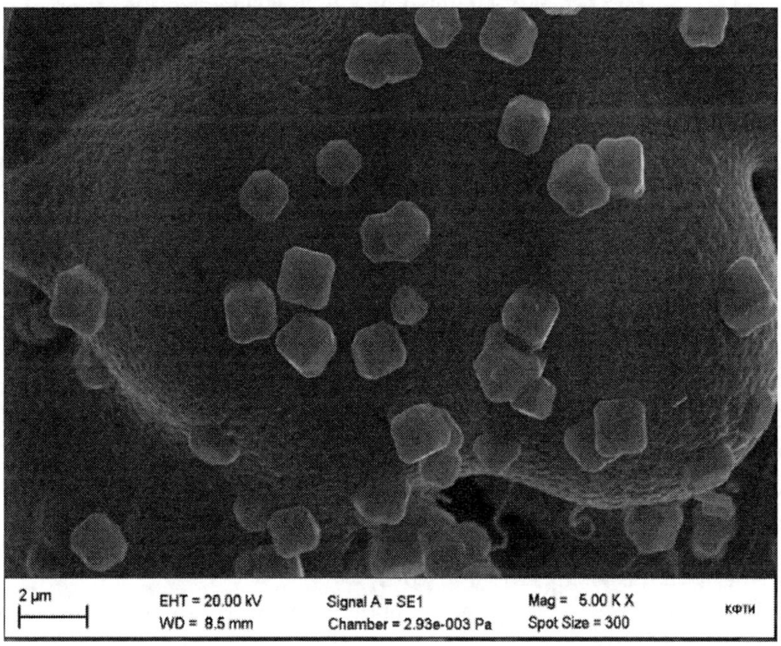

Figure 4. Nanodiamonds of different geometry from the same sample.

Figure 4 shows nanodiamonds of a different geometry. Their vertex squares are much smaller than in the previous figure and average about 0.7 µm. The side hexagons also have irregular shapes. These and other diamonds play with their facets in the light.

4. Conclusion

Thus, germanium nanotubes and microscopic diamonds were synthesized in an argon-germanium arc. Diamonds have 18 facets, 6 of which are squares and 12 are hexagons. The dimensions of all diamonds shown in the figure are strictly the same. Their cross section is about 1.5 μm. Germanium nanotubes have a length of about 2-4 microns and a diameter of about 50 nm. These nanotubes grew, as it were, on nanodiamonds. The sizes of nanodiamonds in this picture are about 800-900 nm and they are like "handles" for these nanotubes. Thus, the presence of such holders, "fasteners" will allow the use of such nanotubes in many applications where it is necessary to deal with individual nanotubes. Low-temperature plasma can be used for the synthesis of various nanostructures and is well suited for the modification of various surfaces. This is shown in many works [10-13].

References

[1] Palyanov Yu.N., Kupriyanov I.N., Khokhryakov A.F., Ralchenko V.G. // Handbook of Crystal Growth (Second Edition). 2015. V. 2a. P. 1418.
[2] Palyanov Yu.N., Kupriyanov I.N., Borzdov Yu.M., Surovtsev N.V. // Scientific reports. 2015. V. 5. P. 14789.
[3] Timerkaev B.A., A.A. Kaleeva, D.B. Timerkaeva, A.I. Saifutdinov Germanium Catalyst for Plasma-Chemical Synthesis of Diamonds. High Energy Chemistry 53(5), 390-395. 2019.
[4] Timerkaev B.A., G.R. Ganieva, A.A. Kaleeva, Z.K. Israfilov, A.O. Sofronitskii. Growing of Carbon Nanotubes from Hydrocarbons in an Arc Plasma. . Journal of Engineering Physics and Thermophysics, 92(5), 1248-1252, 2019. DOI 0.1007/s10891-019-02040-3
[5] Galeev I G, Asadullin T Ya 2016 Obtaining fullerene-containing soot during combustion of gaseous hydrocarbons in an external electric field *J. of Phys. Conf. Ser.* 669 012016
[6] Shakirov A S, Galeev I G and Asadullin T Ya 2020 Features of formation of fullerene-containing soot during combustion gaseous and liquid hydrocarbons *J. of Phys. Conf. Ser.* 1588(1) 012013
[7] Galeev I G, Asadullin T Y 2017 Improving of stability of the volumetric glow discharge in the gas flow *J. of Phys. Conf. Ser.* 789(1) 012012
[8] Asadullin T Y, Galeev I G 2017 Combined method of stabilization of the glow discharge in gas flow *J. of Phys. Conf. Ser.* 927(1) 012003
[9] Galeev I G, Asadullin T Y 2016 Enhancing of the glow discharge stability in chamber with cathode sections coated by a discontinuous dielectric coating *J. of Phys. Conf. Ser.* 669(1) 012015
[10] Saifutdinov, A. I., & Sofronitskii, A. O. (2021). Numerical Study of Breakdown and Formation Dynamics of Arc Discharge Plasma Parameters at Ultrahigh Pressures. High Energy Chemistry, 55(3), 228-232.DOI: 10.1134/S0018143921030115
[11] Shemakhin, A. Y., Zheltukhin, V. S., Shemakhin, E. Y., Terentev, T. N., & Sofronitsky, A. O. (2020, July). Experimental installation to study the RF plasma flow at low pressures with experiment data synchronization. In Journal of Physics: Conference Series (Vol. 1588, No. 1, p. 012018). IOP Publishing. DOI: 10.1088/1742-6596/1588/1/012018
[12] Fairushin, I. I., Saifutdinov, A. I., & Sofronitskiy, A. O. (2020). Numerical and Experimental Studies of the Synthesis of Copper Nanoparticles in a High-Pressure Discharge. High Energy Chemistry, 54(2), 150-153. DOI: 10.1134/S0018143920020071
[13] Sadikov, K. G., Sofronitskiy, A. O., & Larionov, V. M. (2017, November). The effect of electrically conductive additives on the plasma pyrolysis of heavy hydrocarbons. In Journal of Physics: Conference Series (Vol. 927, No. 1, p. 012046). IOP Publishing. DOI: 10.1088/1742-6596/927/1/012046

Universal Structural and Dynamic Features in Metals Near Their Melting Points

R R Khairullina, R M Khusnutdinov*

Kazan (Volga region) Federal University, Kazan, Russia

khrm@mail.ru

Abstract. The results of a comparative analysis of experimental data for the equilibrium properties and characteristics of liquid alkaline earth (magnesium, calcium, strontium), alkaline (lithium, sodium, potassium) and transition metal groups: elements of the subgroups of nickel (Ni, Pd, Pt) and copper (Cu, Ag, Au) near their melting points are presented. Reduced spatial $r/r_m = rk_m/2\pi$ and time $t/t_m = tk_m/\sqrt{m\beta}$ scales, in which k_m is the first peak position of the static structure factor *S(k)* and $\beta = 1/k_BT$ is the inverse temperature, are introduced as the basis for the law of corresponding states. Based on these scale transformations and x-ray diffraction analysis, it was found that the groups of liquid alkaline, alkaline earth, and transition metals are described by universal *r*- and *k*- dependencies. It has been established that the dispersion law of longitudinal polarization $\omega_c(k)$, given in accordance with these scale relations, for elements of groups of liquid alkaline, alkaline earth and transition metals has a single universal character. An analysis of the properties of three groups (alkaline, alkaline earth and transition) liquid metals using scale transformations shows that the law of the corresponding states is valid for these substances.

1. Introduction

The problem of a universal description of the structural and dynamic properties of various groups of liquids is relevant for modern condensed matter physics [1]. On the one hand, this would avoid the difficulties associated with the extraction of experimental information. The fact is that almost until recently, the properties of microscopic collective and single-particle dynamics were experimentally studied mainly by the method of scattering of slow neutrons. In this case, various difficulties arose, firstly, with the separation of collective and single-particle contributions, and, secondly, with the presence of significant experimental errors in some *(k,ω)*-regions, sometimes leading to the impossibility of extracting reliable experimental information. On the other hand, the possibility of describing in a unified way the properties of various groups of liquids would make it possible to create a universal microscopic theory of the liquid state.

In this paper, based on an analysis of experimental data [2-6] and the principle of the corresponding states [1], we consider the possibility of a unified description of the structural features and microscopic dynamics of various groups of metals (alkaline, alkaline earth and transition metals).

2. The law of corresponding states for liquid metal melts

In accordance with the principle of corresponding states, the equilibrium properties of various groups of liquids can be described in a certain universal way [3]. Here, the main problem is only in

Content from this work may be used under the terms of the Creative Commons Attribution 3.0 licence. Any further distribution of this work must maintain attribution to the author(s) and the title of the work, journal citation and DOI.
Published under licence by IOP Publishing Ltd

choosing a certain set of scaling transformations, for example, for such parameters and characteristics of the systems under study as temperature, density, spatial and frequency characteristics [4-6]. To compare the structural and dynamic properties of various groups of metals (alkaline, alkaline earth and transition metals) near their melting point ($T/T_m \approx 1$), we use scale transformations proposed by us in Ref. [3]. Let us choose as the unit of the spatial scale the value $r_m = 2\pi/k_m$, which corresponds to the average interatomic distance in the system; here k_m is the first peak position of the static structure factor *S(k)*. As a unit of the time scale, let us consider the value $t/t_m = tk_m/\sqrt{m\beta}$, where m is the mass of the particle, and $\beta = 1/k_BT$ is the inverse temperature. Note that the value t_m is nothing but the mean free path of a particle on the length scale $l = 2\pi/k_m$. Therefore, the value $\omega_m = 2\pi/t_m$ will represent the frequency scale. We have calculated the characteristics presented above for the systems under study near their melting points on the basis of experimental data on neutron diffraction and X-ray diffraction analysis [2-6]. The calculated values of parameters and characteristics for the studied systems are presented in table 1.

Table 1: A set of reduced physical parameters for the systems under study.

	$T_m(K)$	$\rho_m(g/cm^3)$	$m(10^{-24}g)$	T/T_m	$k_m(Å^{-1})$	r_m (Å)	$t_m(10^{-14}s)$
Alkali metals							
Li	453.70	0.534	11.50	1.021	2.503	2.51	5.29
Na	371.00	0.971	38.20	1.019	2.001	3.14	13.32
K	336.35	0.856	64.90	1.019	1.599	3.93	23.14
Rb	312.64	1.532	141.9	1.002	1.500	4.19	37.79
Cs	301.55	1.873	221.0	1.005	1.441	4.36	50.08
Alkaline earth metals							
Mg	923.0	1.740	40.374	1.032	2.400	2.62	7.30
Ca	1115	1.540	66.575	1.007	1.950	3.22	10.63
Sr	1050	2.640	145.55	1.003	1.780	3.53	17.78
Nickel subgroup elements							
Ni	1728	8.90	97.497	1.026	3.100	2.03	6.45
Pd	1828	12.0	176.18	1.014	2.780	2.26	9.46
Pt	2041	21.5	324.06	1.006	2.750	2.28	12.29
Copper subgroup elements							
Cu	1358	8.96	105.56	1.048	3.00	2.09	7.73
Ag	1235	10.5	179.18	1.031	2.60	2.42	12.28
Au	1337	19.3	327.19	1.064	2.65	2.37	15.40

Experimental data on neutron diffraction and X-ray diffraction analysis for the static structure factor *S(k)* of alkali metallic melts (except for francium), alkaline earth (magnesium, calcium, and strontium), and transition (nickel, palladium, platinum, copper, silver, and gold) metals near their melting points ($T/T_m \approx 1$) [2] are presented in Fig. 1a. Here, *S(k)* is a function of the reduced wave number (in units of k/k_m). As can be seen from the figure, all experimental data can be described by the universal dependence on reduced wave number. In Fig.1b shows the pair correlation functions *g(r)*, given in accordance with the values from Table 1. As can be seen from the figure, the reduced pair correlation functions for the systems under consideration are collapse onto a universal master curve.

Raw experimental and reduced longitudinal dispersion curves for liquid melts of alkali, alkaline earth and transition metals are shown in Figs. 2a and Fig. 2b, respectively. As can be seen from the figure, the dispersion curves in the reduced form for all the systems under study are described by a universal master curve. Thus, the results presented in this work testify to the validity of the law of the corresponding states for various groups of liquid metals. It is gratifying that the equilibrium characteristics for liquid alkali, alkaline earth and transition metals can be described by universal dependencies. This conclusion provides a basis for understanding the mechanisms of microscopic processes in various groups of liquids from a unified standpoint and is some prerequisite to creation of a unified theory of the liquid state.

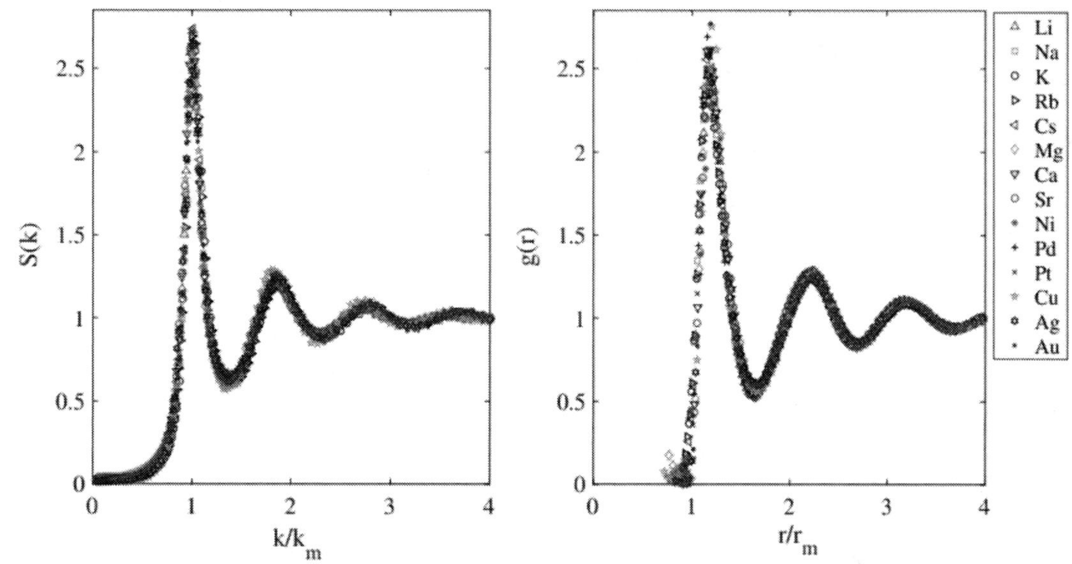

Figure 1. Structural characteristics [*S(k)* and *g(r)*] for liquid alkaline, alkaline earth and transition (elements of nickel and copper subgroups) metals in reduced units: $r_m = 2\pi/k_m$, where k_m is the maximum position in *S(k)* [2].

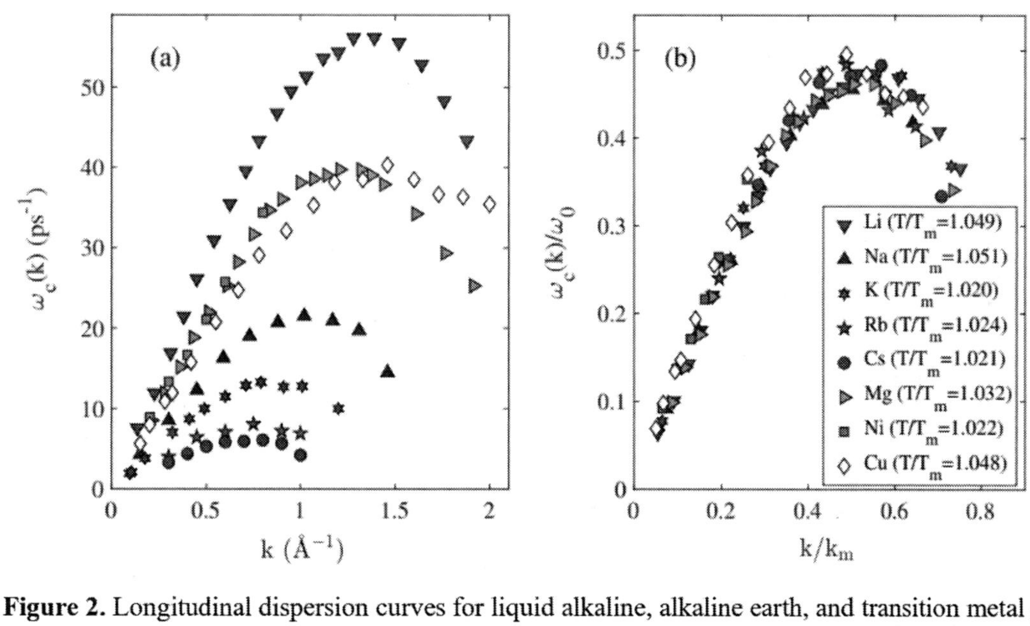

Figure 2. Longitudinal dispersion curves for liquid alkaline, alkaline earth, and transition metal groups in absolute (a) and reduced (b) units (where $\omega_0 = 2\pi k_m\sqrt{k_B T/m}$).

3. Conclusions

An important result of the present work is the possibility of a single description of the structural features and dynamic processes of metals near their melting point (groups of alkaline, alkaline earth and transition metals). Based on experimental data on neutron diffraction, X-ray analysis [2] and scale transformations, it was shown that the structural properties of groups of liquid alkaline, alkaline earth, and transition metals near their melting temperatures are characterized by universal features. Thus, the reduced forms of the pair correlation function *g(r)* and the experimental quantity – the static structural factor *S(k)* for elements of the group of alkali, alkaline earth, and transition metals are described by uniform *r*- and *k*- dependencies. It is shown that the longitudinal dispersion curves, given in accordance with scale transformations, for the systems under consideration have a single universal character. The results presented in this work indicate the validity of the law of the corresponding states for various groups of liquid metals.

Acknowledgments

This article has been supported by the Kazan Federal University Strategic Academic Leadership Program (PRIORITY-2030).

References

[1] Balucani U and Zoppi M 1994 *Dynamics of the Liquid State*, (Oxford: Clarendon)
[2] Waseda Y 1980 *The Structure of Non-Crystalline Materials: Liquids and Amorphous Solids*, (New York: McGraw-Hill)
[3] Mokshin A V, Khusnutdinov R M, Akhmerova A R, et al 2017 *Jetp Lett.* **106** 366
[4] Mousazadeh M H, Khanchi A, Ghanadi Marageh M 2006 *JICS* **3** 22
[5] Leland T W, Chappelear P S 1968 *Ind. Eng. Chem.* **60** 15
[6] Meyer N, Xu H, Wax J-F 2017 *Phys. Rev. B* **96**, 094201

A model of the positive column of longitudinal glow discharge in planar channel with additional gas injection through the side walls

I G Galeev[1], **T Y Asadullin**[1], **D R Valeeva**[1]

[1]KNRTU–KAI, Kazan, Russia

galeev.physics@kstu-kai.ru

Abstract. A model of positive glow discharge column in longitudinal flowing glow discharge with an additional distributed gas injection through the side walls of flat channel is proposed. The main flow is pumped through mesh electrodes located at the beginning and end of the discharge chamber. The solution of an integro-differential system of equations is obtained, expressions for the distribution of the electric field strength, the degree of gas ionization, and the gas temperature in the discharge region are given.

Low-temperature gas-discharge plasma finds numerous applications in various fields of science and technology. It serves as active medium for gas lasers [1], active element of plasma chemical reactors for production of nanoparticles and nanotubes [2-7], used in production of various composite materials and etc. In this regard, the study of the possibilities of influencing the discharge characteristics and maintaining them at the required level is an urgent task. Such influence can be carried out in various ways, for example through different organization of gas flow pumped through the discharge region. In [8-10], supersonic gas pumping is used to create various conditions for maintaining the discharge in separate spatial parts.

In [11] a two-dimensional flat model is presented that takes into account, in addition to the main gas flow, a distributed additional gas injection through one of the walls of channel with constant cross-section.

This paper presents a model of the positive column of longitudinal glow discharge in gas flow in discharge chamber of constant cross-section with gas injection through both walls of the channel. The main gas flow is carried out through mesh electrodes located in the initial and final parts of the channel, and gas is additionally symmetrically supplied through the channel walls.

The positive glow discharge column is described by a system of equations

$$c_a \frac{\partial^2}{\partial y^2}\alpha - NW_x \frac{\partial \alpha}{\partial x} - NW_y \frac{\partial \alpha}{\partial y} + v_i N\alpha = 0,$$

$$I = e\mu_e b_z E \int_{-b}^{b} \alpha dy,$$

$$\frac{\partial}{\partial x} NW_x + \frac{\partial}{\partial y} NW_y = 0,$$

Content from this work may be used under the terms of the Creative Commons Attribution 3.0 licence. Any further distribution of this work must maintain attribution to the author(s) and the title of the work, journal citation and DOI.
Published under licence by IOP Publishing Ltd

$$G = mb_z \int_{-b}^{b} NW_x dy,$$

$$c_p g_x \frac{\partial T}{\partial x} + c_p g_y \frac{\partial T}{\partial y} = \lambda \frac{\partial^2 T}{\partial y^2} + \kappa_q e \mu_e n \frac{E^2}{N},$$

$$P = N\kappa_B T.$$

The boundary conditions are taken as: at the entrance to the discharge zone ($\tilde{x} = 0$)

$$\tilde{\alpha}_e = \frac{\tilde{n}}{\tilde{N}} = \varphi_a(\tilde{y}),\ \tilde{T} = \tilde{\vartheta}(\tilde{y}),$$

on the walls of the discharge chamber

$$\tilde{\alpha}_e = 0, \quad \tilde{T}(\tilde{x}, \pm 1) = \tilde{\theta}_\omega(\tilde{x}),$$

on the plane of symmetry of the discharge chamber ($\tilde{y} = 0$)

$$\tilde{g}_y = 0, \quad \frac{\partial \tilde{\alpha}_e}{\partial \tilde{y}} = 0, \quad \frac{\partial \tilde{T}}{\partial \tilde{y}} = 0,$$

where the discharge chamber circuit is shown in Figure 1.

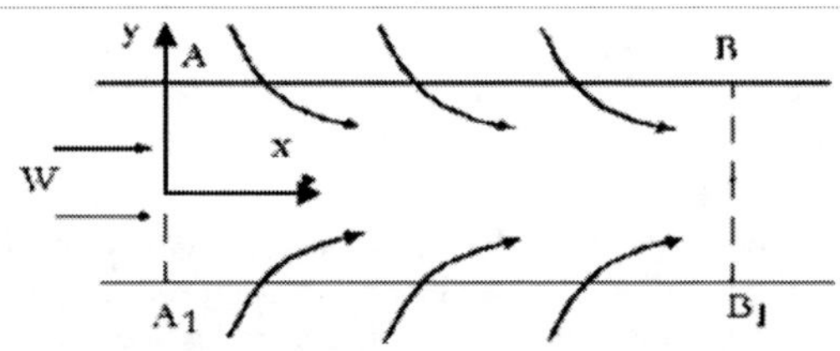

Figure 1. Diagram of a discharge chamber with a distributed gas flow. AA1 and BB1 are mesh electrodes. x0y is a rectangular coordinate system.

As a first approximation, we assume that $g_y = \rho W_x = (\rho W_x)_0 (1+\kappa_w \tilde{x})$, where κ_w is a injection parameter. Then it follows from the equation of continuity of the gas mass that the transverse component of the gas flow rate is determined by the expression

$$\rho W_y = -\kappa_w \tilde{y} (\rho W_x)_0.$$

In dimensionless form, the equation for the gas ionization degree can be represented as

$$\frac{1}{Pe_d} \frac{\partial^2 \tilde{\alpha}_e}{\partial \tilde{y}^2} - (1 + \kappa_w \tilde{x}) \frac{\partial \tilde{\alpha}_e}{\partial \tilde{x}} + \kappa_w \tilde{y} \frac{\partial \tilde{\alpha}_e}{\partial \tilde{y}} + \varsigma_e(\tilde{x}) \tilde{\alpha}_e = 0,$$

where $\tilde{\alpha}_e = \frac{\alpha_e}{\alpha_e(0,0)}$, $Pe_d = \frac{g_0 b}{c_a m}$, $\varsigma_e(\tilde{x}) = c_v \frac{mb}{g_0} E^2$, $g_0 = \frac{G}{2b_z b}$.

With an arbitrary law of change $\varsigma_e(\tilde{x})$, the general solution can be represented as

$$\tilde{\alpha}_e = \exp\left[\int_0^{\tilde{x}} \frac{\zeta_e d\tilde{x}}{1+\kappa_w \tilde{x}}\right] \sum_{m=1}^{\infty} A_m (1+\kappa_w \tilde{x})^{-\frac{\lambda_m}{2\kappa_w Pe_d}} F_m\left(\frac{\lambda_m}{2\kappa_w Pe_d}, \frac{1}{2}, -\frac{Pe_d \kappa_w}{2} \tilde{y}^2\right)$$

where $\varphi_m = F_m\left(\frac{\lambda_m}{2\kappa_w Pe_d}, \frac{1}{2}, -\frac{Pe_d \kappa_w}{2} \tilde{y}^2\right)$ are degenerate hypergeometric functions [12], λ_m are the roots of the equation $F_m\left(\frac{\lambda_m}{2\kappa_w Pe_d}, \frac{1}{2}, -\frac{Pe_d \kappa_w}{2}\right) = 0$.

The constant coefficients A_m are found from the boundary condition at the entrance to the discharge camera

$$\varphi_a(\tilde{y}) = \sum_{m=1}^{\infty} A_m F_m\left(\frac{\lambda_m}{2\kappa_w Pe_d}, \frac{1}{2}, -\frac{Pe_d \kappa_w}{2} \tilde{y}^2\right).$$

The functions are orthogonal with the weight $e^{\frac{\kappa_w Pe_d}{2}\tilde{y}^2}$

$$\int_0^1 e^{\frac{\kappa_w Pe_d}{2}\tilde{y}^2}\varphi_1\varphi_m d\tilde{y} = 0 \quad \text{for } m \neq l.$$

Then

$$A_m = \frac{\int_0^1 \varphi_a(\tilde{y})\exp\left(\frac{\kappa_w Pe_d}{2}\tilde{y}^2\right)F_m(\tilde{y})d\tilde{y}}{\int_0^1 \exp\left(\frac{\kappa_w Pe_d}{2}\tilde{y}^2\right)F_m^2(\tilde{y})d\tilde{y}}.$$

The distribution of the electric field strength satisfying Ohm's law can eventually be written as

$$E(\tilde{x}) = \left[2c_e\Phi^2(\tilde{x})\int_0^{\tilde{x}}\frac{d\tilde{x}}{\Phi^2(\tilde{x})(1+\kappa_w\tilde{x})} + \frac{\Phi^2(\tilde{x})}{E^2(0)\Phi^2(0)}\right]^{-0,5}$$

where $\Phi(\tilde{x}) = \sum_{m=1}^{\infty} A_m(1+\kappa_w\tilde{x})^{-\frac{\lambda_m}{\kappa_w Pe_d}}\gamma_m$, $\gamma_m = \int_0^{\tilde{x}} F_m(\tilde{y})d\tilde{y}$, $c_e = \frac{c_v mb}{g_0}$.

$$\tilde{\alpha}_e = \frac{I\sum_{m=1}^{\infty} A_m(1+\kappa_w\tilde{x})^{-\frac{\lambda_m}{\kappa_w Pe_d}}F_m^*(\tilde{y})}{2e\mu_e b_z b\alpha_e(0,0)E(\tilde{x})\Phi(\tilde{x})}.$$

Taking into account Ohm 's law

$$I = 2e\mu_e b_z b\alpha_e(0,0)E(0)\Phi(0),$$

it turns out

$$\tilde{\alpha}_e = \left[1+2c_e E^2(0)\Phi^2(0)\int_0^{\tilde{x}}\frac{d\tilde{x}}{\Phi^2(\tilde{x})(1+\kappa_w\tilde{x})}\right]^{-0,5}\sum_{m=1}^{\infty} A_m(1+\kappa_w\tilde{x})^{-\frac{\lambda_m}{\kappa_w Pe_d}}F_m(\tilde{y})$$

Knowing the volume distribution of power density of internal heat sources, it is possible to determine the distribution of gas temperatures. In dimensionless form

$$\frac{1}{Pe}\frac{\partial^2\tilde{T}}{\partial\tilde{y}^2} - (1+\kappa_w\tilde{x})\frac{\partial\tilde{T}}{\partial\tilde{x}} + \kappa_w\tilde{y}\frac{\partial\tilde{T}}{\partial\tilde{y}} + q^* = 0,$$

where $q^* = \frac{\kappa_q be\mu_e\alpha_e E^2}{g_0 T_0 c_p}$.

$$\tilde{\alpha}_e = \frac{I\sum_{m=1}^{\infty} A_m(1+\kappa_w\tilde{x})^{-\frac{\lambda_m}{\kappa_w Pe_d}}F_m^*(\tilde{y})}{2e\mu_e b_z b\alpha_e(0,0)E(\tilde{x})\Phi(\tilde{x})}.$$

The temperature distribution can be represented as

$$\tilde{T} = \tilde{\theta}(\tilde{x}) + \sum_{m=1}^{\infty}(1+\kappa_w\tilde{x})^{-\frac{\lambda_m}{Pe b_w}}\left[\int_0^{\tilde{x}}\frac{H_m(\tilde{x})}{Pe}(1+\kappa_w\tilde{x})^{\frac{\lambda_m}{Pe b_w}-1}d\tilde{x} + B_m\right]F_m\left(\frac{\lambda_m}{2\kappa_w Pe},\frac{1}{2},-\frac{\kappa_w Pe}{2}\tilde{y}^2\right),$$

where

$$H_m(\tilde{x}) = \frac{\int_0^1 \exp\left[\frac{\kappa_w Pe}{2}\tilde{y}^2\right]H_m(\tilde{x})F_m\left(\frac{\lambda_m}{2\kappa_w Pe_d},\frac{1}{2},-\frac{\kappa_w Pe}{2}\tilde{y}^2\right)d\tilde{y}}{\int_0^1 \exp\left[\frac{\kappa_w Pe}{2}\tilde{y}^2\right]F_m^2\left(\frac{\lambda_m}{2\kappa_w Pe_d},\frac{1}{2},-\frac{\kappa_w Pe}{2}\tilde{y}^2\right)d\tilde{y}}, \quad B_m = \frac{\int_0^1\left[\tilde{\vartheta}(\tilde{y})-\tilde{\theta}_\omega(\tilde{y})\right]\exp\left[\frac{\kappa_w Pe}{2}\tilde{y}^2\right]F_m^* d\tilde{y}}{\int_0^1 \exp\left[\frac{\kappa_w Pe}{2}\tilde{y}^2\right]F_m^* d\tilde{y}}.$$

In the particular case of setting the initial condition for the degree of ionization of the gas at the entrance to the discharge zone in the form of a function

$$\tilde{\alpha}_e(0,\tilde{y}) = F_m\left(\frac{\lambda_m}{2\kappa_w Pe_d}, \frac{1}{2}, -\frac{\kappa_w Pe_d}{2}\tilde{y}^2\right).$$

after some transformations , you can get

$$E = \frac{E_*}{\sqrt{1-\left[1-\left(\frac{E_*}{E_0}\right)^2\right]\left(1+\kappa_w\tilde{x}\right)^{-\frac{2\lambda_1}{Pe_d \kappa_w}}}},$$

Here $E_*^2 = \lim\limits_{\tilde{x}\to\infty} E^2(\tilde{x}) = \dfrac{\lambda_1}{Pe_d C_E}, \quad \alpha_* = \lim\limits_{\tilde{x}\to\infty} \alpha_e(\tilde{x},0)$

Figure 2 shows the calculated distributions of the profiles of the gas ionization degree at the limit section of the gas injection into positive column. Curves 1 - 3 are plotted for the values of $\kappa_w Pe_d$ = 14.5: 4.39: 0, respectively.

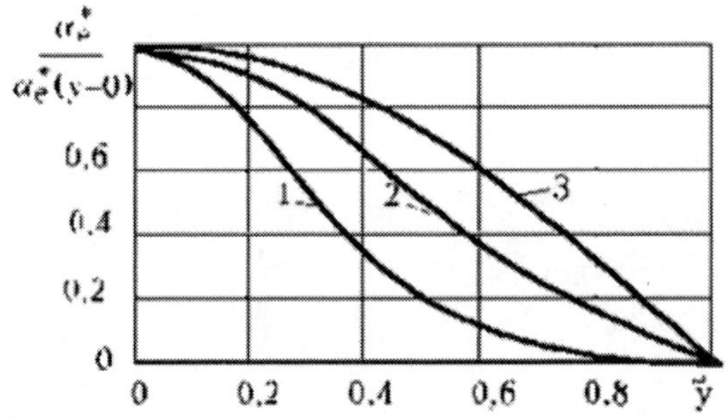

Figure 2. Distribution of profiles of the degree of gas ionization at the limit section of the gas injection into positive column. Curves 1 - 3 are plotted for the values of $\kappa_w Pe_d$ = 14.5: 4.39: 0, respectively.

References

[1] Raizer Yu P 2009 *Physics of gas discharge* (Moscow: Publishing house: Intellect) 736 p
[2] Saifutdinova A A, Sofronitskiy A O, Timerkaev B A, Saifutdinov A I 2020 *Russian Physics Journal* **62(11)** 2132–36
[3] Timerkaev B A, Ganieva G R, Kaleeva A A, Israfilov Z K and Sofronitskii A O 2019 Journal of Engineering Physics and Thermophysics **92(5)** 1248–52
[4] Timerkaev B A, Sofronitskiy A O and Andreeva A A 2016 J. of Phys. Conf. Ser. **669(1)** 012062
[5] Timerkaev B A, Andreeva A A and Sofronitskiy A O 2017 J. ouf Phys. Conf. Ser. **927(1)** 012068
[6] Fairushin I I, Saifutdinov A I, Sofronitskiy A O, Timerkaev B A and Dautov G Yu 2019 J. of Phys. Conf. Ser. **1328(1)** 012088
[7] Shakirov A S, Galeev I G and Asadullin T Ya 2020 *J. of Phys. Conf. Ser.* **1588(1)** 012013
[8] Timerkaev B A and Zalyaliev B R 2014 *High Temperature* **52(4)** 471–4
[9] Timerkaev B A, Zalyaliev B R and Saifutdinov A I 2016 *High Temperature* **54(5)** 632–8
[10] Zalyaliev B R, Timerkaev B A and Amirzyanov I I 2017 *J. of Phys. Conf. Ser.* 927(1) 012079
[11] Galeev I G, Asadullin T Ya and Tsybikov R O 2021 *J. Phys.: Conf. Ser.* **1870** 012016
[12] Korn G and Korn T 1984 Handbook of mathematics (Moscow: Nauka) 832 p

Journal of Physics: Conference Series **2270** (2022) 012033 doi:10.1088/1742-6596/2270/1/012033

On the question of the applicability of the principle of thermodynamic similarity in liquid alkali metals

A A Tsygankov

Kazan Federal University, Kazan, Russia

E-mail: tsigankov.artiom@gmail.com

Abstract. In this paper, we have created an interparticle interaction potential for liquid alkali metals using ab-initio simulations and machine learning. The resulting family of potentials was tested by molecular dynamics simulation, in which the liquid metals were heated to the melting point and the resulting system was then averaged. The simulation results show that near the melting point, the radial distribution function and the static structure factor are reproduced. This results indicates that the principle of corresponding states is fulfilled in liquid alkali metals.

1. Introduction

At present, the structural features and properties of alkali metals are of great interest to researchers. Firstly, this class of metals is the first one after the fully studied inertial gases, which makes it possible to test, generalize and adapt new models to existing experimental data. Secondly, liquid alkali metals have many practical applications, for example, coolants in nuclear reactors, purification of alloys from impurities, the creation of photovoltaic cells. The high chemical activity of this metals is explained by the presence of one electron on the outer shell. Therefore, when studying these metals in their pure form, where the alkali metal atom is represented as a "positive ion and ns^1 electronic structure", on which many different theories of simple metals were tested and well studied experimentally [1, 2, 3, 4]. It was shown in the Ref.[5] [5] that the interaction potential for liquid alkali metals has an oscillating form, due to the periodicity in the arrangement of the electronic structure (Friedel oscillations). For lithium, good agreement was achieved between the simulation results and experimental data [6]. It was also shown the possibility of fulfilling the principle of thermodynamic similarity [7] in liquid alkali metals near the melting point. Authors have shown that this approach is correct and the pair oscillating potential makes it possible to describe the structural characteristics in other liquid alkali metals.

2. Obtaining the coefficients of potentials for alkali metals

The interaction potential is assumed to be paired due to the fact that alkali metal melt is consists of positively charged ions and electrons moving between them, creating oscillations in the structure. To determine the interaction potential, we need a functional dependence of the interaction energy on distance and a set of parameters. In our case, the set of parameters will be chosen on the assumption that for alkali metals the potential will have one dependence:

$$U(r) = A\exp(-\eta(r - r_0)) + \frac{B}{r^n}\cos(kr + \phi), \tag{1}$$

Content from this work may be used under the terms of the Creative Commons Attribution 3.0 licence. Any further distribution of this work must maintain attribution to the author(s) and the title of the work, journal citation and DOI.
Published under licence by IOP Publishing Ltd

where the first term characterizes the short-range repulsion, the second term characterizes periodic oscillations in the electronic structure. The potential parameters for considered systems were obtained by machine learning: the algorithm is a controlled random search with local mutation [8]. The convergence criterion of the machine learning algorithm is the forces acting on the atoms obtained from the earlier ab-initio simulation in the VASP package (100 system configurations) [9, 10]. Initial conditions for ab-initio simulation: 54 particles, bcc lattice type, NVT ensemble, final simulation temperatures are 475 K, 390 K, 343 K, 320 K, 303 K respectively for lithium, sodium, potassium, rubidium, cesium, respectively. The lattice constants are chosen in such a way as to correspond to the quantitative density of the systems in the experiment and are equal to 3.557, 4.282, 5.386, 5.772, 6.215 Åfor lithium, sodium, potassium, rubidium, cesium, respectively. The parameters for the potentials obtained in the course of machine learning are shown in Table 1:

Table 1. Potential parameters for alkali metals.

Element	A	η	r_0	B	n	k	ϕ
Li	3	3.5	1	9.068	4.24	-1.995	3.440
Na	35.73	6.593	1	20.16	4.532	-1.675	3.595
K	2.7527	2.82863	2.47148	2	1.661	-1.358	-0.574
Rb	1.00047	2	2.88956	0.5	2	1.379	4.137
Cs	$1.416*10^{-5}$	1.019	13.66	14.44	2.778	0.883	13.793

3. Structure and collective dynamics of liquid alkali metals

Simulation was performed in the LAMMPS package [11] with the following initial conditions: a bcc lattice of $N = 9826$ atoms with periodic boundary conditions. The lattice constants are same as in the ab-initio simulation case with initially zero velocities and accelerations. A canonical NVT ensemble with an integration time step of $\tau = 10$ fs was used, where the systems were heated to the temperature 800 K over 100 000 time steps and held for 70,000 time steps, after which the liquid system were cooled to the following target temperatures: for lithium is 475 K, sodium is 390 K, potassium is 343 K, rubidium is 320 K, cesium is 303 K. This state is located near the melting curve, and there is a fairly extensive experimental material for it. One of the most common ways to check the correctness of the interaction potential is to calculate the radial distribution function of atoms [12]:

$$g(r) = \frac{V}{4\pi r^2 N} \left\langle \sum_{i=1}^{N-1} \frac{\Delta n_i(r)}{\Delta r} \right\rangle \quad (2)$$

and the static structure factor [12]:

$$S(k) = 1 + \frac{N}{V} \int [g(r) - 1] \exp(i\mathbf{kr})\, d\mathbf{r}, \quad (3)$$

where V is the volume of the system, N is the number of atoms in the system, $\Delta n_i(r)$ is the number of atoms in a spherical layer of thickness Δr, k is the wave vector, and $\mathbf{r}$ is the radius vector of the particle, angular brackets mean averaging over the entire system and over the time sample. The results of calculations of the radial distribution function and the static structure factor are presented below.

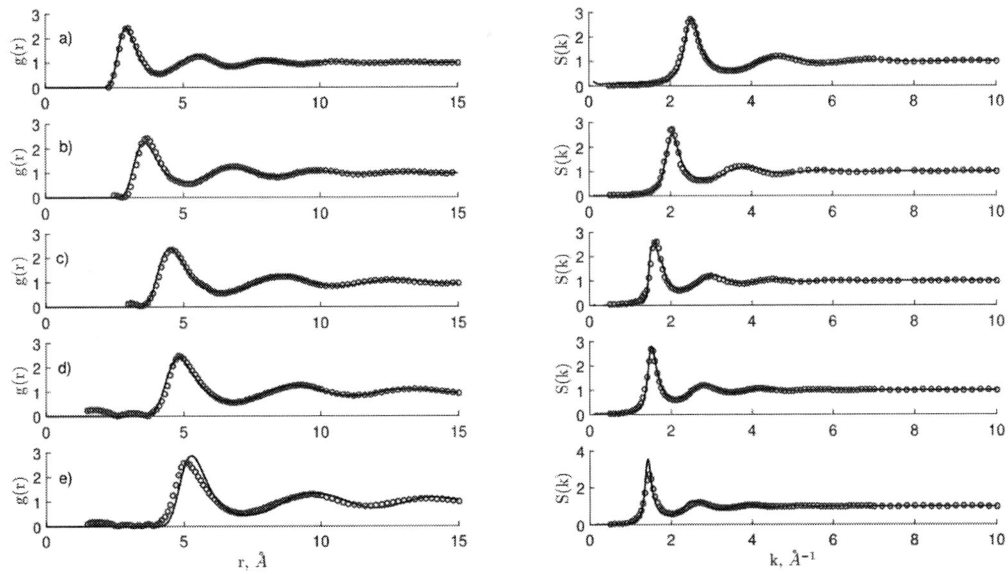

Figure 1. Radial distribution function (left panel) and static structure factor (right panel) for: a) lithium, b) sodium, c) potassium, d) rubidium, e) cesium. Blue line is the simulation results, red circles are the experimental results [13].

In our case, we will consider also dynamical structure factor [14, 15, 16, 17, 18, 19]. This quantity is determined from the experiments of inelastic scattering neutrons and X-rays [12]:

$$S(k,\omega) = \frac{1}{2\pi N}\int_{-\infty}^{+\infty} \exp(-i\omega t)\sum_{i,j}\langle \exp\left[-i\mathbf{k}\mathbf{r}_i(0)\right]\exp\left[-i\mathbf{k}\mathbf{r}_j(t)\right]\rangle\, dt. \tag{4}$$

The angular brackets mean averaging over the entire system and over the time sample. To compare the simulation results with the experimental data, the X-ray scattering intensity spectra $I(k,\omega)$ were determined for different values of the wave number k, which is related to the dynamic structure factor $S(k,\omega)$ by the relations [18]

$$I(k,\omega) = E(k)\int R(k,\omega-\omega')S_q(k,\omega')d\omega', \tag{5}$$

$$S_q(k,\omega) = \frac{\hbar\omega/k_BT}{1-\exp(\hbar\omega/k_BT)}S(k,\omega). \tag{6}$$

4. Conclusions

In the present work, the oscillating potential model (1) is constructed and tested as applied to liquid alkali metals near the melting point. It is shown that such a pair potential accurately reproduces the "structural" features of alkali metal melts and describes experimental data on neutron and X-ray diffraction. It has been established that the resulting model can be used to describe the static structure in a wide range of wave numbers. But in the dynamical structure factor, there is some mismatch between the simulation results, especially in sodium.

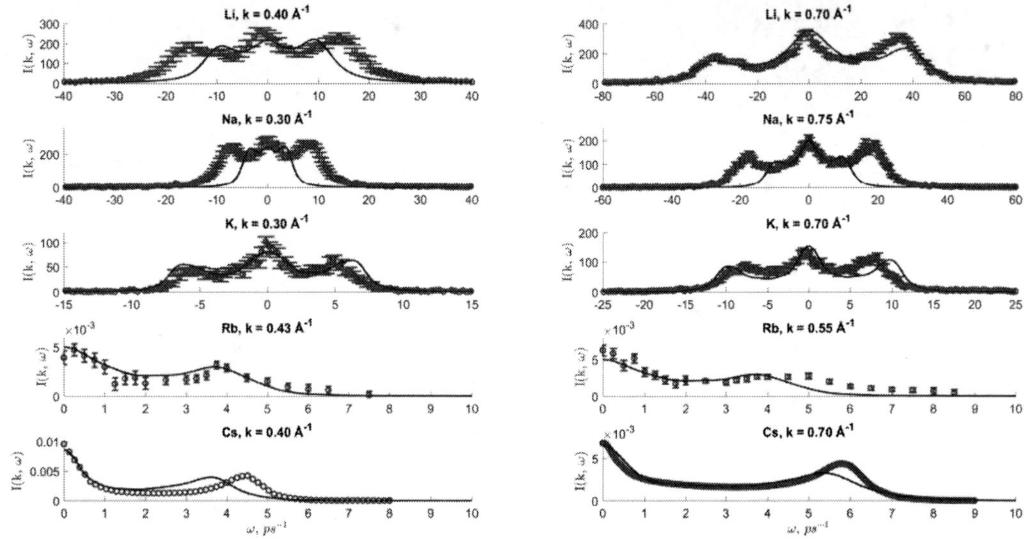

Figure 2. Spectra of inelastic scattering of X-rays and neutrons (red circles) for liquid lithium at a temperature of $T = 475$ K [20], sodium $T = 390$ K [21], potassium $T = 343$ K [22], rubidium $T = 320$ K [23], cesium $T = 303$ K [24]. The experimental errors of the intensity are shown for all liquid metals, except cesium. The results of simulation with model potential are marked as blue line.

5. Acknowledgments
This study is supported by the Kazan Federal University Strategic Academic Leadership Program.

References
[1] Nichol A and Ackland G J 2016 *Phys. Rev. B* **93** 6
[2] Hoshino K and Joarder R N 1986 *J. Non-Cryst. Sol.* **16** 1659
[3] Das T and Joarder R N 1990 *J. Non-Cryst. Sol* **117** 583
[4] Belashenko D K 2012 *Inorg. Mater.* **48** 79
[5] Egelstaff P 1967 *An Introduction to the Liquid State* (New York: Academic Press London)
[6] Gonzalez L E, Gonzalez D J and Lopez J M 2001 *Phys.: Condens. Matter* **13** 7801
[7] Mokshin A V, Khusnutdinov R M, Akhmerova A R and Musabirova A R 2017 *JETP Letters 106* 343
[8] Stukowski A, Fransson E, Mock M and Erhart P 2017 *Modelling Simul. Mater. Sci. Eng.* **25** 055003
[9] Kresse G and Furthmiller J 1996 *Phys. Rev. B* **54** 11169
[10] Kresse G and Hafner J 1993 *Phys. Rev. B* **47** 558
[11] Plimpton S 1995 *Phys. Rev. B* **117** 1
[12] Khusnutdinoff R M, Galimzyanov B N and Mokshin A V 2018 *JETP* **126** 83
[13] Waseda Y 1980 *The Structure of Non-Crystalline Materials: Liquids and Amorphous Solids* (New York: McGraw-Hill)
[14] Yulmetyev R M, Mokshin A V, Hanggi P and Shurygin V Y 2002 *JETP Letters* **76** 147
[15] Mokshin A V, Yulmetyev R M and Hanggi P 2004 *J. Chem. Phys.* **121** 7341
[16] Mokshin A V, Yulmetyev R M, Khusnutdinov R M and Hanggi P 2007 *J. Phys.: Condens. Matter.* **19** 046209(1)
[17] Mokshin A V 2015 *Theoret. and Math. Phys.* **183** 449
[18] Mokshin A V and Galimzyanov B N 2018 *J. Phys.: Condens. Matter.* **30** 085102
[19] Khusnutdinoff R M et al. 2020 *Phys. Rev. B.* **101**(21) 214312-1

[20] Scopigno T, Balucani U, Ruocco G and Sette F 2000 *J. Phys.: Condens. Matter* **12** 8009
[21] Scopigno T, Balucani U, Ruocco G and Sette F 2002 *Phys. Rev. E.* **65** 31205
[22] Monaco A and Scopigno T 2004 *J. Chem. Phys.* **120** 8089
[23] Copley J R D and Rowe J M 1974 *Phys. Rev. Lett.* **32** 49
[24] Bodensteiner T, Morkel C, Glaser W and Dorner B 1992 *Phys. Rev. A.* **45** 5709

Structural transformations in binary $Ni_{62}Nb_{38}$ alloy at ultrahigh temperatures

Dinar T. Yarullin, Maria A. Doronina

Kazan Federal University, Kazan, Russia

E-mail: YarullinDT@gmail.com

Abstract. In this study, we performed the molecular dynamics simulation of the binary $Ni_{62}Nb_{38}$ alloy at heating from 300 K to 9000 K. The temperature dependences of the density and the fraction of liquid atoms were calculated. The boiling temperature of the system is determined. It has been established that near the temperature 7000 K the system undergoes phase transition from liquid to gas phase. It is shown that near the boiling temperature, the molten metal forms a branched structure. The obtained results made it possible to estimate the region of the liquid phase and to determine the probable scenario for the occurrence of structural changes near the boiling point.

1. Introduction

The plasma-arc cutting is the most popular method for processing of refractory metals, which include the Ni-Nb system [1-3]. This system has the liquidus temperature ~1450 K at the eutectic point, which exceeds the eutectic temperature of steel [4]. In addition, the Ni-Nb binary system is capable of forming bulk metallic glass at room temperature [5]. The glass-forming ability is most pronounced in the $Ni_{62}Nb_{38}$ alloy [6, 7].

Most modern studies of the microscopic structure of $Ni_{62}Nb_{38}$ alloy cover the temperature range below the liquidus temperature and do not consider ultrahigh temperatures [7-10]. For the correct processing of this alloy, for example, by ionized gas, it is necessary to know the critical temperatures as the boiling temperature [11]. The melting process of the system occurs through the formation and growth of liquid phase nuclei, which contain only a few tens of atoms. The growth laws of these nuclei will depend on the temperature of the ionized gas and on the rate of heat removal from the melting zone. Here the molecular dynamics simulation method is the most suitable tool for studying structural-phase transformations on small spatial scales. In this work, the structure of the $Ni_{62}Nb_{38}$ alloy is calculated at heating from 300 K to 9000 K at the constant pressure 1 atm. This study is aimed to determine the boiling temperature of this alloy.

2. Simulation procedure and results

The considered system consists of 11664 atoms (7229 atoms for Ni and 4435 atoms for Nb). The interaction of atoms is given by the Finnis-Sinclair type potential [12]. The initial samples have a

Content from this work may be used under the terms of the Creative Commons Attribution 3.0 licence. Any further distribution of this work must maintain attribution to the author(s) and the title of the work, journal citation and DOI.
Published under licence by IOP Publishing Ltd

crystal structure equilibrated at the temperature 300 K and at the pressure 1 atm. The crystalline system is heated with the rate 10^{11} K/s to the temperature 9000 K.

Figure 1 shows decreasing density from 8.71 g/cm³ to 4.1 g/cm³ at heating to the temperature ~7000 K. With a further increase in temperature, a rapid decrease in density is observed due to the transition to the gas phase. Hence we obtain that the temperature $T_b \approx (7000 \pm 300)$K is the boiling point of the system. The obtained ratio between the liquidus and boiling temperatures is $T_b/T_l \approx 4.5$, which is within reasonable range for metallic systems [13, 14]: the value of T_b/T_l can vary widely from 1.7 to 5.5. For example, for tin the ratio is $T_b/T_l \approx 5.5$, while for nickel this ratio is only $T_b/T_l \approx 1.78$ [15, 16].

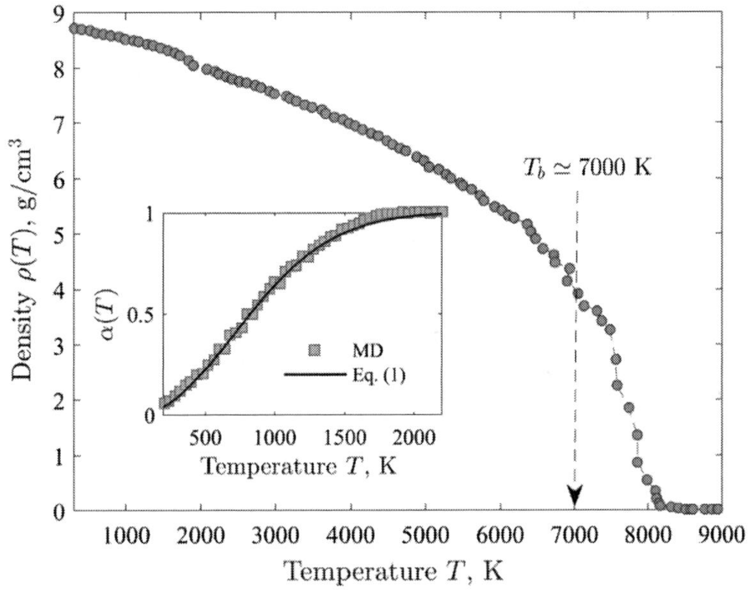

Figure 1. System density ρ as function of the temperature T. Inset: temperature dependence of fraction of the liquid phase atoms.

Figure 1 (see inset) shows the T-dependence of the fraction of liquid phase atoms with respect to the atoms of the entire system. The form of this dependence is typical for activation-type processes, such as the crystal nucleation and the formation of liquid droplets in saturated vapor [17–20], and this dependence is well reproduced by the exponential expression of the following form

$$\alpha(T) = 1 - \exp(-[\gamma T]^n). \quad (1)$$

Note that Eq. (1) is similar to the Kolmogorov-Johnson-Mehl-Avrami equation, which is used to analyze the kinetics of isothermal crystallization based on time-dependent size of crystal nuclei [21]. In Eq. (1), the parameters γ and n are adjustable and these parameters take the values $\gamma = 0.001$ K^{-1} and $n = 2$, respectively. This small value of the parameter n may indicate that the growth of liquid nuclei occurs as a result of coalescence of small melt droplets. It follows from Figure 2 that, at temperatures above 6000 K the coalescence leads to formation of the branched structure consisting molten metal. The complete transition of the system to the liquid phase occurs at the temperature $T_l \approx$ 1550K. For comparison, the experimentally calculated value of the liquidus temperature of the $Ni_{62}Nb_{38}$ alloy is 1543 K [22].

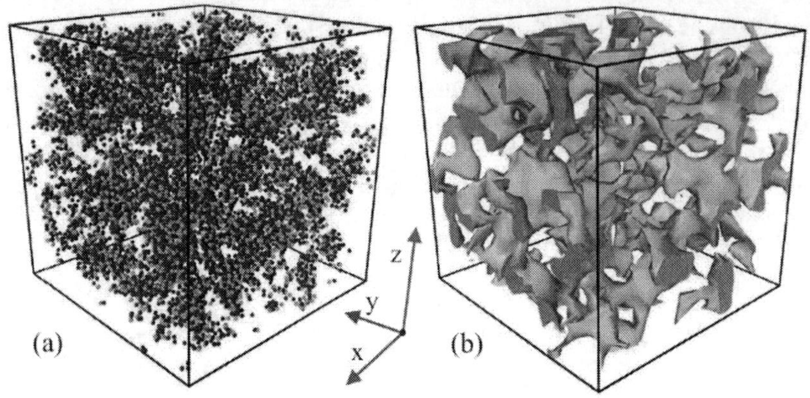

Figure 2. Snapshots of the system at the temperature 7000 K: (a) Ni atoms (blue circles) and Nb atoms (green circles); (b) liquid branched structure (blue areas).

3. Conclusions

Thus, the density of the $Ni_{62}Nb_{38}$ alloy at heating from 300 K to 9000 K was calculated by the method of molecular dynamics simulation. The boiling temperature of this alloy was also determined. The obtained results show that the liquid melt forms the branched structure near the boiling point. The results of this study made it possible to determine the temperature range of the liquid phase and predict a probable scenario for the occurrence of structural changes near the boiling point of the Ni62Nb38 system.

Acknowledgments

This study is supported by the Kazan Federal University Strategic Academic Leadership Program. D.T.Y. acknowledges the Foundation for the Development of Theoretical Physics and Mathematics "Basis" (project No. 20-1-2-38-1).

References

[1] Wang J, Hoang T, Floyd E L, Regens L J 2017 *Annals of Work Exposures and Health* **61** 311
[2] Fan W, Ji W, Wang L, Zheng L, Wang Y 2020 *The International Journal of Advanced Manufacturing Technology* **110** 2863
[3] Fairushin I I, Vasiliev M M, Petrov O F 2021 *Molecules* **26** 6974
[4] Bhatia D N, Rao M N, Taneja A K 2009 *Mat.-wiss. u. Werkstofftech.* **40** 634
[5] Xia L, Li W H, Fang S S, Wei B C, Dong Y D 2006 *J. Appl. Phys.* **99** 026103
[6] Zeng Y Q, Yu J S, Tian Y, Hirata A, Fujita T, Zhang H Z, Nishiyama N, Kato H, Jiang J Q, Inoue A, Chen M W 2020 *Acta Mater.* **200** 710
[7] Jeon J, Kim G, Seo N, Choi H, Kim H-J, Lee M-H, Lim H-K, Son S B, Lee S-J 2022 *J. Mater. Research and Technology* **16** 129
[8] Galimzyanov B N, Doronina M A, Mokshin A V 2021 *J. Non-Cryst. Solids* **572** 121102
[9] Galimzyanov B N, Yarullin D T, Mokshin A V 2018 *JETP Letters* **107** 629
[10] Galimzyanov B N, Mokshin A V 2021 *Journal of Non-Crystalline Solids* **570** 121009
[11] Fairushin I I, Khrapak S A, Mokshin A V 2020 *Results in Physics* **19** 103359
[12] Zhang Y, Ashcraft R, Mendelev M I, Wang C Z, Kelton K F 2016 *J. Chem. Phys.* **145** 204505

[13] Mokshin A V, Khusnutdinoff R M, Galimzyanov B N 2020 *Physical Chemistry Chemical Physics* **22** 4122
[14] Mokshin A V, Galimzyanov B N 2015 *J. Chem. Phys.* **142** 104502
[15] Molodets A M, Nabatov S S 2000 *High Temperature* **38** 715
[16] Kerfoot, Derek G E 2000 *Nickel. Ullmann's Encyclopedia of Industrial Chemistry* (Weinheim: Wiley-VCH)
[17] Mokshin A V, Galimzyanov B N 2014 *J. Chem. Phys.* **140** 024104
[18] Mokshin A V, Galimzyanov B N, Yarullin D T 2020 *European Physical Journal: Special Topics* **229** 427
[19] Yarullin D T, Galimzyanov B N, Mokshin A V 2020 *J. Chem. Phys.* **152** 224501
[20] Galimzyanov B N, Mokshin A V 2021 *International Journal of Solids and Structures* **224** 111047
[21] Kashchiev D 2000 *Nucleation: Basic Theory with Applications* (Butterworth-Heinemann, Oxford)
[22] Lesz S, Dercz G 2016 *J. Therm. Anal. Calorim.* **126** 19

On the accuracy of analytical solution of the Poisson-Boltzmann equation in the linear approximation

I. I. Fairushin[1, 2]

[1]Kazan Federal University, Kazan, Russia
[2]Joint Institute for High Temperatures, Russian Academy of Sciences, Moscow, Russia

fairushin_ilnaz@mail.ru

Abstract. In this paper, it is shown that the choice of the reference point of the potential for the Poisson-Boltzmann equation can affect the accuracy of the solution in the linear approximation.

1. Introduction

As is known, the Poisson-Boltzmann equation for the potential φ of the electric field finds application in many areas of physics. In particular, in plasma physics this equation is an integral part of the system of equations describing the processes in a gas discharge in the fluid approximation [1, 2].

In the case of a spherically symmetric setting, the Poisson-Boltzmann equation can be written as follows [3-6]:

$$\frac{d^2\varphi}{dr^2}+\frac{2}{r}\frac{d\varphi}{dr}=\frac{q}{\varepsilon\varepsilon_0}(n_e-n_i), \tag{1}$$

where r is the spatial coordinate, q is the absolute value of the electron charge, ε is the dielectric constant, ε_0 is the electric constant, n_e is the electron concentration, and n_i is the ion concentration. The last two quantities in the general case can be represented as functions of the potential:

$$n_e=n_{e0}f_e(\varphi), \qquad n_i=n_{i0}f_i(\varphi), \tag{2}$$

where n_{e0} and n_{i0} are, respectively, the concentrations of electrons and ions at $\varphi = 0$. In this case:

$$f_e(\varphi)=1+k_1\varphi+k_2\varphi^2+\cdots+k_n\varphi^n+\cdots \tag{3a}$$

$$f_i(\varphi)=1+m_1\varphi+m_2\varphi^2+\cdots+m_n\varphi^n+\cdots. \tag{3b}$$

2. Theoretical background

In this works devoted to the study of physical processes in thermal dusty plasma [3-9], expressions for the functions $f_e(\varphi)$ and $f_i(\varphi)$ can be given in the following form:

$$f_e(\varphi)=\exp\left(\frac{q\varphi}{k_BT}\right), \tag{4a}$$

Content from this work may be used under the terms of the Creative Commons Attribution 3.0 licence. Any further distribution of this work must maintain attribution to the author(s) and the title of the work, journal citation and DOI.
Published under licence by IOP Publishing Ltd

$$f_i(\varphi) = \begin{cases} 1, \text{ at } 0 \le r \le R, \\ 0, \text{ at } r > R. \end{cases} \tag{4b}$$

Here R – is the dust particle radius, T – is the absolute temperature of the system, k_B – is the Boltzmann constant. Further, reducing equation (1) to a dimensionless form and taking into account (2) and (3) we obtain:

$$\psi_1'' + \frac{2}{x}\psi_1' = a^2(\bar{n}_{e0}\exp(\psi_1) - 1), \tag{5a}$$

$$\psi_2'' + \frac{2}{x}\psi_2' = a^2\bar{n}_{e0}\exp(\psi_2). \tag{5b}$$

Here

$$\psi = \frac{q\varphi}{k_B T}, \quad x = \frac{r}{R}, \quad a^2 = \frac{q^2R^2n_i}{\varepsilon_0 k_B T}, \quad \bar{n}_{e0} = \frac{n_{e0}}{n_i}, \tag{6}$$

Moreover, the boundary conditions for equations (5a) and (5b) are set as follows:

$$\psi_1'(0) = 0; \ \psi_2'(\lambda) = 0; \ \psi_1(0) = 0; \ \psi_1(1) = \psi_2(1); \ \psi_1'(1) = \psi_2'(1).$$

Here $\lambda = \frac{l}{R}$.

3. Results

Equations (5a) and (5b) can only be solved numerically, for example, by the Runge-Kutta method. For the analytical solution of the system of equations (5a) and (5b), one can use the linear approximation for the exponential on the right-hand side, namely:

$$\exp(\psi_{1,2}) \approx 1 + \psi_{1,2}. \tag{7}$$

Then the solution of the system of equations (5a) and (5b) is given by the following expressions:

$$\psi_1(x) = \frac{C}{\bar{n}_{e0}}\frac{\exp(a\sqrt{\bar{n}_{e0}}x) - \exp(-a\sqrt{\bar{n}_{e0}}x)}{x} + \frac{1}{\bar{n}_{e0}} - 1, \tag{8a}$$

$$\psi_2(x) = \frac{D}{\bar{n}_{e0}}\frac{\exp(a\sqrt{\bar{n}_{e0}}x) + F\exp(-a\sqrt{\bar{n}_{e0}}x)}{x} - 1. \tag{8b}$$

Here

$$C = \frac{(a\sqrt{\bar{n}_{e0}} - 1)\exp(a\sqrt{\bar{n}_{e0}}) - F(a\sqrt{\bar{n}_{e0}} - 1)\exp(-a\sqrt{\bar{n}_{e0}})}{2(F+1)a\sqrt{\bar{n}_{e0}}};$$

$$D = \frac{C(\exp(a\sqrt{\bar{n}_{e0}}) - \exp(-a\sqrt{\bar{n}_{e0}})) + 1}{\exp(a\sqrt{\bar{n}_{e0}}) + F\exp(-a\sqrt{\bar{n}_{e0}})};$$

$$F = \frac{a\sqrt{\bar{n}_{e0}}\lambda - 1}{a\sqrt{\bar{n}_{e0}}\lambda + 1}\exp(2a\sqrt{\bar{n}_{e0}}\lambda);$$

$$\bar{n}_{e0} = 1 + 2Ca\sqrt{\bar{n}_{e0}}.$$

Pay attention to the fact that reference point of the potential can be chosen arbitrarily. If we choose the point $x = 1$ as the reference point of the potential, then the absolute value of the potential will be the sum of the positive contribution and the modulus of the negative part of the potential for $x > 1$. Moreover, each term is obviously less than the sum of the contributions in absolute value, this makes it possible to expand the range of parameters where the analytical the solution is carried out with acceptable accuracy. In the case of choosing the first initial condition in the form:

$$\psi_{1,2}(1) = 0, \tag{9}$$

we obtain the following expressions for the analytical solution of the system of equations (5a) and (5b) in the linear approximation:

$$\psi_1(x) = \frac{C'}{\bar{n}'_{e0}} \frac{\exp\left(a\sqrt{\bar{n}'_{e0}}x\right) - \exp\left(-a\sqrt{\bar{n}'_{e0}}x\right)}{x} + \frac{1}{\bar{n}'_{e0}} - 1, \tag{10a}$$

$$\psi_2(x) = \frac{D'}{\bar{n}'_{e0}} \frac{\exp\left(a\sqrt{\bar{n}'_{e0}}x\right) + F' \exp\left(-a\sqrt{\bar{n}'_{e0}}x\right)}{x} - 1. \tag{10b}$$

Here

$$C' = \frac{\left(a\sqrt{\bar{n}'_{e0}} - 1\right)\exp\left(a\sqrt{\bar{n}'_{e0}}\right) - F'\left(a\sqrt{\bar{n}'_{e0}} - 1\right)\exp\left(-a\sqrt{\bar{n}'_{e0}}\right)}{2(F' + 1)a\sqrt{\bar{n}'_{e0}}};$$

$$D' = \frac{\bar{n}'_{e0} + 1}{\exp\left(a\sqrt{\bar{n}'_{e0}}\right) + F' \exp\left(-a\sqrt{\bar{n}'_{e0}}\right)};$$

$$F' = \frac{a\sqrt{\bar{n}'_{e0}}\lambda - 1}{a\sqrt{\bar{n}'_{e0}}\lambda + 1} \exp\left(2a\sqrt{\bar{n}'_{e0}}\lambda\right);$$

$$\bar{n}'_{e0} = C'\left(\exp\left(a\sqrt{\bar{n}'_{e0}}\right) - \exp\left(-a\sqrt{\bar{n}'_{e0}}\right)\right) + 1.$$

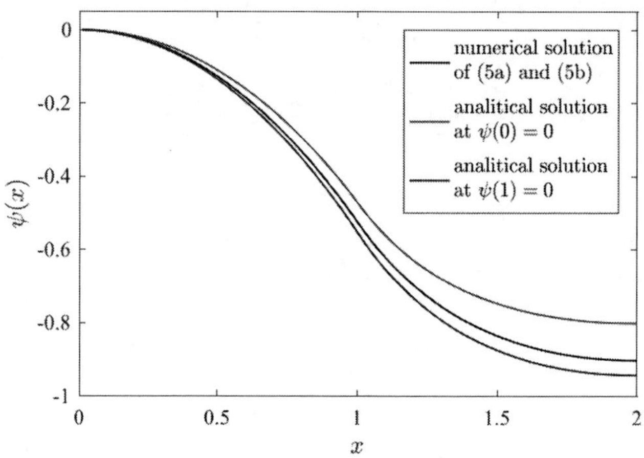

Fig. 1. Potential distribution inside and outside the macroparticle in thermal dusty plasma for the following system parameters: $R = 10^{-7}$m, $T = 1700$ K, $n_i = 3{,}5 \cdot 10^{21}\text{m}^{-3}$, $\lambda = 2$.

In fig. 1 shows the graphs of potential distributions obtained using the numerical solution and different analytical ways for solving equations (5a) and (5b).

4. Conclusion
In this work, it is shown that, the use of the point $x = 1$ as a initial point of the potential makes it possible to obtain an analytical solution of the linear Poisson-Boltzmann equation that is closer to the numerical solution of the nonlinear equation, than the solution that is obtained using the zero potential in the point $x = 0$. The maximum deviation in the first case is 4.5%., in the second case the maximum deviation is 11.2%.

5. Acknowledgements
This paper has been supported by the Kazan Federal University Strategic Academic Leadership Program.

References
[1] Benilov M S 2019 *Journal of Physics D: Applied Physics* **53** (1) 013002
[2] Saifutdinov A I, Fairushin I I and Kashapov N F 2016 *JETP letters* **104** (3) 180-185
[3] Vishnyakov V I 2012 *Physical Review E* **85** 026402.
[4] Vishnyakov V I and Dragan G S 2006 *Physical Review E* **73** (2) 026403
[5] Vishnyakov V I and Dragan G S 2006 *Physical Review E* **74** (3) 036404
[6] Vishnyakov V I, Dragan G S and Evtuhov V M 2007 *Physical Review E* **76** (3) 036402
[7] Khrapak S A, Morfill G E, Fortov V E, D'yachkov L G, Khrapak A G, and Petrov O F 2007 *Phys. Rev. Lett.* **99** 055003
[8] Fairushin I I, Dautov I G, Kashapov N F and Shamsutdinov A R 2017 *Technical Physics Letters* **43** (1) 27-30
[9] Fayrushin I I, Dautov I G and Kashapov N F 2017 *International Journal of Environmental Science and Technology* **14** (12) 2555-2560

Creation of a flexible thin-film thermoelectric converter for the range -50...150 degrees Celsius

O P Plotnikova[1], D V Kirillov[2], N A Tumanov[1, 2], D S Manegin[1, 2], V A Riazanov[2]
[1]"TTEG" LLC, Moscow, Russia
[2]Bauman Moscow State Technical University, Moscow, Russia

plotnikovaolgap97@gmail.com

Abstract. The paper presents the results of a project to create highly sensitive thermoelectric converters for small temperature differences on a flexible film base, consisting of a large number of compact thermocouples connected in series. Such sensors may be used, for example, as the part of fire alarm systems without any power supply. The sensors were produced as multilayer structure coating of thermoelectric alloys on the flexible polymer substrate through photolithographic masks. A pair of chromel-T and kopel was taken as thermoelectric alloys. The preliminary operating parameters of the obtained samples are presented.

1. Introduction

Differential temperature sensors are used everywhere, including fire or coolant leakage warning systems. However, in the range of low temperatures (-50...150°C) measurements via passive sensors are difficult. The smaller the temperature difference, the greater the measurement error, the more accurate and expensive equipment is required to interpret the signal received from the sensor. For this reason, thermal resistances are more often used for this temperature range, but their operation requires additional power supplies [1]. A multiple increase in the specific thermoelectromotive force of a passive thermal sensor would make it possible to significantly simplify and reduce the cost of measuring temperature differences in the specified range.

Vacuum-plasma technologies make it possible to create compact and efficient electronic components, including sensitive elements with a much higher output signal level without a significant increase in price, dimensions, weight and material consumption relative to existing passive thermal sensors.

The aim of the work was to develop technology, manufacture and confirm the operability of prototypes of thin-film thermoelectric converters.

2. Materials and Methods

The manufacturing technology of thin film multilayer thermoelectric converters is based on the method of magnetron sputtering [2, 3] of thermoelectric alloys to form the two-layer coating on a polymer substrate. This method makes it possible to form thin films of metals, including alloys, with almost no change in their composition.

Before deposition, the substrate was cleaned with ion beam produced by anode layer accelerator. The exposure and energy of the ion beam were selected to provide the mask retention. The thickness of

Content from this work may be used under the terms of the Creative Commons Attribution 3.0 licence. Any further distribution of this work must maintain attribution to the author(s) and the title of the work, journal citation and DOI.
Published under licence by IOP Publishing Ltd

the deposited layers was 280 nm, since it is difficult to remove the photoresist with the layer thickness more than 300 nm. Chromel-T and kopel alloys were chosen due to the combination of their high specific thermopower and manufacturability. A polyimide film was used as the substrate material, since the substrate required mechanical strength, heat resistance, solvent resistance, and minimal shrinkage upon heating.

The best way to arrange the maximum number of junctions pairs per unit area and separate the areas of hot and cold junctions is a meander pattern. This pattern was created from two layers of alloys, providing an overlap between the layers in the hot and cold side junction areas. Conventional photolithography, based on chemical etching of the required pattern on a continuous metallization layer, is not suitable for a two-layer element due to damage of the lower deposited layer during the etching the upper one. Therefore, the method of templateless reverse photolithography [4, 5] was chosen, and it was used twice for successive deposition of chromel-T and kopel films. The photolithography method required a reliable fixation of the film on a solid base, which was a silicon wafer. Two methods were considered: the deposition of liquid polyimide on a tungsten sublayer and the application of a polyimide film with an adhesive layer on a solid silicon wafer. The first method supposed the chemical deposition of polyimide film on a silicon wafer with tungsten sublayer coating. After the process the tungsten sublayer was chemically dissolved but polyimide film was damaged during the process. So, it was decided to use a ready-made film 20 µm thick with an adhesive layer fixed uniformly and without defects as the base for metal thin film coatings on a silicon substrate. The magneton sputtering was used as the coating method.

3. Results

A number of flexible sensitive elements with dimensions of 20x24 mm on a polyimide base were obtained. The resolution of the structures in the experimental batch ranged from 1 mm to 50 µm (Figure 1). The calculated resistances of such elements are from 0.95 to 91.7 kOhm, the calculated specific thermoelectromotive force is from 0.68 to 6.57 mV/K. The dimensions of the elements and the resolution of the structures are determined by the equipment used for photolithography.

Flexible polyimide base was fixed on a silicon wafer with the adhesive. The film was taken off by various methods, but the separation with heating of silicon wafer showed the best result. However, some samples were damaged and lost their conductivity. The elements before separation from the silicon wafer are shown in Figure 1.

Figure 1. Thin-film thermoelectric sensors on a silicon wafer

As a result, prototypes of sensitive elements were obtained with a resolution of 0.5 mm and 0.3 mm with the number of junction pairs 11 to 18 respectively. The resistance was from 1.4 kOhm to 7.8 kOhm. The elements produced thermoelectromotive force when the "hot" junctions were heated up to

ΔT 100 degrees, ranging was 0.57 and 0.93 mV/K, which in terms of a pair of junctions corresponds to 52 μV/K.

4. Discussion

Theoretical resistance values of operable samples range from 0.95 and 2.59 kOhm for a track width of 0.5 mm and 0.3 mm, respectively. The actual resistance values are 1.5…3 times higher than the calculated values. Such a significant difference is most likely associated with the appearance of microcracks during the separation of the samples from the solid substrate.

The theoretical values of the specific thermoelectromotive force of operable samples range from 0.68 mV/K to 1.12 mV/K at the rate of 62 μV/K from one pair of junctions. The lower value of the actual specific thermoelectromotive force in comparison with the theoretical values is presumably due to the change in the stoichiometric compositions of the alloys during magnetron deposition on the substrate. It is also possible that chemical contamination of the junction areas may take place during the masking process.

The work performed has shown the operability of the method to obtain thin-film thermoelectric converters with layer-by-layer magnetron sputtering technology on a polymer substrate. The obtained samples showed an average specific thermoelectromotive force in terms of a junction pair of 52 μV/K, which is close to the theoretical value of 62 μV/K.

However, a serious technological problem was the separation of the finished samples from the solid base required for applying masks by photolithography. The solution of this problem will make it possible to obtain operable samples with a resolution of 50 μm with a thermoelectromotive force of the element up to 6.57 mV/K, which at a temperature difference of 100 K corresponds to a generated voltage of 0.66 V.

Acknowledgements

The work was carried out with the financial support of the Federal State Budgetary Institution "Fund for the Promotion of Small Forms of Enterprises in the Scientific and Technical Sphere" (Fund for the Promotion of Innovations) - Agreement No. 4005GS1 / 65541

References

[1] Ash ZH. Datchiki izmeritel'nyh sistem: V 2-h knigah. Kn. 1 – per. s franc. – M.: Mir, 1992. – 480 s.

[2] Marahtanov M.K., Duhopel'nikov D.V., Vorob'ev E.V., Kirillov D.V. Tribologicheskie svojstva serebryanyh pokrytij, poluchennyh metodom magnetronnogo raspyleniya // Nanoinzheneriya. – 2014. – № 9 (39). – S. 30-32.

[3] Duhopel'nikov D.V., Ryazanov V.A., Vorob'ev E.V., Abgaryan V.K., Popov G.A., Hartov S.A. Effektivnye koefficienty raspyleniya titana, nitrida titana i molibdena ionami azota i smesi azot-kislorod // Poverhnost'. Rentgenovskie, sinhrotronnye i nejtronnye issledovaniya. – 2016. – № 12. – S. 15-21.

[4] Bobrova YU.S., Andronik M., SHCHerbakov V.I., Kitaev I.V. Fotolitografiya v tekhnologii izgotovleniya plat silovyh modulej special'nogo naznacheniya // Tekhnologii v elektronnoj promyshlennosti. – 2016. – №2 (86). – S. 31-34.

[5] Avakov S.M., Plebanovich V.I. Bezmaskovaya litografiya // Nanoindustriya. – 2018. – №82. – S. 200-202.

Local density dynamics in a supercritical Lennard-Jones fluid

R A Khabibullin

Department of Computational Physics, Institute of Physics, Kazan Federal University, 420008 Kazan, Russia

E-mail: roman.khabibullin@outlook.com

Abstract. The collective particle dynamics of supercritical Lennard-Jones fluid is investigated on the basis of molecular dynamic modeling data. The intermediate scattering functions and dynamic structure factor spectra for the wavenumber range $k \in [0.18; 3.26]\ \sigma^{-1}$ were calculated. The characteristics of dynamic structure factor spectra such as the thermal diffusivity and the sound velocity were estimated.

1. Introduction

A detailed description of the collective particle dynamics in a condensed disordered system can be obtained, on the one hand, on the basis of experimental INS/IXS spectroscopy data [1–3]. On the other hand, the study of the dynamics of such a system can be carried out by molecular dynamics simulation with some given potential [4]. In this case, the frequency spectra of time correlation functions of dynamical variables are of the greatest interest [5]. As such a variable, it is convenient to choose the quantity

$$\rho_{\mathbf{k}}(t) = \sum_{j=1}^{N} \exp(-i\mathbf{k} \cdot \mathbf{r}_j(t)), \tag{1}$$

which is a spatial Fourier image of the local density function

$$\rho(\mathbf{r}, t) = \sum_{j=1}^{N} \delta(\mathbf{r} - \mathbf{r}_j(t)). \tag{2}$$

Here, N is the number of particles in the system, $\mathbf{r}_j(t)$ is the position vector of the j-th particle at time t and $\mathbf{k}$ is the wavevector; dot denotes scalar product. Next, we consider the temporal correlation function of local density fluctuations, the intermediate scattering function

$$F(\mathbf{k}, t) = \frac{1}{N}\langle \rho_{\mathbf{k}}(t)\rho_{-\mathbf{k}}(0)\rangle, \tag{3}$$

and its frequency spectrum, the dynamic structure factor [6]

$$S(\mathbf{k}, \omega) = \int_{-\infty}^{\infty} \mathrm{d}t\, F(\mathbf{k}, t) \exp(i\omega t). \tag{4}$$

The angle brackets $\langle \ldots \rangle$ in the expression (3) denote the ensemble averaging, which can be replaced by time averaging in the case of an ergodic system.

In this paper, we present the results of molecular dynamics simulation of a supercritical Lennard-Jones fluid. To investigate the collective microscopic dynamics of this system, the dynamic structure factor and sound dispersion were calculated.

Content from this work may be used under the terms of the Creative Commons Attribution 3.0 licence. Any further distribution of this work must maintain attribution to the author(s) and the title of the work, journal citation and DOI.
Published under licence by IOP Publishing Ltd

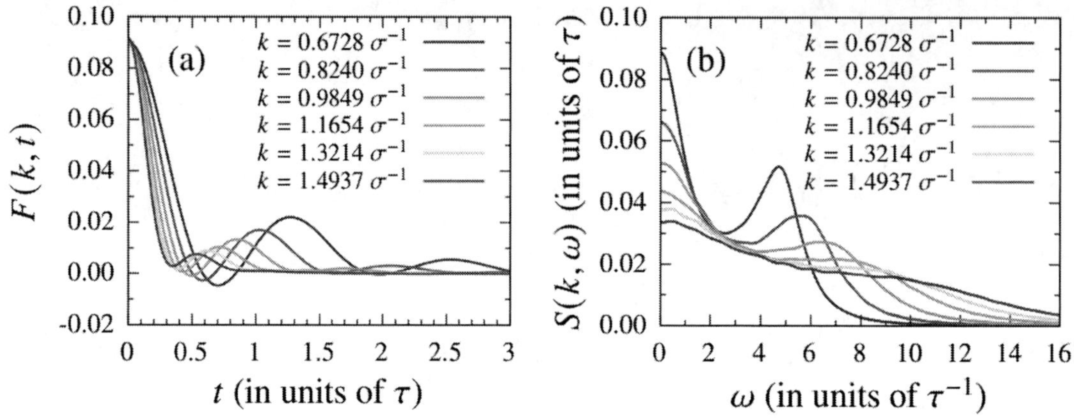

Figure 1: (a) Intermediate scattering function $F(k,t)$ computed by means molecular dynamics simulation with the Lennard-Jones potential for six various wavenumber k values. (b) Dynamic structure factor $S(k,\omega)$ for the same six wavenumber k values. All the functions correspond to the same supercritical [7] state with the temperature $T = 2.8\,\epsilon/k_B$ and the pressure $p = 6.6\,\epsilon/\sigma^3$.

2. Simulation details and results

The procedure for calculating the dynamic structure factor consists of three stages. At the first stage, the MD simulation is performed in order to obtain time series of particles coordinates $\mathbf{r}_j(t)$. For this purpose we used the LAMMPS simulator [8] with the GPU acceleration package [9]. The system consists of $N = 32\,000$ atoms placed in a cubic box with periodic boundary conditions. Pairwise interactions were computed using the Lennard-Jones potential with a $5.0\,\sigma$ cutoff radius. As the initial configuration of the system we used a crystal with a FCC lattice with the density $\rho = 0.75\,\sigma^{-3}$. To bring the system to an equilibrium state with the temperature $T = 2.8\,\epsilon/k_B$ and the pressure $p = 6.6\,\epsilon/\sigma^3$ we used a Nose–Hoover thermostat and barostat with relaxation times of $0.5\,\tau$ and $5.0\,\tau$ respectively. Next, we simulated the equilibrium system for 5 000 000 steps. To exclude manipulations of the thermostat and barostat on the dynamics, we performed that MD run in the NVE ensemble. While $0.005\,\tau$ timestep was used for the isothermal-isobaric simulation, the microcanonical simulation was performed with the smaller timestep of $0.001\,\tau$ for better energy conservation. Thus, the simulation time in the NVE ensemble was $5\,000\,\tau$.

At the second stage the Fourier components of the local density function (1) were calculated. Acceptable values of wavevector $\mathbf{k}$ components for the case when particle trajectories $\mathbf{r}_j(t)$ are obtained from MD simulations in a cubic box with periodic boundary conditions are determined from the expression $\mathbf{k} = (2\pi/L)(a,b,c)$, where a, b, c are integers and L is the box length. It is important to note that for isotropic disordered systems, the spectral characteristics can be averaged over all possible directions of the wavevector $\mathbf{k}_1, \mathbf{k}_2, \ldots, \mathbf{k}_n$ with the fixed magnitude $k = |\mathbf{k}_1| = |\mathbf{k}_2| = \ldots = |\mathbf{k}_n|$, where n is the number of such directions. In terms of calculations, at this stage each stored MD snapshot is turned into one complex number $\rho_{\mathbf{k}}(t_m)$, where m is the snapshot index. We calculated $\rho_{\mathbf{k}}(t_m)$ time series with a $0.005\,\tau$ timestep for 830 wavevectors corresponding to 22 wavenumbers in the $k \in [0.18; 3.26]\ \sigma^{-1}$ range.

Finally, consider the third and final stage of the calculation. The intermediate scatter function $F(k,t)$ was calculated on the basis of time series $\rho_{\mathbf{k}}(t_m)$ with time averaging. The averaging window is an array of 1 501 elements, which corresponds to a $7.5\,\tau$ time interval. Fig. 1(a) shows the obtained scattering functions for 6 wavenumbers. Then, using Filon's numerical integration method [10], we turned these $F(k,t)$ functions into spectra of the dynamic structure factor $S(k,\omega)$, some of which are shown in Fig. 2(b). All computational procedures corresponding to the second and third stages were performed using the original parallel code.

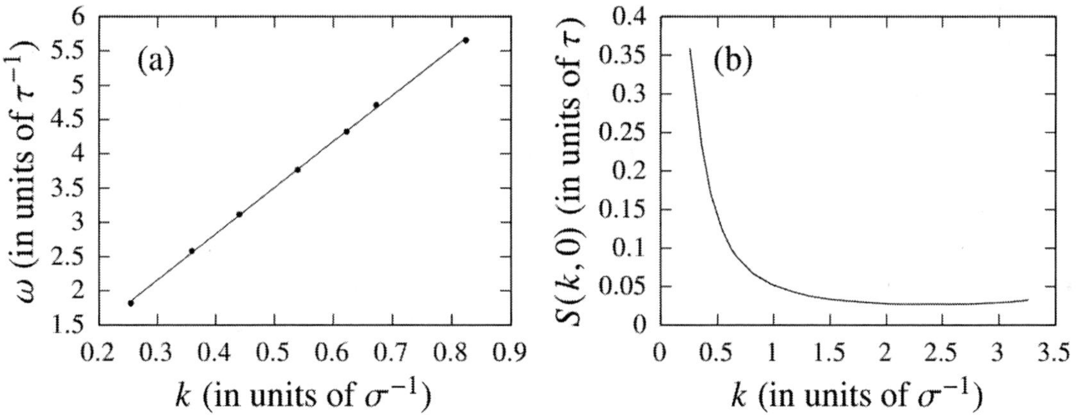

Figure 2: (a) Sound dispersion: filled dots represent the data obtained from molecular dynamics simulation, the solid line corresponds to a linear fit with the slope $v_s = 6.74\sqrt{\epsilon/m}$. (b) Zeroth frequency spectral component $S(k, \omega = 0)$.

As can be seen from Fig. 1(b), the high-frequency (Brillouin) components of the spectrum of the dynamic structure factor decay quite rapidly with increasing wave number k and at $k \approx 1.5\,\sigma^{-1}$ become almost indistinguishable. We plotted the wavenumber dependence of the side peak maximum position for such a range of wavenumbers, when the position of the maximum of the Brillouin peak can be accurately determined. This relationship represents the law of dispersion for the system under study and is shown in Fig. 2(a). Further, the speed of sound v_s can be estimated using the expression

$$\omega_c = \lim_{k \to 0} v_s k \tag{5}$$

from the linear low-k asymptotic of sound dispersion [11]. To do this, we fit the points from Figure 2(a) with a line whose slope coefficient is the desired speed of sound and is equal to $v_s = 6.74\sqrt{\epsilon/m}$. The thermal diffusivity D_T was calculated using the following relationship [12]

$$\Delta\omega_R(k) = D_T k^2, \tag{6}$$

where $\Delta\omega_R(k)$ is the half-width at half-height of the Rayleigh peak. From the spectrum of the dynamic structure factor $S(k, \omega)$ at $k = 0.6728\,\sigma^{-1}$, we determined that the thermal diffusivity of the studied system is $D_T = 2.82\,\sigma\sqrt{\epsilon/m}$. We also determined the dependence of the intensity of the zero-frequency spectral component $S(k, \omega = 0)$ on the wave number k, which is shown in Figure 2(b). This dependence can be used, for example, to determine the sound attenuation coefficient Γ [11].

3. Conclusion

The results of a long-time molecular dynamic simulation of a supercritical Lennard-Jones fluid at the temperature $T = 2.8\,\epsilon/k_B$ and the pressure $p = 6.6\,\epsilon/\sigma^3$ were presented in this paper. For the considered state of the system it is shown that as the number of waves increases, the side peak of the dynamic structure factor $S(k, \omega)$ becomes blurred and shifts to the region of high frequencies.

Acknowledgements

The author is grateful to A. V. Mokshin for useful discussions and recommendations. This paper has been supported by the Kazan Federal University Strategic Academic Leadership Program. The author acknowledges the Foundation for the Development of Theoretical Physics and Mathematics "BASIS" (project No. 20-1-2-38-1).

References

[1] Yulmetyev R M, Mokshin A V, Hanggi P and Shurygin V Yu 2002 *Jetp Lett.* **76** 147–151
[2] Mokshin A V, Yulmetyev R M and Hanggi P 2004 *J. Chem. Phys.* **121** 7341
[3] Mokshin A V, Khusnutdinov R M, Akhmerova A R and Musabirova A R 2017 *Jetp Lett.* **106** 366–370
[4] Mokshin A V, Yulmetyev R M, Khusnutdinoff R M and Hanggi P 2007 *J. Phys.: Condens. Matter* **19** 046209
[5] Mokshin A V, Khusnutdinoff R M, Vilf Y Z and Galimzyanov B N 2021 *Theor. Math. Phys.* **206** 216–235
[6] Hansen J P and McDonald I R 2006 *Theory of Simple Liquids* (London: Academic Press)
[7] Stephan S, Thol M, Vrabec J and Hasse H 2019 *J. Chem. Inf. Model.* **59** 10
[8] Plimpton S 1995 *J. Comp. Phys.* **117** 1–19
[9] Brown W M *et al* 2011 *Comput. Phys. Commun.* **182** 898–911
[10] Allen M P and D. J. Tildesley D J 1987 *Computer Simulation of Liquids* (Oxford: Oxford University Press)
[11] Mokshin A V and Galimzyanov B N 2018 *J. Phys.: Condens. Matter* **31** 209501
[12] Mokshin A V 2015 *Theor. Math. Phys.* **183** 449–477

The development of statistical analysis methods for the study of correlations and statistical memory effects in the recorded data of physical experiments

S A Demin[1], V A Yunusov[1]

[1]Kazan Federal University, Kazan, Russian Federation

serge_demin@mail.ru

Abstract. In this paper, we discuss the prospects for the use of statistical analysis methods in the study of correlations and statistical memory effects in the experimental data of physical experiments. Developed methods under consideration: Memory Functions Formalism and Flicker-Noise Spectroscopy allow obtaining a large set of quantitative parameters and qualitative characteristics directly from temporal signals generated by complex physical systems. Additionally, for the study of collective phenomena and effects, one- and two-parameter cross-correlation functions are proposed, which allow studying cross-correlations between simultaneously recorded signals in spatially separated areas of the object under study. The introduced analytical relations and numerical algorithms can be applied to solve the problems of metrology of surface structures at the nano- and microlevels, to study collective phenomena in the plasma of astrophysical objects, the structure of molecules, and their complexes.

1. Introduction. Physics of complex systems

In the second half of the 20th century, interdisciplinary areas that can be described as "sciences of complexity" acquired relevance. Beginning with the works of Henri Poincaré, the theory of dynamical systems has been actively used in the study of nonlinear physical systems that are sensitive to even small changes in the initial parameters. Later, the theory of dynamic chaos, the theory of catastrophes, the theory of strange attractors appeared in this area, developing a mathematical apparatus for describing a wide range of natural phenomena [1, 2]. Starting with the work of Ludwig von Bertalanffy, general systems theory outlined methods and approaches to the study of processes, regardless of their nature, based on the discovery of relationships between parts of complex systems, as well as their reactions to external influences [3]. "Black box" cybernetics and neurocybernetics contributed to the development of artificial intelligence and machine learning systems [4, 5]. Due to the mutual penetration of these areas and the wide dissemination of the results obtained in the scientific community, it became possible to study the unique properties of complex both physical and non-physical systems.

The scientific literature highlights such properties of complex systems as complexity, openness, nonlinearity, adaptability, emergence, self-organization, and critical transitions [6, 7]. A complex system is understood as an object consisting of a significant number of interacting elements. The interaction between the parts of the whole leads to the enumerated unique properties that make systems "difficult" to study by classical methods. When studying such an interaction, special attention is paid

Content from this work may be used under the terms of the Creative Commons Attribution 3.0 licence. Any further distribution of this work must maintain attribution to the author(s) and the title of the work, journal citation and DOI.
Published under licence by IOP Publishing Ltd

to collective effects, statistical memory, and correlations. Collective phenomena have a significant impact on the evolution of complex systems: from the physics of accelerators (beam intensity limitation) and condensed matter physics (collective effects) to collective behavior in wildlife [8].

2. The application of Memory Functions Formalism and Flicker-Noise Spectroscopy in the analysis of autocorrelations and statistical memory effects

In this paper, we discuss the prospects for using the author's methods: Memory Functions Formalism (MFF) and Flicker-Noise Spectroscopy (FNS) for the analysis of correlations and statistical memory effects in the experimental data of physical experiments. In addition, cross-correlators are discussed, which allow analyzing cross-correlations in simultaneously fixed sets of experimental data.

Figure 1 shows the theoretical methods and numerical algorithms developed by the team of authors, as well as the results of applying these methods in the analysis of the temporal dynamics of complex physical and non-physical systems.

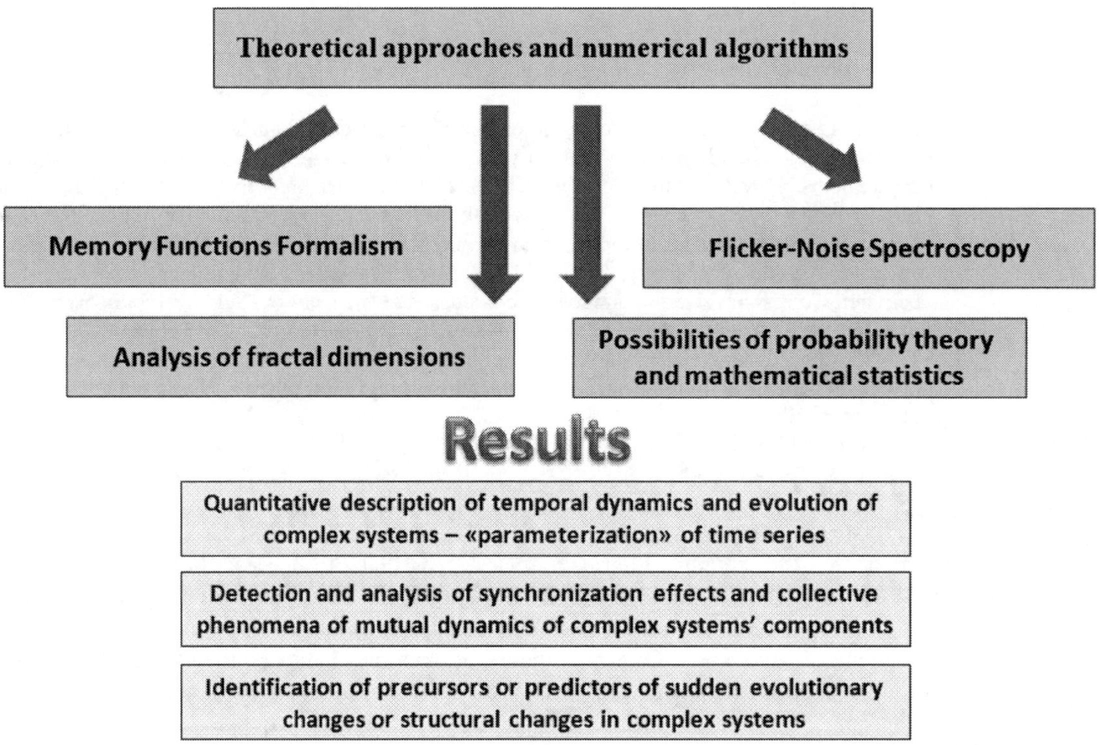

Figure 1. Developed methods for the analysis of experimental data sets of physical experiments and the results obtained using them.

2.1. Memory Functions Formalism

Memory Functions Formalism is a finite-difference generalization of the Zwanzig-Mori projection formalism [9, 10] for analyzing the discrete dynamics of complex systems. The method is based on the representation of the temporal dynamics of the process under study in the form of a multidimensional state vector that obeys the equation of motion written in a discrete form. The use of the Zwanzig-Mori projection technique and the Gram-Schmidt orthogonalization procedure allows shortening the description. Within the framework of the method, for the studied time series, a chain of finite-difference linked kinetic equations of the Zwanzig-Mori type is constructed for the time correlation function and statistical memory functions for interrelated variables [11–13]. Due to the presented concept, it becomes possible to describe the aftereffects at different relaxation levels. The method provides a large set of characteristics and dependences: time dependences of orthogonal dynamic variables, phase portraits of combinations of dynamic variables, relaxation and kinetic parameters, statistical memory functions and their power spectra, frequency dependences of statistical memory measures calculated directly from sequences of dynamic variables of complex systems.

In [14, 15] correlations and statistical memory effects were studied in the sequence of Wolf numbers as one of the quantitative characteristics of solar activity, as well as the X-ray intensity of the GRS 1915+105 microquasar. The results obtained reveal individual physical mechanisms of X-ray generation, the energy released during accretion in the GRS 1915+105 binary system, and the formation of sunspots in the solar photosphere.

Subsequently, the MFF was adapted to the study of the statistical memory effects and cross-correlations in the dynamics of time series of simultaneously recorded experimental parameters characterizing the evolution of complex systems [16].

2.2. Flicker-Noise Spectroscopy

Flicker-Noise Spectroscopy is a method based on the representation of time signals as a combination of jump irregularities (low-frequency components) and burst irregularities (high-frequency chaotic components) [17–19]. These irregularities manifest themselves in correlation relationships at different levels of the spatio-temporal hierarchy. For this, the corresponding procedures and algorithms for parameterization are formulated in the form of general phenomenological expressions for the difference moments and power spectra. The method allows obtaining information about both low-frequency resonant and high-frequency chaotic components of the considered signals using power spectra and transient "structural" functions, as well as a set of parameters characterizing the duration of correlation times, the loss of correlations in sequences of irregularities of the studied signals.

The use of the FNS to introduce "passport parameters" of chaotic surface structures, i.e. roughness profiles studied by atomic force microscopy, makes it possible to approach the solution of problems of surface metrology at nanoscales [20]. In [21] the "certification" of sufficiently homogeneous surfaces of lithium fluoride single crystals grown from a melt was carried out. In [22] a study was made of inhomogeneous surfaces, which can be formed, for example, during the destruction of the material, in the processes of crystallization on the surface, chemical modification.

The FNS approach allows obtaining direct information about the dynamics of correlation relationships between simultaneously recorded signals – dynamic variables of the same essence, measured at spatially separated points of a complex system. In [23, 24] authors study the cross-correlations of the time signals of the radio emission of quasars recorded in different frequency ranges, as well as various indicators of solar activity. An analysis of the X-ray emission of microquasars, which is produced mainly by the inner hot layers of accreting disks, allows obtaining information about the dynamics of the processes occurring in these regions. The study of the dynamics of solar activity indicators allows us to establish the effects of frequency-phase synchronization and periodic patterns at different time intervals.

3. Conclusion

The methods of statistical analysis being developed: the Memory Functions Formalism and Flicker-Noise Spectroscopy, as well as the numerical algorithms and procedures obtained within their framework for calculating parameters, characteristics, and graphical dependencies, are based on the author's theoretical concepts and applied results in the field of nonequilibrium statistical physics, as well as physics of complex systems [25, 26]. The physics of complex systems is aimed at finding physical patterns in describing the behavior and properties of complex systems.

The Memory Functions Formalism allows introducing a set of statistical markers – information measures of memory, kinetic and relaxation parameters, linear measures of synchronization, which are determined directly from experimental signals, and to describe the behavior of these characteristics at different time intervals (scales). The results obtained contribute to the formation of detailed ideas about the role of statistical memory effects, correlations, and fluctuations of the initial time sequences in the description of the dynamic states of complex physical systems.

To extract sufficient quantitative information to make it possible to judge the nature of frequency-phase synchronization, the Flicker-Noise Spectroscopy method is used with two-parameter dependences of cross-correlators calculated based on digitized signals from experimental data sets recorded simultaneously at different frequencies. In addition, parameters introduced in the framework of the FNS allow detecting chaotic and resonant components of the initial discrete time series.

The proposed methods can be effectively applied to the study of autocorrelations, cross-correlations, and statistical memory effects in time sequences and spatial maps of experimentally recorded parameters of physical experiments (see for example [27, 28]), including the study of collective effects in plasma [29], the study of surface structures on nano- and microscales [30], the structure of molecules and their complexes according to the characteristic absorption bands.

Acknowledgments
This paper has been supported by the Kazan Federal University Strategic Academic Leadership Program (PRIORITY-2030). The authors express gratitude to D. Sc. (Physics and Mathematics), Prof. S. F. Timashev for discussing some of the results of the present work.

References

[1] Boccaletti S, Latora V, Moreno Y, Chavez M and Hwang D U 2006 *Phys. Rep.* **424** 175
[2] Boccaletti S, Grebogi C, Lai Y C, Mancini H and Maza D 2000 *Phys. Rep.* **329** 103
[3] Diebner H H, Zerjatke T, Griehl M and Roeder I 2018 *BioSystems* **167** 1
[4] Mehta P, Bukov M, Wang C H, Day A G R, Richardson C, Fisher C K and Schwab D J 2019 *Phys. Rep.* **810** 1
[5] Mokshin A V, Mokshin V V and Sharnin L M 2019 *Commun. Nonlinear Sci. Numer. Simul.* **71** 174
[6] Kwapień J and Drożdż S 2012 *Phys. Rep.* **515** 115
[7] Speth J, Drożdż S and Grümmer F 2010 *Nucl. Phys.* A **844** 30c
[8] Fiegna F, Yu Y T N, Kadam S V and Velicer G J 2006 *Nature* **441** 310
[9] Zwanzig R 1961 *Phys. Rev.* **124** 983
[10] Mori H 1965 *Prog. Theor. Phys.* **34** 399
[11] Demin S A, Yulmetyev R M, Panischev O Y and Hänggi P 2008 *Physica* A **387** 2100
[12] Yulmetyev R, Demin S, Emelyanova N, Gafarov F and Hänggi P 2003 *Physica* A **319** 432
[13] Yulmetyev R M, Demin S A, Panischev O Y and Hänggi P 2005 *Physica* A **353** 336
[14] Yulmetyev R M, Demin S A, Khusnutdinov R M, Panischev O Y and Hänggi P 2006 *Nonlinear Phenom. Complex Syst.* **9** 313
[15] Yulmetyev R M, Demin S A, Hänggi P and Galeev A I 2004 *Nonlinear Phenom. Complex Syst.* **7** 210
[16] Panischev O Y, Demin S A and Bhattacharya J 2010 *Physica* A **389** 4958
[17] Timashev S F 2007 Flicker-Noise Spectroscopy: Information in Chaotic Signals (Moscow: Fizmatlit) [in Russian]
[18] Timashev S F, Polyakov Y S, Yulmetyev R M, Demin S A, Panischev O Y, Shimojo S and Bhattacharya J 2009 *Laser Phys.* **19** 836
[19] Timashev S F, Polyakov Y S, Yulmetyev R M, Demin S A, Panischev O Y, Shimojo S and Bhattacharya J 2010 *Laser Phys.* **20** 604

[20] Timashev S F 2006 *Russ. J. Electrochem.* **42** 424
[21] Letnikova S F, Solovieva A B and Timashev S F 2001 *J. Mater. Sci.* **7** 98
[22] Timashev S F, Belyaev V E, Timashev P S and Solov'eva A B 2006 *Colloid J.* **68** 481
[23] Demin S A, Panischev O Y and Nefedyev Y A 2015 *J. Phys. Conf. Ser.* **661** 012003
[24] Demin S A, Nefedyev Y A, Andreev A O, Demina N Y and Timashev S F 2018 *Adv. Space Res.* **61** 639
[25] Demin S A, Panischev O Y, Timashev S F and Latypov R R 2020 *Bull. Russ. Acad. Sci.: Phys.* **84** 1349
[26] Demin S, Panischev O, Yunusov V and Timashev S 2021 *Stud. Syst. Decis. Control* **350** 27
[27] Ferrari L 2019 *Chem. Phys.* **523** 42
[28] Mokshin A V, Yulmetyev R M and Hänggi P 2005 *Phys. Rev. Lett.* **95** 200601
[29] Suttorp L G and Schooldерman A J 1987 *Physica* A **143** 494
[30] Zayats A V, Smolyaninov I I and Maradudin A A 2005 *Phys. Rep.* **408** 131

Spatial distributions of X-ray radiation during ferrite surface breakdown

S N Andreev[1], A V Bernatskiy[1], G A Dushkin[1, 2], S N Tskhai[1], I N Tilikin[1], S A Pikuz[1], S Y Savinov[1]

[1]P.N. Lebedev Physical Institute of the Russian Academy of Sciences, Moscow, Russia
[2]Moscow Institute of Physics and Technology, Moscow, Russia

bernatskiyav@lebedev.ru

Abstract. Spatial properties of the radiation induced by a ferrite surface flashover supported by a 300 kV and ~1 ns rise time pulsed voltage source were studied. Peak value and rise time were obtained for the discharge current.

1. Introduction

Radiation above ferrite surface in pulse discharge along it was researched in many works, such as, for example, [1-5]. However, only in [6] properties of radiation directed along the discharge on BIN experimental setup [1, 6] were studied. In [6] a short (~2 ns) directional (~3°-5°) X-ray (> 1 keV) pulse was observed during the pre-discharge stage, with pulse intensity increasing as square of the discharge gap length. Discharge parameters were as follows: ~300 kV voltage (~20 ns rise time) and ~270 kA current. Results shown in [6] present great interest for creating coherent X-ray sources.

Radiation appearing in pre-discharge state suggests that phenomenon observed in [6] is not a consequence to discharge current. This article studies spatial properties of radiation induced by a ferrite surface flashover supported by a low power pulsed voltage generator [7] with peak voltage up to 300 kV of ~1 ns rise time.

2. Experimental setup

Schematic 3D layout of the experiment is presented in Fig. 1, and an in-depth description of the setup and measuring techniques are outlined in [7]. An original block (a heavily modified X-ray device «Pamir-300», peak voltage ~300 kV, voltage rise time ~1 ns) was used to power the discharge. High negative voltage was applied through a copper electrode to a wide surface of a rectangular ferrite prism (brand M1000NN) with cross section of 10x20 mm^2. Anode, made also of copper plate, was mounted at a distance of 45 mm from the cathode in a deepening in the ferrite so that it didn't stand out above ferrite surface. Two independent methods were used to measure the discharge current: using a 0.018 Ω shunt and a single-turn Rogowski coil. Both these components were mounted between anode *3* and ground.

Content from this work may be used under the terms of the Creative Commons Attribution 3.0 licence. Any further distribution of this work must maintain attribution to the author(s) and the title of the work, journal citation and DOI.
Published under licence by IOP Publishing Ltd

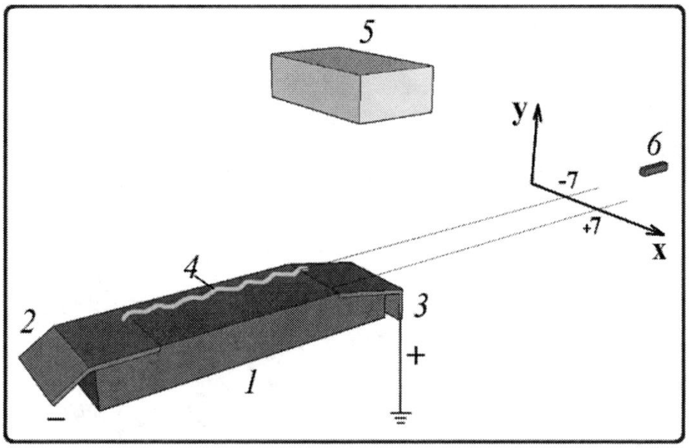

Figure 1. Experiment layout: *1* – ferrite of brand M1000NN; *2* – cathode; *3* – anode; *4* – plasma filament; *5* – neodymium magnet; *6* – scintillator.

The part of the setup responsible for the discharge as well as recording equipment and partly the high voltage source were placed in a spherical stainless steel vacuum chamber 50 cm in diameter. Pumping system consisting of a Leybold Turbovac 350 turbomolecular pump, an Anest Iwata ISP90 spiral vacuum pump and a Leybold IONIVAC ITR 90 combined pressure sensor provided pressure of $\sim 10^{-5}$ Torr in the chamber.

The radiation along ferrite surface was registered from the anode side (XY plane, see Fig. 1) using a circular quartz optic fiber ~1 mm in diameter and ~60 m in length, connected to a FEU-87 photomultiplier, which provided ~2 ns of time resolution. Photomultiplier signal was observed on a 300 MHz bandwidth TDS 3032 oscilloscope. To register X-ray radiation the free fiber end was connected to a scintillator (*6* in Fig. 1) placed inside a lead tube collimator. The 10x2x2 mm^3 piece made of polystyrene doped with π-terphenyl and POPOP had its response time measured separately with use of a phase shift method utilizing a MFLI (Zurich Instruments) lock-in amplifier. It was established to be 1.8 ± 0.1 ns (at the excitation by 365 nm LED emission), well consistent with the time resolution of the receiving path. Visible light from discharge was prevented from reaching the photomultiplier by placing a 4 μm filter made of polypropylene with aluminum cover in front of scintillator.

An original two-dimensional motorized translator controlled programmatically from a personal computer was used to obtain a spatial distribution for the studied radiation. Optical fiber end with the scintillator was attached to its moving part at a distance of 76 mm from the anode side of the ferrite prism. The system allowed to position the radiation receiver in an area of 10x10 cm^2 with an accuracy of 0.1 mm for each axis.

Electron beam formed in discharge was separated from X-ray radiation by placing a 0.3 T 4x2x1 cm^3 neodymium magnet (*5* in Fig. 1) between ferrite and scintillator. Magnet location and orientation were chosen so that the electron beam would be diverted considerably «down» relative to the ferrite surface. A simple estimate was made for beam image shift in the measurement plane using a self-written integrator. A modified Euler method was implemented to calculate electron trajectory with account for relativism and using a varying time step ($\tau i \propto 1/vi$-1, where vi-1 is electron velocity at the previous step) to ensure sufficient accuracy. A set of coils with current was used to roughly model magnet's field. The integrator was implemented inside Godot engine using built-in scripting tools in order to simplify real-time visualization process and problem parameters setup. Estimates made with it for electrons with energies of 150 keV, 300 keV and 600 keV show that in measurement plane the beam image center in each case would be shifted by more than 15 mm in the negative Y direction.

This, as will be shown later, is right on the edge of measurement area and should not interfere with main obtained results.

The ferrite surface was polished, and then copper paste was applied to it to create a conductive path close to one of its edges (*4* in Fig. 1) which corresponded to the X coordinate of -7 cm in the plane of translator. The discharge was photographed during the breakdown, one of the photos can be seen in Fig. 2. The photos demonstrated that the path didn't deform throughout all of the experiments described in this work and that the discharge always stayed right on it. More in depth description of the discharge propagation mechanism along the conductive path over ferrite surface was described in [2].

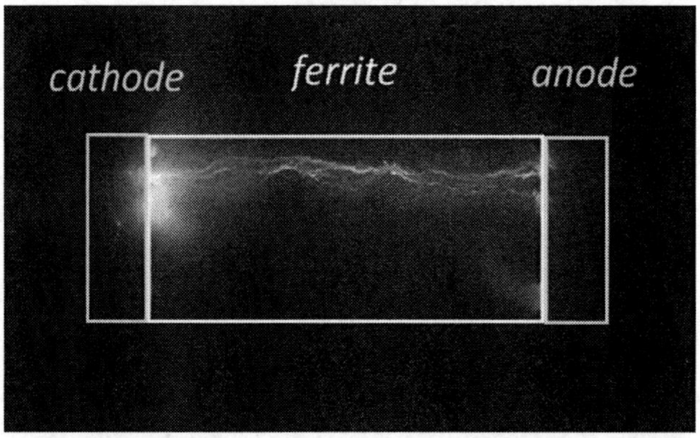

Figure 2. Photo of a ferrite surface discharge along a conductive path.

3. Experimental results and discussion

In the first stage, discharge current dynamics and peak value were studied. This is necessary to determine the income power and to later establish relations between radiation processes and high-voltage discharge dynamics. As described earlier, current values were obtained through two independent means: a shunt, and a Rogowski coil. The results obtained by both are consistent between each other (see Fig. 3), peak current measured is ~1 kA.

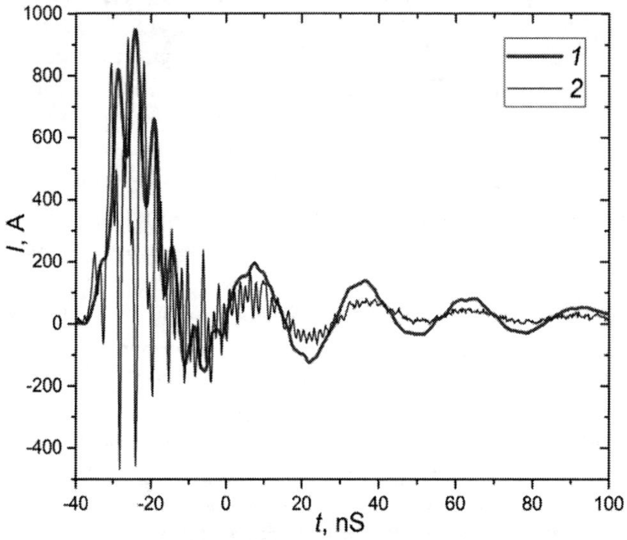

Figure 3. Ferrite surface flashover current measurements: *1* — using a Rogowski coil; *2* — using a shunt.

The next step were spatial radiation distribution measurements. In Fig. 3 intensity distributions along *y* axis (perpendicular to ferrite surface, see Fig. 1) opposite to the plasma channel (x=-7 cm) and to the side of it (x=7 cm) are shown. Each data point on the graph represents an average of at least 3 independent measurements. These results allow, similar to [6], angular divergence of observed radiation. At x=7 cm it is ~4°, and a little wider in from of discharge. It is probably due to structural damage to ferrite surface (as described in [2], it becomes amorphous) during discharge along a specified path. Change in structure as well as increased roughness lead to increased dispersion. Peak intensities differ by ~20%, which may be related to a higher electron density in front of the discharge path. If angular distribution of radiation along the y axis corresponds to the results of [6], then along the ferrite surface (the x axis) this distribution does not have a narrow direction [7].

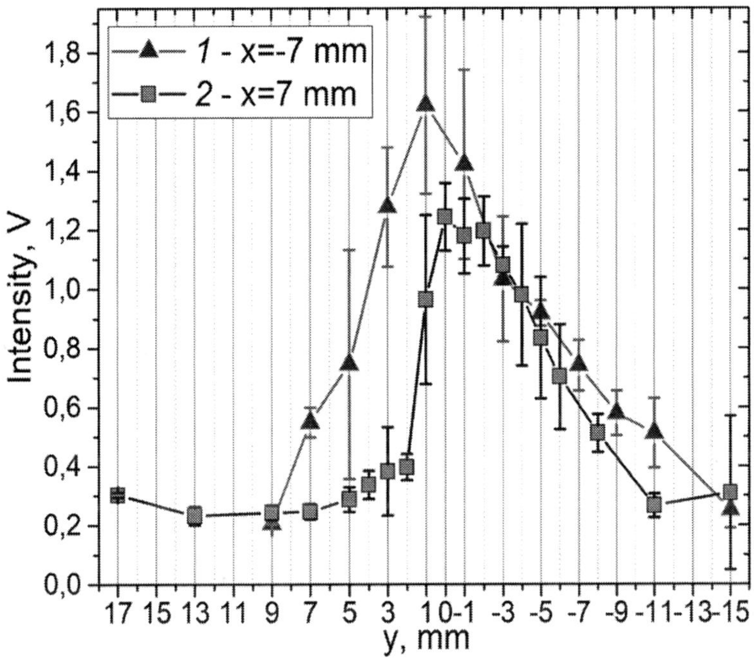

Figure 4. Spatial distributions of radiation along the ferrite surface.

4. Conclusion

Current dynamics in a discharge were studied with rise time of ~1 ns and up to 1 kA peak. Spatial distribution for radiation, induced by a ferrite surface flashover, is obtained in the direction along discharge axis. Angular diversion of the radiation to the side from discharge is ~4°. Obtained results correspond well with the results from the high-current experiments with a similar geometry [6].

Acknowledgments

This work was supported by the Russian Science Foundation (project no. 19-79-30086).

References

[1] Tilikin I N, Tskhai S N, Shelkovenko T A, Savinov S Yu, Pikuz S A 2018 *Plasma Physics Reports* **44** 600 https://doi.org/10.1134/S1063780X18060107

[2] Watanabe K, Kashiwabara S, Fujimoto R 1987 *Applied Physics Letters* **50** 629 https://doi.org/10.1063/1.98102

[3] Kashiwabara S, Watanabe K, Fujimoto R 1998 *Journal of Applied Physics* **63** 2514 https://doi.org/10.1063/1.341031

[4] Watanabe K, Furusho K, Kashiwabara S, Fujimoto R 1990 *Journal of Applied Physics* **68** 5059 https://doi.org/10.1063/1.347068

[5] Sentis M L, Chazaud F, Delaporte Ph C, Marine W I 1993 *Applied Physics Letters* **63** 2333 https://doi.org/10.1063/1.110517

[6] Tilikin I N, Pestovskii N V, Pikuz S A, Shelkovenko T A, Tskhai S N, Savinov S Yu 2021 Coherent x-ray radiation induced by high-current breakdown on a ferrite surface. *arXiv: 1910.13851v2* https://arxiv.org/abs/1910.13851v2

[7] Andreev S N, Bernatskiy A V, Dushkin G A, Tskhai S N, Pikuz S A, Tilikin I N, Shelkovenko T A, Savinov S Yu 2021 *Bulletin of the Lebedev Physics Institute* **48** 321 https://doi.org/10.3103/S1068335621100031

Influence of ion-plasma treatment on the surface morphology of epitaxial lead-tin telluride films

Y D Belov[1], S P Zimin[1, 2], I I Amirov[1], V V Naumov[1]

[1]Valiev Institute of Physics and Technology of Russian Academy of Sciences, Yaroslavl Branch, Yaroslavl, Russia
[2]P.G. Demidov Yaroslavl State University, Yaroslavl, Russia

yadbelov@gmail.com

Abstract. The paper discusses the physical aspects of surface modification of $Pb_{1-x}Sn_xTe$ (x = 0.00-0.80) epitaxial films during ion-plasma treatment in argon plasma. Lead–tin telluride films 1–2 μm thick were grown on (111) BaF_2 substrates by molecular beam epitaxy. The ion-plasma treatment of the samples was carried out in a high-density low-pressure radio frequency (RF) inductively coupled plasma at an ion energy of 75 eV. The duration of the process is 240 s. The evolution of the surface morphology of the films and the formation of micro- and nanostructures at different ratios of lead and tin are studied.

1. Introduction

A solid solution of lead-tin telluride $Pb_{1-x}Sn_xTe$ is actively used in the manufacture of infrared photodiodes, lasers, optoelectronic pairs, thermoelectric devices, multilayer systems with quantum wells and superlattices [1,2, etc.]. Recently, much attention has been paid to studying the properties of a material during the transition from a trivial state to a topological one [3], which makes the development of surface nanostructuring processes relevant [4, 5]. In our previous works [6, etc.] it was shown that effective nanostructuring of the surface of lead-tin telluride is possible by ion-plasma treatment of the surface with argon ions with an energy of 25 and 200 eV. The purpose of this work was to continue research on the processes of micro- and nanostructuring of the surfaces of $Pb_{1-x}Sn_xTe$ epitaxial films under the conditions of using argon ions with an energy of 75 eV.

2. Experimental details

(111) $Pb_{1-x}Sn_xTe$ films with x=0.00; 0.40; 0.80 were grown by molecular beam epitaxy on a Riber 32 P setup (INPE, Brazil) on (111) BaF_2 substrates by the authors of [7]. The film thickness was 1–2 μm; the properties of the layers in the initial state are described in detail in [6]. The surface of the films was treated in argon plasma in a high-density low-pressure radio frequency (RF) inductively coupled plasma, which is described in [6]. Argon plasma was ignited by applying an RF power of 800 W (frequency 13.56 MHz) to the inductor. An average Ar+ ion energy (E_i) was of 75 eV. The ion current density was 5.2 mA cm^{-2}. Argon flow was 20 sccm, and the operating gas pressure was 0.15 Pa. The plasma treatment duration (t) was of 240 s. The surface morphology was studied with scanning electron microscopy (SEM) using a Supra 40 Carl Zeiss microscope. The local chemical composition of the crystal surface was determined using energy-dispersive X-ray (EDX) microanalysis using an INCAx-act spectrometer (Oxford Instruments).

Content from this work may be used under the terms of the Creative Commons Attribution 3.0 licence. Any further distribution of this work must maintain attribution to the author(s) and the title of the work, journal citation and DOI.
Published under licence by IOP Publishing Ltd

3. Results and discussion

The original samples had a flat surface and were characterized by the presence of triangular pits of dislocation exit. The surface density of dislocation exit pits was in the range (7–80) × 10^8 cm^{-2}, increasing in this range as the Sn content in the sample increased [6]. Carrying out plasma treatment with ions with an energy of 75 eV for 240 s led to a significant modification of the surface. Estimation of the sputtering rates showed that they decrease from 5.5 nm/s to 0.4 nm/s as x increases from 0.00 to 0.80. Comparison with data from previous studies indicates that the sputtering rate under these conditions is higher than at an ion energy of 25 eV, but less than at E_i=200 eV.

When treated with ions with an energy of 75 eV for 240 s for x=0 (PbTe), against the background of small-scale roughness, the formation of identically oriented truncated triangular pyramids with triangle sides at the base of 450 nm on average was observed (Fig. 1). The height of the pyramids was 200–300 nm. The surface density of such truncated pyramids was 6 × 10^8 cm^{-2}, which is very close to the surface density of the dislocation exit regions, which was 7 × 10^8 cm^{-2}. This suggests that the formation of pyramids is associated with the well-known fact of micromasking in the places where dislocations emerge on the surface [8]. The shape of the triangular pyramids in this case is explained by the (111) orientation of lead telluride films, while the (100) orientation is characterized by the appearance of tetrahedral pyramids [9]. The pyramids in this experiment are truncated due to partial sputtering of the vertices and at the same time formation of the nanorelief.

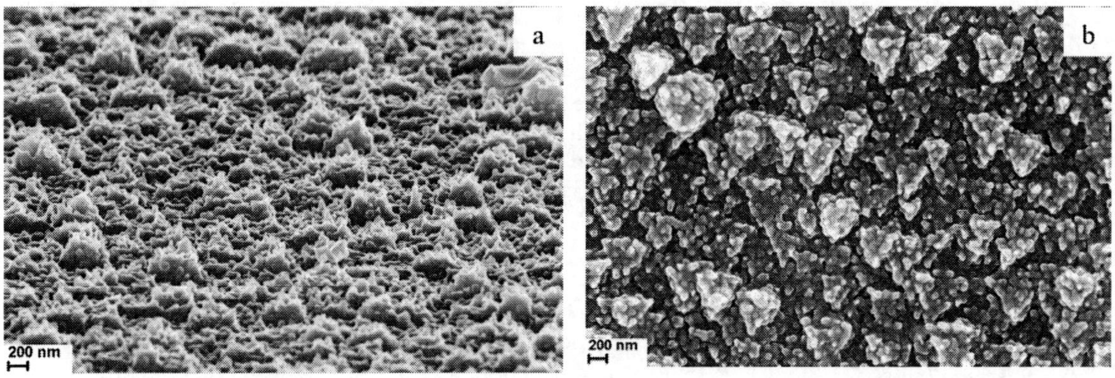

Figure 1. Surface modification of PbTe films at an ion energy of 75 eV in 240 s. View at an angle of 70° (a) and from above (b).

The surface morphology of $Pb_{0.60}Sn_{0.40}Te$ films after plasma treatment (E_i=75 eV, t=240 s) is shown in Fig.2. Under these conditions, cone-shaped structures with a quasi-spherical cap at the top were formed on the surface. The surface density of these structures was (2–3) × 10^{10} cm^{-2}, and the dimensions of the quasi-spherical cap were on average 65 nm. The inclination of the structures at an angle same structures of up to 55° from the normal is traced.

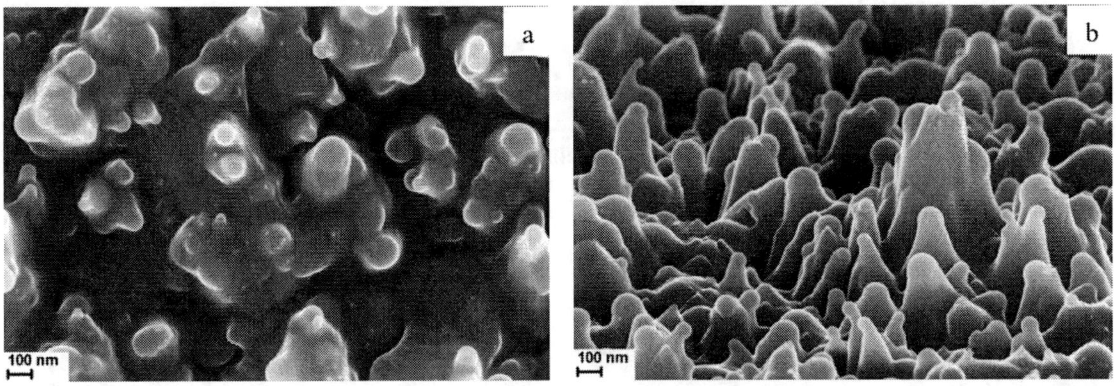

Figure 2. Surface modification of $Pb_{0.60}Sn_{0.40}Te$ films after plasma treatment (E_i=75 eV, t=240 s). (a) and (b) - view at an angle of 0° and 70° respectively.

To elucidate the mechanism of formation of such structures, an additional experiment was carried out, which consisted in the fact that a mask was placed on the surface, between which and the surface there was a microgap of ~1 μm (Fig. 3). During plasma treatment, cone-shaped structures with a quasi-spherical cap at the top appeared on the free surface. In the microgap under the mask, inclined columnar structures with a spherical cap were formed. The inclination of the columnar structures by 55° from the vertical is associated with growth in the <100> directions, which corresponds to the crystal growth in the directions, which have minimal free energy [6]. These structures grow by the vapor-liquid-solid (VLS) mechanism [10]. It is possible that in an open area, structures also grow according to the VLS mechanism. As a result ion bombardment, the formed nanocones have specific shape. Same structures described earlier in [6]. The use of methods of local EDX analysis and electron microscopy in backscattered electrons has shown that quasi-spherical caps are two-phase and consist of separate phases of lead and tin. These experiments will be described in detail in our next paper.

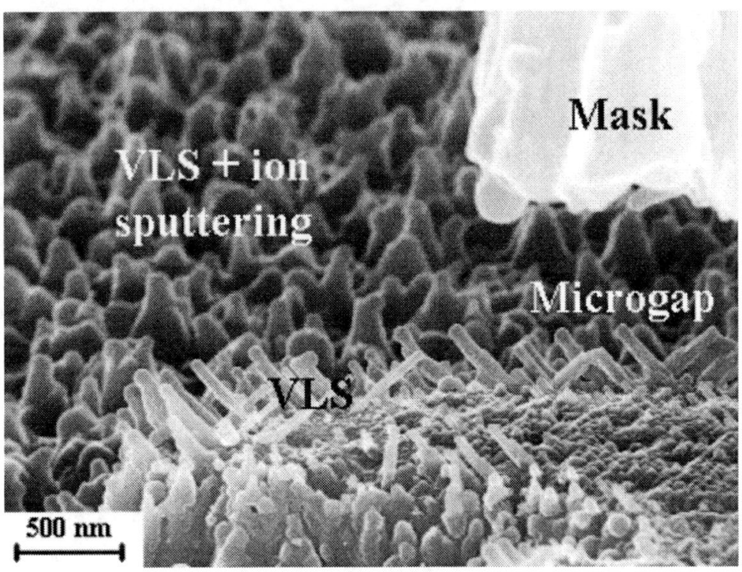

Figure 3. Surface nanostructuring processes in the microgap under the mask and on the open surface (E_i=75 eV, t=240 s).

The surface morphology of the $Pb_{0.20}Sn_{0.80}Te$ films after plasma treatment is shown in Fig. 4. At a high content of tin (x=0.80), quasi-spherical nanodroplet structures with sizes of 30–60 nm were formed on the surface. The surface density of all quasi-spherical structures was $\sim 3.5 \times 10^{10}$ cm^{-2}. EDX analysis showed that the surface under these plasma treatment conditions is enriched in tin and depleted in tellurium.

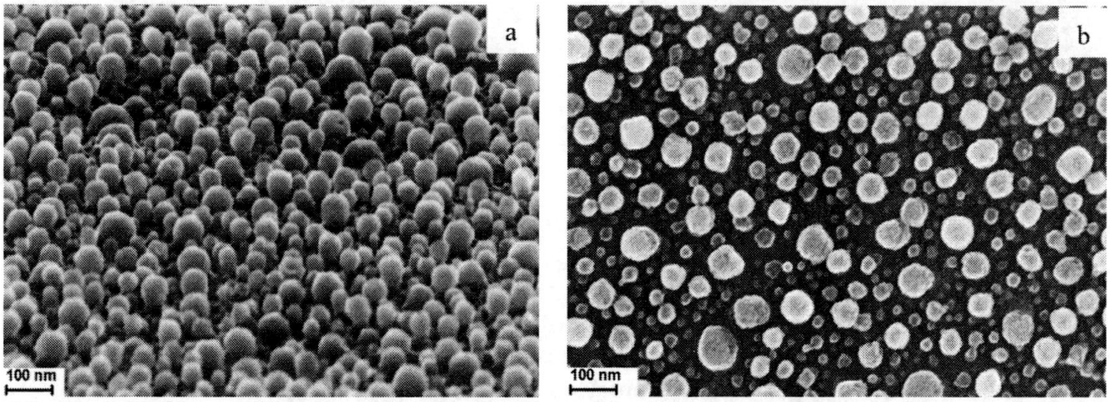

Figure 4. Surface modification of $Pb_{0.2}Sn_{0.8}Te$ films after plasma treatment at an ion energy of 75 eV for 240 s. (a) and (b) - view at an angle of 70° and 0°.

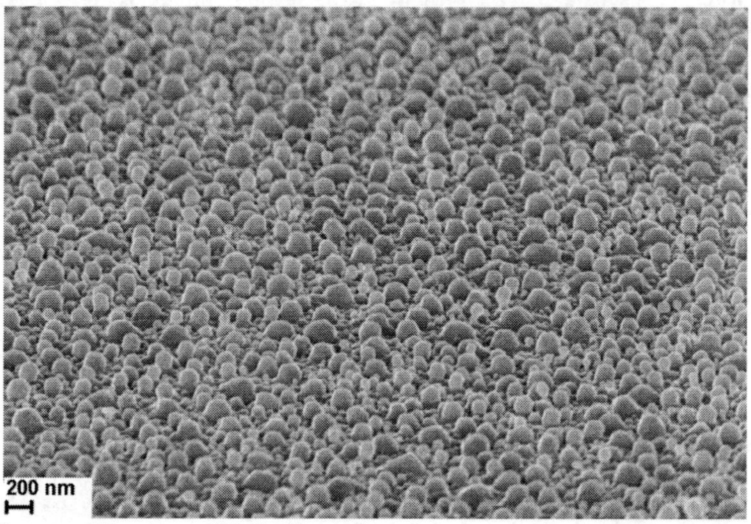

Figure 5. Surface modification of PbTe films after plasma treatment at an ion energy of 25 eV for 240s.

The mechanism of formation of the observed structures on the surface of $Pb_{0.2}Sn_{0.8}Te$ films requires further study. Similar structures have been described by us when the surface of PbTe was treated with ions with an energy of 25 eV (Fig.5). Now, by analogy with the results of [11], it can be assumed that under these conditions, the processes of nucleation of metallic quasi-spherical particles can take place, which can become hollow with an increase in the process time.

4. Conclusions

The results of this work showed that during ion-plasma treatment of $Pb_{1-x}Sn_xTe$ epitaxial films at an ion energy of 75 eV, nano- and microstructures of various types appear on their surface, the formation of which depends on the ratio of lead and tin in the solid solution films. If at an ion energy of 25 eV self-formation of faceted and quasi-spherical structures occurred on the surface, as well as the formation of nanodrops and nanorods with a drop-like "cap", then during plasma treatment of the surface with ions with an energy of 75 eV at a similar duration of the process, the surface morphology of $Pb_{1-x}Sn_xTe$ was different. Under these technological conditions, depending on the tin content on the surface, identically oriented truncated triangular pyramids, conical and quasi-conical structures with spherical formations at the top, or ensembles of quasi-spherical structures were formed. The obtained results of the surface nanostructuring during ion-plasma treatment at an ion energy of 75 eV, together with the data for E_i 25 and 200 eV [6, etc.], indicate a variety of shapes and geometric dimensions of the resulting nanostructures, which allows them to be used in instrument devices for various purposes.

Acknowledgments

The investigation was supported by the Program no. FFNN-2022-0017 of the Ministry of Science and Higher Education of Russia for Valiev Institute of Physics and Technology of RAS and within the framework of the initiative research work of the P.G. Demidov Yaroslavl State University. SEM studies were carried out at the Center for Collective Use "Diagnostics of Micro- and Nanostructures" with the financial support of the Ministry of Education and Science of the Russian Federation. The authors are grateful to Abramof E. and Rappl P.H.O. (Brazil) for providing $Pb_{1-x}Sn_xTe$ samples for the study.

Reference

[1] Nimtz G and Schlicht B 1983 *Springer Tracts in Modern Physics.* **98** 1
[2] Xu E, Li Z, Acosta J A, Li N, Swartzentruber B, Zheng S, Sinitsyn N, Htoon H, Wang J and Zhang S 2016 *Nano Research.* **9** 820
[3] Xu S-Y, Liu C, Alidoust N, Neupane M, Qian D, Belopolski I, Denlinger J D, Wang Y J, Lin H, Wray L A, Landolt G, Slomski B, Dil J H, Marcinkova A, Morosan E, Gibson Q, Sankar R, Chou F C, Cava R J, Bansil A and Hasan M Z 2012 *Nature communications.* **3** 1192
[4] Tarasov A S, Ishchenko D V, Akhundov I O, Golyashov V A, Klimov A E, Suprun S P, Fedosenko E V, Sherstyakova V N, Rybkin A G, Vilkov O Y and Tereshchenko O E 2021 *Applied Surface Science.* **569** 150930
[5] Saghir M, Sanchez A M, Hindmarsh S A, York S J and Balakrishnan G 2015 *Cryst. Growth Des.* **15** 5202
[6] Zimin S P, Gorlachev E S, Amirov I I, Naumov V V, Juskenas R, Skapas M, Abramof E and Rappl P H O 2019 *Semicond. Sci. Technol.* **34** 095001
[7] Rappl P H O, Closs H, Ferreira S O, Abramof E, Boschetti C, Motisuke P, Ueta A Y and Bandeira I N 1998 *Cryst. Growth.* **191** 466
[8] Markov O I, Khripunov Yu V 2020 Russian Physics Journal 62 11
[9] Zayachuk D M, Zayachuk Y D, Hunyadi M, Slynko V E and Csík A 2021 *Journal of Alloys and Compounds* **883** 160978
[10] Charalampous A, Zervos M, Kioseoglou J, Tsagaraki K, Androulidaki M, Konstantinidis G, Tanasa E and Vasile E 2019 *Appl. Energy Mater.* **6** 4274
[11] Zimin S P, Amirov I I, Naumov V V and Guseva K E 2018 *Tech. Phys. Lett.* **44** 518

Classical and quantum warm dense electron gas dynamic characteristics: analytic predictions

J. Ara[1], A.V. Filinov[2], I.M. Tkachenko[3,4]

[1] Instituto de Tecnología Química, Universitat Politècnica de València-Consejo Superior de Investigaciones Científicas, Valencia, Spain
[2] Institut für Theoretische Physik und Astrophysik, Christian-Albrechts-Universität zu Kiel, Germany
[3] Departament de Matemàtica Aplicada, Universitat Poliècnica de València, Valencia, Spain
[4] Al-Farabi Kazakh National University, Almaty, Kazakhstan

E-mail: `imtk@mat.upv.es`

Abstract.
We apply a novel 9-moment variational version of the self-consistent non-perturbative method of moments to study how the temperature affects the dynamic response of the electron gas in thermodynamic equilibrium. The theoretical results are obtained with the only input being the static structure factor. Comparison is carried out with the data obtained in the random-phase and the effective static local-field (ESA) (1) approximations. A quite satisfactory agreement is achieved with the system dynamic structure factor evaluated within the ESA interpolation scheme. We analyze the system properties for the temperature values ($1 \leq T/T_F \leq 4$) chosen in the range where the electron gas starts to undergo a transition from the degenerate to classical behaviour. The extension of the method to a broader range of temperatures and densities is straightforward and is left for future studies. Nevertheless, we demonstrate a systematic way to investigate the gradual transition from degenerate to classical systems.

1. Introduction

A warm dense gas of electrons (see (2) and references therein) is an important particular case of one-component fluids of charged particles and a key-model for a broad class of warm dense matter systems which (i) is widely used to testify different theoretical approaches, and (ii) is of significant practical importance for the interpretation and development of real experimental studies. Development of novel theoretical simulation techniques for warm dense matter is also of utmost interest and importance.

Until recently only *ab initio* path-integral Monte-Carlo (PIMC) simulation results on different characteristics of a dense warm uniform electron gas (UEG) (see (3) and references therein) were available. As any simulation data, these results are available in a limited realm of physical parameters, e.g. the smallest wavenumber being accessible is specified by the system size, $k = 2\pi/L$. Lately, we have presented (4; 5; 6) preliminary theoretical results on the static and dynamic properties of the UEG obtained within the self-consistent approach based on the non-perturbative method of moments capable of expressing the system (inverse) dielectric function, $\epsilon^{-1}(q,\omega)$, its static value, the dynamic structure factor, $S(q,\omega)$, and other dynamic characteristics exclusively in terms of the UEG static structure factor (SSF), $S(q)$, and using

Content from this work may be used under the terms of the Creative Commons Attribution 3.0 licence. Any further distribution of this work must maintain attribution to the author(s) and the title of the work, journal citation and DOI.
Published under licence by IOP Publishing Ltd

up to nine sum rules and other exact properties (7; 8; 5). The input SSF has been taken from the recent PIMC simulations (9) by T. Dornheim, S. Groth and M. Bonitz. The obtained interrelations produce results which are in a good agreement with the available PIMC data (3).

In the present work, we extend the self-consistent method of moments (SCMM) onto a broader domain of thermodynamic parameters (r_s, θ) and wavenumbers (k/k_F), and focus on the analysis of dynamic characteristics vs. the degeneracy (controlled by the temperature).

2. Parameters and sum rules

The thermodynamic state of the UEG can be uniquely specified by the standard coupling and degeneracy parameters ($\{\Gamma, D\}$ or $\{r_s, \theta\}$)

$$\Gamma = \beta e^2/a = \beta e^2 \sqrt[3]{4\pi n/3}\ , \quad D = \beta E_F = \theta^{-1}\ , \quad r_s = 1.842\Gamma\theta\ , \tag{1}$$

where β is the inverse temperature, n is the number density, and a is the average interparticle spacing. In the following analysis we will consider the warm dense matter regime (10) with $r_s, \theta \sim 1$. Nevertheless, the approach we suggest is non-perturbative so that the region of parameter values can be extended as soon as new simulation data (on the SSF) become available. The wavenumber-dependence will be expressed via the dimensionless parameter $q = ka$ or k/k_F.

In the linear response regime, the loss function (or the spectral density) is the central quantity for modeling of the inverse dielectric function $\epsilon^{-1}(q, \omega)$ and the dynamic structure factor $S(q, \omega)$

$$\mathcal{L}(q,\omega) = -\mathrm{Im}\,\epsilon^{-1}\left(q,\omega\right)/\pi\omega, \quad S(q,\omega) = \frac{q^2 n}{3\pi\Gamma}\beta\hbar\omega(1-\exp(-\beta\hbar\omega))^{-1}\mathcal{L}(q,\omega). \tag{2}$$

The method of moments is based on a direct relation between the above dynamical properties and the sum rules introduced as the loss function frequency power moments

$$C_\ell\left(q\right) = \int\limits_{-\infty}^{\infty} \omega^\ell\, \mathcal{L}(q,\omega)\,\mathrm{d}\omega\ , \quad \ell = 0, 1, 2, ...8, \tag{3}$$

or the characteristic frequencies

$$\omega_j\left(q\right) = \sqrt{C_{2j}/C_{2j-2}\left(q\right)}\ , \quad j = 1, 2, 3, 4 \tag{4}$$

The odd-order moments vanish due to the symmetry of the loss function. The zero-th moment, due to the Kramers-Kronig relations, specifies the static limit of the inverse dielectric function, $C_0\left(q\right) = 1 - \epsilon^{-1}\left(q, 0\right)$, and the plasma frequency enters in the f-sum rule, $C_2 = \omega_p^2$. The knowledge of the fourth moment, by virtue of the detailed balance condition (see (5) and references therein),

$$C_4\left(q\right) = \frac{6e^2}{aq^2\hbar n}\int_{-\infty}^{\infty}\omega^3 S\left(k,\omega\right)d\omega =$$
$$= \omega_p^4\left\{1 + \frac{q^2}{\Gamma}\theta^{3/2}I_{3/2}\left(\eta\right) + \frac{q^4}{12r_s} + \frac{1}{3\pi}\int_0^{\infty}\left[S\left(p\right)-1\right]f\left(p,q\right)p^2dp\right\}\ , \tag{5}$$

is reduced to the UEG static structure factor, $S\left(q\right)$. Here

$$f\left(p,q\right) = \frac{5}{6} - \frac{p^2}{2q^2} + \frac{\left(p^2-q^2\right)^2}{4q^3p}\ln\left|\frac{q+p}{q-p}\right|\ , \tag{6}$$

and $I_{3/2}\left(\eta\right)$ is the Fermi integral, $I_{1/2}(\eta) = (2/3)/\sqrt{\theta^3}$ is the normalization condition. In a classical system the third term in (5) vanishes and the second contribution takes the Vlasov form, q^2/Γ.

Journal of Physics: Conference Series **2270** (2022) 012041 doi:10.1088/1742-6596/2270/1/012041

3. The self-consistent solution of the five- and nine-moment problems

It was demonstrated in our previous works (7; 8) that the dynamic properties of classical Coulomb and Yukawa systems can be quite accurately described by the method of moments in the *five-moment approximation*. The corresponding Hamburger problem $\{C_0(q), 0, C_2, 0, C_4(q)\}$ and the Nevanlinna theorem (11; 12; 13) results in the following ansatz for the inverse dielectric function and the dynamic structure factor

$$\epsilon^{-1}(q,\omega;Q_2) = 1 + \frac{\omega_p^2\left(\omega + Q_2(q,\omega;\omega_1,\omega_2)\right)}{\omega\left(\omega^2-\omega_2^2(q)\right) + Q_2(q,\omega;\omega_1,\omega_2)\left(\omega^2-\omega_1^2(q)\right)}, \tag{7}$$

$$S(q,\omega) = \frac{q^2 n}{3\pi\Gamma}\left(\frac{\beta\hbar\omega}{1-\exp(-\beta\hbar\omega)}\right)\left(-\frac{\mathrm{Im}\epsilon^{-1}(q,\omega;Q_2)}{\omega}\right). \tag{8}$$

Both characteristics are determined by the frequencies $\omega_{1(2)}(q)$ and a non-phenomenological function $Q_2(q,\omega;\omega_1,\omega_2)$. The quantitative agreement with the OCP molecular-dynamics data (both on Coulomb and Yukawa systems) could be achieved (7; 8) even within the static approximation of the Nevanlinna (response) function $Q_2(q,z)$ such that $\lim_{z\to\infty}(Q_2(q,z)/z)$, $\mathrm{Im}z > 0$ (14),

$$Q_2(q,z;\omega_1,\omega_2) = Q_2\left(q,0^+;\omega_1,\omega_2\right) = ih_2(q;\omega_1,\omega_2)\ . \tag{9}$$

The key physical assumption made and validated (7; 8) was that the zero-frequency Rayleigh mode is generally absent in the above classical systems. This permits to determine the static parameter as $h_2(q;\omega_1,\omega_2) = \omega_2^2(q)/\left(\sqrt{2}\omega_1(q)\right)$.

Such a relatively simple approach proves to be insufficient for quantum Fermi fluids of charged particles, as it is demonstrated in the analysis below (see Fig. 1 and the corresponding discussion). Our interpretation of this result is the existence in the spectrum of the second harmonic of the collective plasmon mode. Some indications of this mode, however, quite weak have been recently reported in (3).

The three-mode structure of the system spectrum can be adequately considered taking into account the sixth and the eighth moments or the frequencies $\omega_3(q)$ and $\omega_4(q)$. To this end we applied the nine-moment Nevanlinna formula (11; 12; 13), and equalized the five- and nine-moment expressions for the dielectric function. As a result the response-function $Q_2(q,z;\omega_1,\omega_2)$ introduced above is substituted with a new Nevanlinna parameter $Q_4(q,z;\omega_1,\omega_2,\omega_3,\omega_4)$. The absence of the zero-frequency Rayleigh mode (7) also in quantum fluids by the quantum-classical correspondence principle, allows us to employ the same static approximation to the function $Q_4(q,z;\omega_1,\omega_2,\omega_3,\omega_4)$ and find a new parameter $h_4(q;\omega_1,\omega_2,\omega_3,\omega_4)$. Thus we relate the dynamic five-moment Nevanlinna function,

$$\begin{aligned} Q_2(q,\omega;\omega_1,\omega_2,\omega_3,\omega_4) &= -\frac{\omega_2^2\left(\omega_3^2-\omega_2^2\right)}{\omega_2^2-\omega_1^2}\frac{(\omega+ih_4)}{\omega(\omega+ih_4)+\left(\frac{\omega_1^2\left(\omega_3^2-\omega_2^2\right)}{\left(\omega_2^2-\omega_1^2\right)}-\frac{\omega_3^2\left(\omega_4^2-\omega_3^2\right)}{\left(\omega_3^2-\omega_2^2\right)}\right)}, \\ h_4(q;\omega_1,\omega_2,\omega_3,\omega_4) &= \frac{\omega_3^2\left(\omega_4^2-\omega_3^2\right)\left(\omega_2^2-\omega_1^2\right)}{\omega_1\sqrt{2\left(\omega_3^2-\omega_1^2\right)\left(\omega_3^2-\omega_2^2\right)^3}}, \end{aligned} \tag{10}$$

to the new parameters $\omega_3(q)$ and $\omega_4(q)$. Certainly, the above expressions simplify into the *five-moment* case as soon as we consider two successive limiting transitions: $\omega_4(q)\to\infty$ and then $\omega_3(q)\to\infty$ (with $\omega_3(q) < \omega_4(q)$ following from the Cauchy-Schwarz-Bunyakovsky inequalities).

The characteristic frequency $\omega_1(q) = \omega_p\sqrt{1-\epsilon^{-1}(q,0)}$ and the static dielectric function were determined by direct analytical and numerical integration in the five- and nine-moment cases, respectively (5), without any significant difference being observed. The frequency $\omega_2(q)$ was determined by $S(q)$, see Eqs. (4),(5). Thus we reduced the knowledge of the dynamic structure factor, see Fig. 1, to that of the SSF.

4. The Shannon entropy maximization technique

The frequencies $\omega_3(q)$ and $\omega_4(q)$ formally introduced in terms of the higher-order moments, see Eq. (4), are determined by the virtually unknown three- and four-point static correlators. On the other hand, the frequencies $\omega_1(q)$ and $\omega_2(q)$ are uniquely determined by the SSF (5). Taking into account that the spectral density (here the loss function $\mathcal{L}(q,\omega)$) is formed by different combinations of quasi-particle excitations – the microstates in the sense of statistical ensemble, we decided that the optimal solution (most probable) should satisfy the principle of maximum entropy. This allows to determine the unknown parameters $\omega_{3(4)}$ by means of the Shannon information-entropy maximization procedure (15; 16; 17; 18), see also (13). Precisely, we suggest and resolve the maximization equations for the two-parameter Shannon entropy functional:

$$S(q;\omega_3,\omega_4) = -\int_{-\infty}^{\infty} \mathcal{L}(\omega_3,\omega_4,q,\omega)\ln\left[\mathcal{L}(\omega_3,\omega_4,q,\omega)\right]\mathrm{d}\omega, \tag{11}$$

$$\begin{cases} -\int\limits_{-\infty}^{\infty}\left\{\frac{\partial\mathcal{L}(\omega_3,\omega_4;q,\omega)}{\partial\omega_3}\ln\left[\mathcal{L}(\omega_3,\omega_4;q,\omega)\right]\right\}d\omega = 0, \\ -\int\limits_{-\infty}^{\infty}\left\{\frac{\partial\mathcal{L}(\omega_3,\omega_4;q,\omega)}{\partial\omega_4}\ln\left[\mathcal{L}(\omega_3,\omega_4;q,\omega)\right]\right\}d\omega = 0. \end{cases} \tag{12}$$

The Hessian of the entropy (11) was studied to warrant its maximization at the solution of the system (12).

5. Numerical results

To validate our model solution with the higher order moments included (to allow for a second harmonic) and the principle of maximum entropy for the spectral density, we have performed comparison of the 9-moment solution (9MA) against the effective static local-field approximation (ESA) (1) data and the 5-moment solution (5MA), i.e. the static approximation for the Nevanlinna function, see Eq. (9), being accurate in the classical limit. The corresponding dynamical structure factor (DSF) from three different models and also the RPA-result is presented in Fig. 1. A systematic analysis is left for a future work, while here we concentrate on the intermediate coupling, $r_s = 2$, where the RPA is already insufficient to accurately reproduce the exchange-correlation effects in the dynamical properties.

First, we observe a very good agreement between the "9MA" and the "ESA". Both independent approaches reproduce a very similar theoretical solution for the DSF, its variation with the wavenumber and even with the temperature varied in the range $1 \le \theta \le 4$. The half-width of the DSF around the maximum (the plasmon mode) and the position of the latter demonstrate a remarkable agreement. Some noticeable discrepancies appear only at small wavenumbers (i.e. $k/k_F = 0.60$). However, this result could be expected as the principle of entropy maximization tends to smooth any sharp (resonance) features present in the spectral density.

Both theoretical approximations (ESA and 9MA), where the correlations effects are included via the power moments $C_0(q)$ and $C_4(q)$, significantly deviate from the RPA-result, in particular by lowering the temperature and the wavenumber (see the plots of DSF in Fig. 1 for $q \le 2.3$ and $\theta \le 2$). On the other hand, all three theories agree for large wavenumbers ($k/k_F \gtrsim 2.98$) where a broad multiexcitation continuum is present.

Now we turn to the discussion of the DSF in the "classical" limit, see the "5MA"-solution. The term "classical" applies only to the analytical form used for the inverse dielectric function (and the dynamic structure factor) similar to the one considered earlier for the description of the dynamic properties of the classical OCP systems. In contrast, here the power moments C_0, C_4

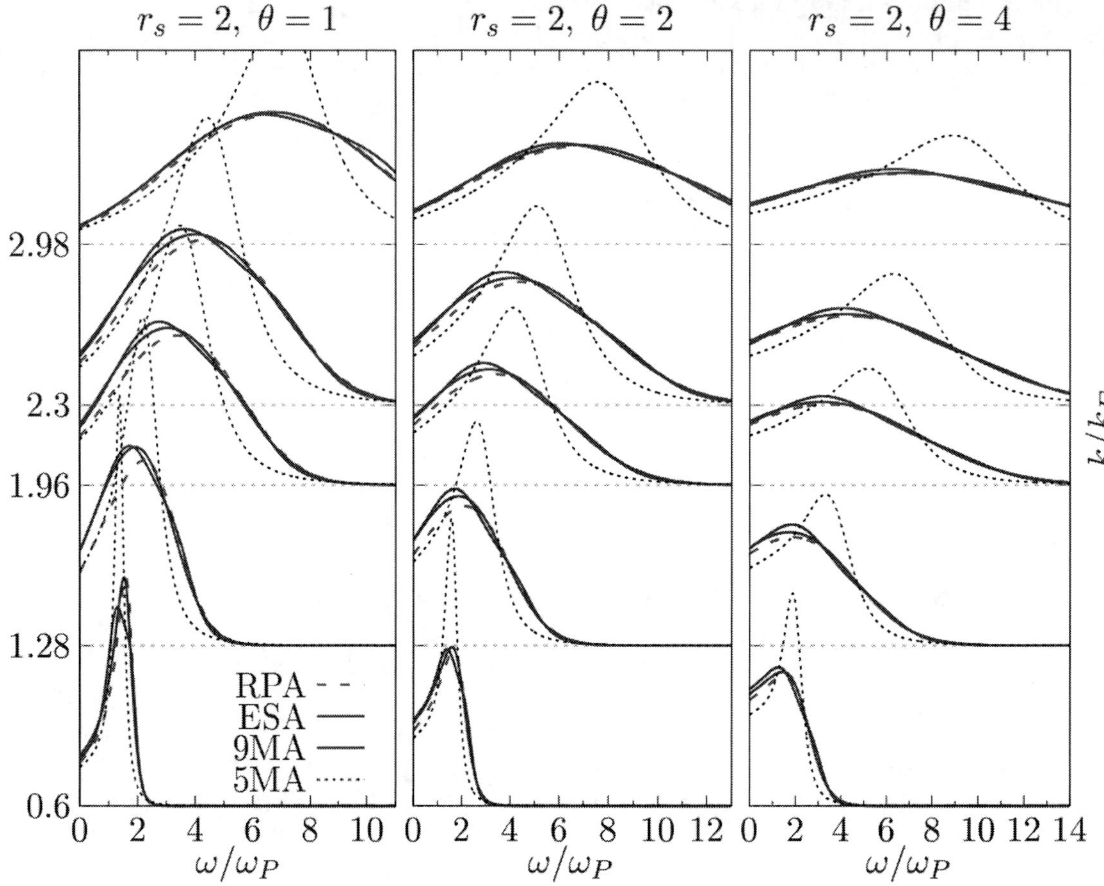

Figure 1. The dynamic structure factor $S(k, \omega)$ at three temperatures $\{\theta = 1, 2, 4\}$ and $r_s = 2.0$. The DSF plots are shifted by the value of k/k_F (horizontal dotted lines). Compared are the results of the random-phase approximation (RPA), the *effective static approximation* (ESA) (1) and the present self-consistent method of moments (SCMM) in the nine-moment ("9MA") and five-moment ("5MA") cases. Note that the five-moment approximation corresponds to the static approximation, Eq. (9), for the Nevanlinna parameter function Q_2. Discrepancies observed for $k/k_F = 0.60$ are due the Shannon entropy maximization which tends to smooth sharp energy resonances and is much better suited for description of a broad multi-excitation continuum. In this case a slight modification of the frequency $\omega_3(k)$ was enough to achieve the level of agreement observed for other wavenumbers.

go beyond the traditional classical description and include the correlation and degeneracy effects via the static structure factor (SSF) provided by the QMC simulations.

For the lowest temperature considered ($\theta = 1$) the 5MA-model provides the correct position of the main plasmon mode, however, it significantly underestimates its broadening due to the damping effects and presence of the second harmonic. The absence of a second harmonic in this solution also at other temperatures ($\theta = 2, 4$) shifts the main peak position in the "5MA"-solution to higher frequencies. Simultaneously, the peak amplitude is reduced and the overall shape approaches that of the 9MA-solution. In the limit of higher temperatures ($\theta \gtrsim 16$) both solutions are excepted to coincide. This gradual transition can be systematically investigated by the analysis of the temperature dependence of the higher-moments (C_6 and C_8) or the frequencies

$\omega_{3(4)}$.

While the high-temperature case is left for the future work, as a preliminary result we demonstrate in Fig. 2 the temperature and the wavenumber dependence of the characteristic frequencies. One can clearly see, that the frequencies $\omega_{3(4)}(q)$ grow faster with the temperature in contrast to $\omega_{1(2)}(q)$. As a result, in the limiting case ($\theta \gg 1$) the dynamic frequency-dependent Nevanlinna function (10) should reduce to its static limiting form $ih_2(q;\omega_1,\omega_2)$. In other words, when the system effectively loses its quantum character, we return to the five-moment case which proved to be valid for classical systems.

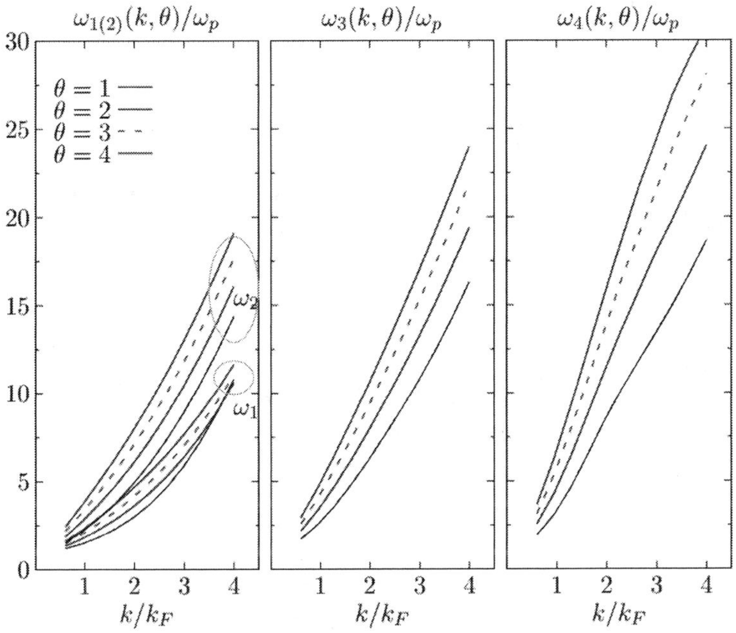

Figure 2. The temperature and wavenumber dependence of the characteristic frequencies $\omega_i\,(i = 1,\dots 4)$ (in units of plasma frequency) defined via the power moments, see Eq. (4). The frequencies $\omega_{3(4)}$ have been obtained via the Shannon entropy maximization. Simulation parameters: $r_s = 2$ and $\theta = 1, 2, 3, 4$.

Finally, we note that the properties of the collective modes (and the presence of a second harmonic) within our approach can be studied by the solution of the dispersion equation or as the poles of the inverse dielectric function. The corresponding algebraic equation is of the fifth-order and can be written in a closed analytical form. But this analysis goes beyond of the scope of the present short paper.

6. Conclusions

The dynamic structure factor of a warm and dense uniform electron gas is successfully studied within two sequential versions of the self-consistent method of moments based on the exact sum rules, mathematical theorems, and other exact relations. The novel analytical expression for the loss-function is introduced based on the nine-moment approximation and is complemented by the Shannon information entropy maximization technique. As a result, we reduce the knowledge of the dynamic properties of quantum plasmas to the knowledge of a single quantity, i.e. the static structure factor.

Special attention is paid to the temperature dependence and transition between the degenerate ($\theta = 1$) and non-degenerate regime ($\theta = 4$). We demonstrate the superior accuracy and validity of our approach both for low and high temperatures by a direct comparison with the effective static local-field approximation (1) data based on the ab-initio PIMC simulations. For all cases considered both theoretical results are found to be in a nice agreement.

The comparison between the more general 9-moment approximation and the 5-moment solution can be used in the further analysis of the transition from the quantum to classical cases.

We stress that in the 5- and 9-moment calculations the only input was the static structure factor calculated independently using, e.g., the effective static local-field correction (1) or direct quantum Monte-Carlo simulations. The dielectric function we obtain simultaneously can be also used in a variety of problems, see e.g. Ref. (19) and references therein.

6.1. Acknowledgments

A.V. Filinov and I.M. Tkachenko acknowledge support provided by the German Science Foundation (DFG) (grant BO1366-15) and by the Ministry of Education and Science of Kazakhstan (grant AP09260349), respectively.

References

[1] Dornheim T, Cangi A, Ramakrishna K, Böhme M, Tanaka S and Vorberger J 2020 *Phys. Rev. Lett.* **125**(23) 235001 URL https://link.aps.org/doi/10.1103/PhysRevLett.125.235001

[2] Dornheim T, Moldabekov Z A and Vorberger J 2021 *Contributions to Plasma Physics* **61** e202100098

[3] Hamann P, Dornheim T, Vorberger J, Moldabekov Z A and Bonitz M 2020 *Phys. Rev. B* **102**(12) 125150 URL https://link.aps.org/doi/10.1103/PhysRevB.102.125150

[4] Ara J, Coloma L and Tkachenko I 2022 *17th Intl. Conf. on the Physics of Non-Ideal Plasmas, September 20 - 24, 2021, Dresden, Alemania, Book of Abstracts, p. 26*

[5] Ara J, Coloma L and Tkachenko I M 2021 *Physics of Plasmas* **28** 112704 (*Preprint* https://doi.org/10.1063/5.0062259) URL https://doi.org/10.1063/5.0062259

[6] Ara J, Coloma L, Filinov A and Tkachenko I 2022 *42-nd International Workshop on High-Energy-Density Physics with Intense Ion and Laser Beams, Book of abstracts, p. 22*

[7] Arkhipov Y V, Askaruly A, Davletov A E, Dubovtsev D Y, Donkó Z, Hartmann P, Korolov I, Conde L and Tkachenko I M 2017 *Phys. Rev. Lett.* **119**(4) 045001 URL https://link.aps.org/doi/10.1103/PhysRevLett.119.045001

[8] Arkhipov Y V, Ashikbayeva A, Askaruly A, Davletov A E, Dubovtsev D Y, Santybayev K S, Syzganbayeva S A, Conde L and Tkachenko I M 2020 *Phys. Rev. E* **102**(5) 053215 URL https://link.aps.org/doi/10.1103/PhysRevE.102.053215

[9] Dornheim T, Groth S and Bonitz M 2017 *Contributions to Plasma Physics* **57** 468–478

[10] Dornheim T, Groth S and Bonitz M 2018 *Phys. Rep.* **744** 1 – 86 ISSN 0370-1573 URL http://www.sciencedirect.com/science/article/pii/S0370157318300516

[11] Nevanlinna R 1922 *Asymptotische Entwicklungen beschränkter Funktionen und das Stieltjessche Momentenproblem* STK (Helsinki)

[12] Shohat J and Tamarkin J 1943 *The Problem of Moments* Amer. Math. Soc. (Providence, R.I.)

[13] Tkachenko I, Arkhipov Y and Askaruly A 2012 *The Method of Moments and its Applications in Plasma Physics* (Lambert)

[14] Krein M and Nudel'man A 1977 *The Markov moment problem and extremal problems* Trans. of Math. Monographs 50, Amer. Math. Soc. (Providence, R.I.)

[15] Shannon C E 1948 *Bell System Technical Journal* **27** 379–423

[16] Khinchin A 1953 *Uspekhi Mat. Nauk* **8** 3

[17] Zubarev D 1974 *Nonequilibrium Statistical Mechanics* (Consultants Bureau, London)

[18] Jaynes E T 1957 *Phys. Rev.* **106**(4) 620–630 URL `https://link.aps.org/doi/10.1103/PhysRev.106.620`

[19] Zan X, Lin C, Hou Y and Yuan J 2021 *Phys. Rev. E* **104**(2) 025203 URL `https://link.aps.org/doi/10.1103/PhysRevE.104.025203`

Journal of Physics: Conference Series **2270** (2022) 012042 doi:10.1088/1742-6596/2270/1/012042

The study of statistical features of the evolution of complex physical systems using adaptive machine learning methods

V A Yunusov[1], S A Demin[1], A A Elenev[1]

[1]Kazan Federal University, Kazan, Russian Federation

valentin.yunusov@gmail.com

Abstract. In this work, we discuss various machine learning methods and their implementation in the field of complex physical systems for the analysis of experimental data. These methods: classical machine learning, neural nets and deep learning allow greatly outperforming classical analysis methods by giving the algorithm the ability to "learn" and perform tasks adapting to the data provided and search. Neural nets and deep learning approaches are used to search for hidden patterns of the suggested input data that can't be analyzed using common methods. This variety of methods can be applied to study collective phenomena in plasma and thermonuclear fusion on the basis of experimental data of physical experiments with a higher level of performance than classical approaches.

1. Introduction

Nowadays, one of the most relevant and remarkable areas in physics is the physics of complex systems. Such systems are widely distributed in nature on scales from a group of cells to plasma in stars and stellar cluster systems. A complex system is a composite object, the parts of which relate to each other by certain relationships, as a result of which such a system acquires new properties that cannot be reduced to the properties of its parts. Complex systems are distinguished by the following properties: non-linearity, limited predictability, evolutionary dynamics, self-organization, openness, and adaptability [1, 2].

Machine learning is one of the new and actively developing methods of analysis, combining approaches that can "learn" based on the received data, which allows you to perform a wide range of different tasks. Machine learning can be used to solve problems of detection, recognition, prediction, prediction, diagnostics, and optimization. In the area of physics of complex systems machine learning methods are widely used in the study of complex systems structure and for analysis of the dynamic behavior of nonlinear physical complex systems: forecasting of the future evolution of the systems and establishing a causal relationship [3].

2. Application of machine learning methods in the analysis of the dynamic behavior of complex physical systems

In this paper, we discuss the prospects for applying the latest methods of machine learning, neural networks and deep learning for the analysis of experimental data from physical experiments in modern science.

Content from this work may be used under the terms of the Creative Commons Attribution 3.0 licence. Any further distribution of this work must maintain attribution to the author(s) and the title of the work, journal citation and DOI.
Published under licence by IOP Publishing Ltd

2.1. Machine Learning methods

Figure 1 shows the relationship between machine learning and other learning methods within artificial intelligence technologies [4].

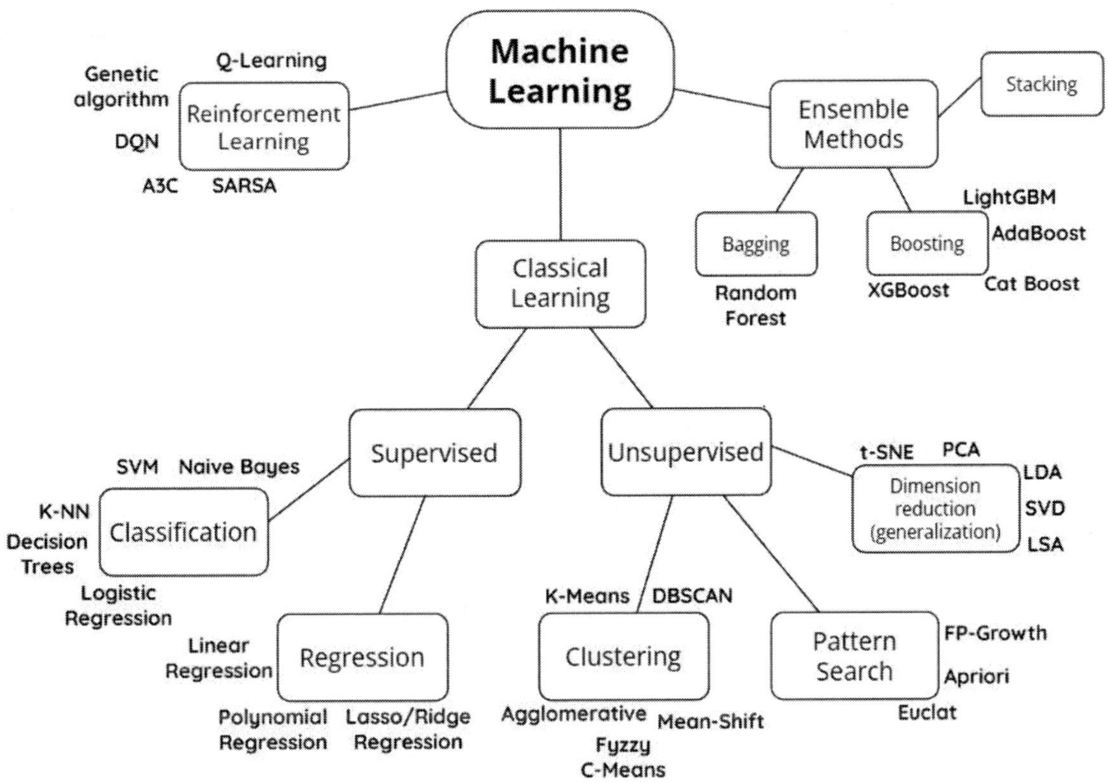

Figure 1. The relationship among different types of machine learning methods.

Machine learning is a rapidly growing area and one of the latest technologies being used in the modern information technology field. Machine learning is a set of techniques that will allow computer algorithms to be able to learn [5]. It is based on the input and required output of the algorithms, some of which are based on the way how humans can carry out a task [6]. Machine learning is usually classified into the following categories: reinforcement learning, ensemble methods, supervised and unsupervised learning. In reinforcement learning the algorithm needs not just to analyze data, but to act independently in real conditions. The task is to minimize errors, for which it gets the opportunity to continue working without obstacles and failures [7]. Ensemble methods are groups of algorithms that use several machine learning methods at once and correct each other's errors. Supervised learning is a type of algorithm where the method is supplied with example inputs along with the required output, which then allows it to learn a rule that maps inputs to outputs [8]. In unsupervised learning, on the contrary, only the inputs are supplied, and the learning algorithm is required to determine the structure of the input and perform according to unknown characteristics [9].

These algorithms are widely used by scientists in different areas. For example, in the area of construction of accurate density functionals for realistic molecular systems [10]. Figure 2 shows the principle of the authors' method to directly learn the Hohenberg-Kohn map.

Journal of Physics: Conference Series **2270** (2022) 012042 doi:10.1088/1742-6596/2270/1/012042

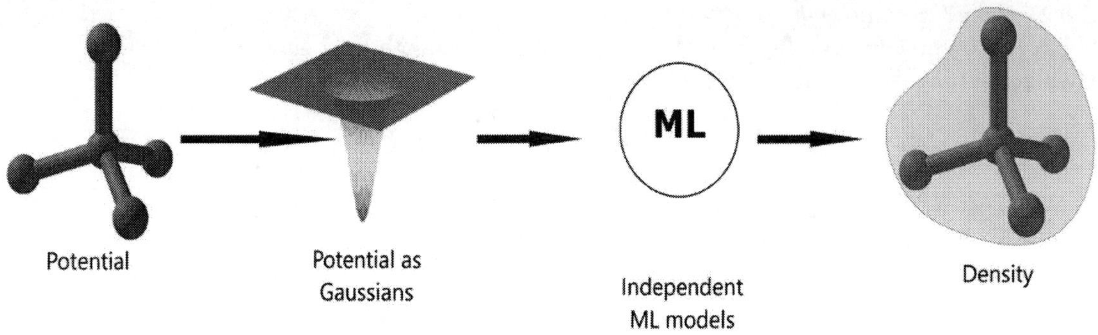

Figure 2. The principle of forecasting machine learning Hohenberg-Kohn map.

Besides, in the field of complex physical systems, machine learning methods are used in plasma physics. In [11] the supervised and unsupervised learning methods are used to predict matrix effects severity and analyte recovery prediction in plasma optical emission spectrometry. The study used non-analyte signals as inputs. The authors declare that the efficiency of collecting and interpreting responses from plasma species may be based on this analysis workflow. Moreover, machine learning methods are also widely used in the tasks of thermonuclear fusion. In the study [12] the authors compare two machine learning tools: Gaussian Mixture Models and Support Vector Machine to carry out the classification task – distinguishing neutrons and gamma-rays in thermonuclear fusion. As a result, the authors declare that the approaches are in very good agreement and these methods greatly outperform previously used classification algorithms, by providing the probability of each example being a neutron or a gamma-ray.

2.2. Neural Nets and Deep Learning methods

The next group of methods is neural nets and deep learning approaches. It is a more complex level of learning algorithm than machine learning. Neural nets are designed in a way that resembles the way the neurons work in the human brain [13]. Neurons form layers through which the signal passes consistently. All this is connected by neural connections – channels for which data is transmitted. Each channel has its own "weight" – a parameter that affects the data it transmits. The input layer takes input, which then is processed in the hidden layer and the output layer sends out the calculated output. Neural nets are a powerful tool for image, speech and signal processing and are widely used in modern science [14]. Deep learning is a set of learning algorithms that can be used to learn complex forecasting models, e.g., multi-layer neural networks with many hidden layers. Figure 3 shows the variety of neural nets and deep learning methods.

Both sets of methods are recently being used in the field of physics of complex systems for the analysis of experimental data. For example, in [15] the authors develop a new method to produce a predictive model for turbulent fluxes in drift-wave turbulence systems, which can be applied to forecasting of turbulence behavior in plasma. Authors use a supervised deep learning methodology that infers a mean-field model for the cross-phases from direct numerical simulation. As a result, the authors develop a model for the drift-wave/zonal flow system directly from numerical solution data. The method reproduces analytical results for the Reynolds stress, such as negative viscosity effects and terms regularizing it.

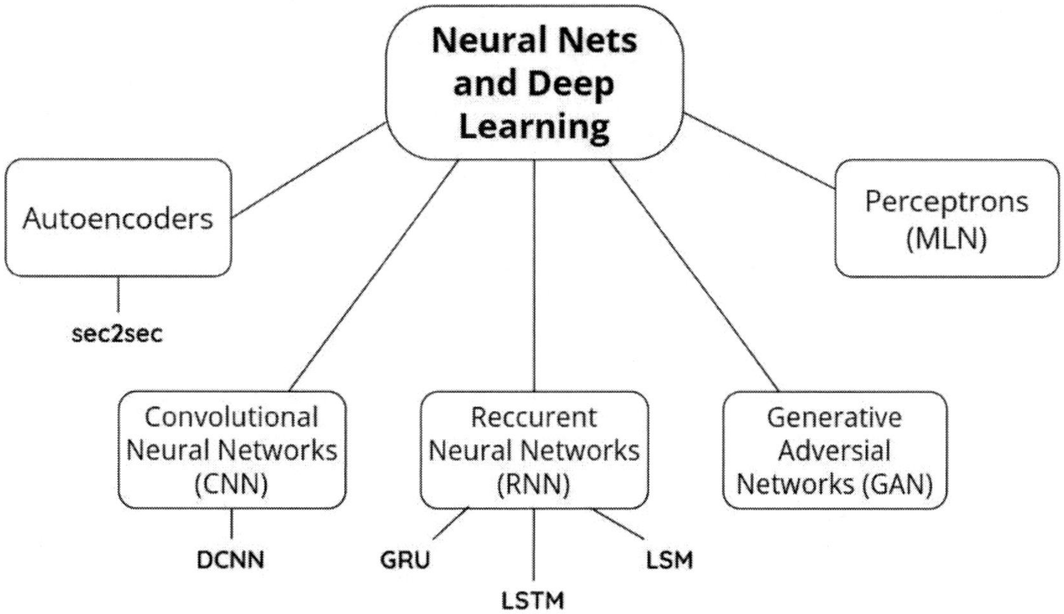

Figure 3. Set of the neural nets and deep learning methods.

3. Conclusion

In this paper, we discuss recently developed methods of analyzing data of various nature: machine learning [16], neural nets and deep learning approaches. We describe different approaches, their meaning and their implementation in modern science. These methods are applied in different fields, including information technologies and physics. In the area of physics of complex systems, the methods are actively used to study experimental data of experiments, e.g., in plasma physics and thermonuclear fusion, in order to replace classical analysis methods due to the higher level of performance.

These approaches are not the only ones that can outperform classical analysis methods: in the study of collective phenomena in physical complex systems, Memory Functions Formalism [17] and Flicker-Noise Spectroscopy [18] are also can show significant results. These methods can be applied to analyze autocorrelations, cross-correlations and statistical memory effects in the recorded experimental data [19], which reflect the manifestation of collective phenomena in plasma and during thermonuclear fusion [20, 21].

Acknowledgments

This paper has been supported by the Kazan Federal University Strategic Academic Leadership Program (PRIORITY–2030). The authors express gratitude to D.Sc. (Physics and Mathematics), Prof. S.F. Timashev for discussing some of the results of the present work.

References

[1] Northrop R B 2011 Introduction to complexity and complex systems (Boca Raton: CRC Press)
[2] Siegenfeld A F and Bar-Yam Y 2020 *Complexity* **2020** 1
[3] Tang Y, Kurths J, Lin W, Ott E and Kocarev L 2020 *Chaos* **30** 063151
[4] Isonkobong C 2020 *Semicond. Sci. Inf. Devices* **2** 5

[5] Hastie T, Tibshirani R and Friedman J 2009 Overview of Supervised Learning *The Elements of Statistical Learning. Springer Series in Statistics* (New York: Springer) p 9
[6] Mnih V, Kavukcuoglu K, Silver D, Rusu A A, Veness J, Bellemare M G, Graves A, Riedmiller M, Fidjeland A K, Ostrovski G, Petersen S, Beattie C, Sadik A, Antonoglou I, King H, Kumaran D, Wierstra D, Legg S and Hassabis D 2015 *Nature* **518** 529
[7] Foerster J, Assael I A, de Freitas N and Whiteson S 2016 *Adv. Neural Inf. Process. Syst.* **29** 2137
[8] Bostanabad R, Bui A, Xie W, Apley D W and Chen W 2016 *Acta Mater.* **103** 89
[9] Bengio Y, Courville A and Vincent P 2013 *IEEE Trans. Pattern Anal. Mach.* **35** 1798
[10] Brockherde F, Vogt L, Li L, Tuckerman M E, Burke K and Muller K R 2017 *Nat. Commun.* **8** 872
[11] Carter J A, O'Brien L M, Harville T, Jones B T and Donati G L 2021 *Talanta* **223** 121665
[12] Gelfusa M, Rossi R, Lungaroni M, Belli F and Murari A 2020 *Nucl. Instr. Meth.* A **974** 164198
[13] Guastoni L, Encinar M P, Schlatter P, Azizpour H and Vinuesa R 2020 *J. Phys. Conf. Ser.* **1522** 012022
[14] Pan S 2021 *J. Phys. Conf. Ser.* **1802** 042055
[15] Heinonen R A and Diamond P H 2020 *Phys. Rev.* E **101** 061201
[16] Mokshin A V, Mokshin V V and Sharnin L M 2019 *Commun. Nonlinear Sci. Numer. Simul.* **71** 174
[17] Yulmetyev R M, Demin S A and Hänggi P 2006 Manifestation of chaos in real complex systems: Case of Parkinson's disease *The logistic map and the route to chaos: From the beginnings to modern applications, Understanding Complex Systems* ed M. Ausloos and M. Dirickx (Berlin/Heidelberg: Springer Science & Business Media) p 175
[18] Demin S A, Nefedyev Y A, Andreev A O, Demina N Y and Timashev S F 2018 *Adv. Space Res.* **61** 639
[19] Ferrari L 2019 *Chem. Phys.* **523** 42
[20] Suttorp L G and Schoolderman A J 1987 *Physica* A **143** 494
[21] Vladimirov S V and Ostrikov K 2004 *Phys. Rep.* **393** 175

Pore formation in simple systems simulated by molecular dynamics

R A Khabibullin

Department of Computational Physics, Institute of Physics, Kazan Federal University, 420008 Kazan, Russia

E-mail: roman.khabibullin@outlook.com

Abstract. Simulation of the pore formation process in a simple one-component system was performed. As the initial configuration of the system a crystal with FCC lattice was used. Pore formation and evolution were observed during isochoric melting of the crystal and subsequent fluid simulation.

1. Introduction

In this paper, we consider pore formation in a simple one-component system in which the interaction between particles is determined by the Mie potential

$$\frac{u(r)}{\epsilon} = \frac{n}{m-n}\left(\frac{r_\mathrm{e}}{r}\right)^m - \frac{m}{m-n}\left(\frac{r_\mathrm{e}}{r}\right)^n , \tag{1}$$

where r is the distance between particle centers, ϵ is the potential well depth, m and n are free parameters. The equilibrium interparticle distance r_e is given by

$$r_\mathrm{e} = \sigma\left(\frac{m}{n}\right)^{1/(m-n)} , \tag{2}$$

where σ is the effective particle diameter. When $m = 12$ and $n = 6$, expression (1) corresponds to the well-known Lennard-Jones potential [1–8].

2. Simulation details and results

At the initial moment of time the system of $N = 16\,384$ particles was a crystal with a face-centered cubic lattice. The density of the system was $\rho = 0.75\,\sigma^{-3}$ and the velocity distribution of the atoms corresponded to temperature $T_0 = 0.1\,\epsilon/k_\mathrm{B}$. Next, we performed a constant-volume simulation of the melting of this crystal for 20 000 timesteps. To control the system temperature, we used a Nose–Hoover thermostat with a relaxation time of $0.5\,\tau$ and an external temperature $T_\mathrm{ext} = 2.8\,\epsilon/k_\mathrm{B}$. The interparticle interaction is given by the Mie potential (1) with parameters $m = 10$, $n = 3$, which are the power indices in the repulsive and attractive contributions, respectively. In this case, the attractive branch of the potential decreases quite slowly with distance, forcing the use of a relatively large cutoff radius $r_\mathrm{cut} = 8.8\,\sigma$. Simulations with such long-range interactions is computationally challenging, so we used the LAMMPS software [9] with the GPU acceleration package [10], which provides significantly better performance than the legacy CPU-based simulation approach.

The results of the simulation are summarized in Figure 1. Based on the visual representation of the configuration of the system under consideration, we can distinguish four stages of pore evolution. For each

Content from this work may be used under the terms of the Creative Commons Attribution 3.0 licence. Any further distribution of this work must maintain attribution to the author(s) and the title of the work, journal citation and DOI.
Published under licence by IOP Publishing Ltd

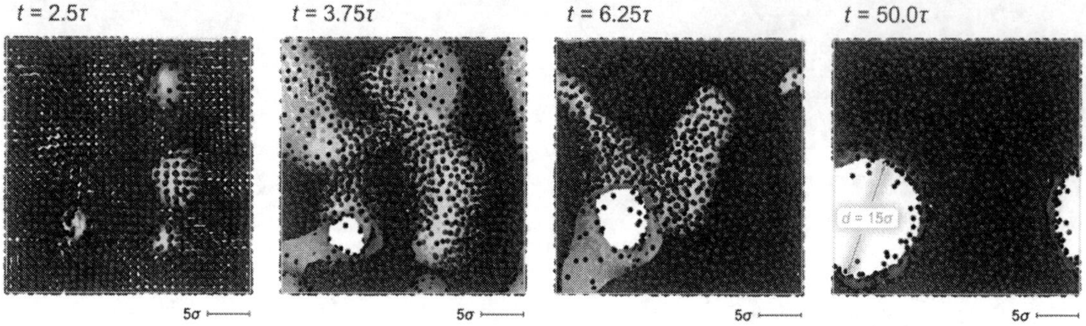

Figure 1: Snapshots showing the time evolution of pores in a fluid with Mie interparticle potential. Blue and red spheres represent bulk and surface atoms, respectively. Visualization of snapshots and construction of pore surfaces were performed using the OVITO software [11].

of these stages, Figure 1 shows a snapshot. Thus, the first snapshot represents the pore nucleation stage, which corresponds approximately to a time interval of 2–2.5 τ. Here we can still observe ordering in the structure, which indicates that the melting process of the crystal is not yet complete. Next, we observe the pore growth stage, which reaches its largest volume by the time $t \approx 3.75\,\tau$. The process of pore growth is followed by the process of pore coalescence, which is shown in the third frame. We can see how the pores merge into one large pore, taking the shape of a tube. At the last stage, the formed pore acquires a stable configuration, which does not change until the end of the simulation. The system temperature by the time $t = 50\,\tau$ stabilizes around $2.8\,\epsilon/k_B$. In the case of the Lennard-Jones system, such a state corresponds to a supercritical fluid [12]. The diameter of the tube-shaped pore can be estimated by approximating its shape with a cylindrical surface:

$$d = \frac{S}{\pi L}, \tag{3}$$

where S is the pore surface area and L is the simulation box length. Construction of the pore surfaces and calculation of the surface area S was performed using the alpha-shape method [13].

3. Conclusion

In this paper, the results of molecular dynamic simulation of pore formation in a simple system were presented. The simulation was performed along an isochore with the density $\rho = 0.75\sigma^{-3}$, the initial temperature $T_0 = 0.1\,\epsilon/k_B$, and the external temperature $T_{ext} = 2.8\,\epsilon/k_B$. This results can be used to develop methods of computer modeling of porous systems.

Acknowledgements

The author is grateful to A. V. Mokshin for useful discussions and recommendations. This paper has been supported by the Kazan Federal University Strategic Academic Leadership Program. The author acknowledges the Foundation for the Development of Theoretical Physics and Mathematics "BASIS" (project No. 20-1-2-38-1).

References

[1] Mokshin A V, Khusnutdinoff R M, Novikov A G *et al* 2015 *J. Exp. Theor. Phys.* **121** 828–843
[2] Mokshin A V 2015 *Theor. Math. Phys.* **183** 449–477
[3] Khusnutdinoff R M and Mokshin A V 2010 *Bull. Russ. Acad. Sci. Phys.* **74** 640–643
[4] Mokshin A V and Galimzyanov B N 2014 *J. Chem. Phys.* **140** 024104
[5] Khusnutdinoff R M, Galimzyanov B N and Mokshin A V 2018 *J. Exp. Theor. Phys.* **126** 83–89
[6] Mokshin A V, Chvanova A V and Khusnutdinoff R M 2012 *Theor. Math. Phys.* **171** 541–552

[7] Khusnutdinov R M, Mokshin A V and Yul'met'ev R M 2009 *J. Exp. Theor. Phys.* **108** 417–427
[8] Mokshin A V and Galimzyanov B N 2018 *J. Phys.: Condens. Matter* **30** 085102
[9] Thompson A P, Aktulga H M, Berger R *et al.* 2022 *Comput. Phys. Commun.* **271** 108171
[10] Brown W M, Wang P, Plimpton S J and Tharrington A N 2011 *Comput. Phys. Commun.* **182** 898–911
[11] Stukowski A 2010 *Modelling Simul. Mater. Sci. Eng.* **18** 015012
[12] Stephan S, Thol M, Vrabec J and Hasse H 2019 *J. Chem. Inf. Model.* **59** 10
[13] Stukowski A 2014 *JOM* **66** 399–407

Investigation of the structure of an aluminum nitride film for a Bragg reflector obtained by magnetron sputtering

L Baranova[1, 2], B Baysova[1, 2], V Strunin[1, 2]

[1]Dostoevsky Omsk State University, Omsk, Russian Federation
[2]Institute of Radiophysics and Physical Electronics OSC SB RAS, Omsk, Russian Federation

baranova@omsu.ru

Abstract. The phase composition of the aluminum nitride film obtained through magnetron sputtering was determined with the help of X-ray phase analysis. The analysis revealed the AlN, Al and V phases deposited during the synthesis of the film. The structural parameters of the primitive unit cell of the AlN and Al phases were determined. The effect of texturing the resulting film was detected.

1. Introduction.

Microwave resonators with a Bragg reflector on bulk acoustic waves (BAW) are widely used in modern electronics due to their good reproducibility [1]. At the same time, piezoelectric aluminum nitride films can be used as a resonant element in thin film resonators [2]. Since the efficiency of a Bragg reflector directly depends on the structure, phase composition and the surface morphology of thin films [3], it is important to study the characteristics of thin film coatings.

The aim of the paper is to investigate the phase composition of aluminum nitride thin films obtained through magnetron sputtering.

2. The experimental procedure.

The phase composition of aluminum nitride films was studied with the help of X-ray phase analysis done on the Bruker D8 Advance device. In this case, the characteristic radiation of the Cu-K $_{\alpha}$ X-ray tube, with a wavelength of 1.541 Å, was used. Bragg-Brentano geometry for reflection was used to photograph the samples of thin films (≈ 1 microns), with the range of 2θ 25-75 °angles. Scanning was performed symmetrically (θ - 2θ scan), the angle between the sample surface and the X-ray tube was equal to the same angle betw een the surface and the detector. The magnitude of the step 2θ in the angle was 0.02 °, and the accumulation time was 3s at each point.

A Goebel mirror was additionally used to reduce the effect of defocusing caused by the height of the sample, the beam spread and the influence of possible inhomogeneities. The use of this device makes it possible to form a quasi-monochromatic parallel beam of primary X-ray radiation. The X-ray diffractogram for the film substance was obtained by subtracting the diffractogram of the pure substrate from the "film+substrate" diffractogram. The phase analysis of the film, as well as the

Content from this work may be used under the terms of the Creative Commons Attribution 3.0 licence. Any further distribution of this work must maintain attribution to the author(s) and the title of the work, journal citation and DOI.
Published under licence by IOP Publishing Ltd

calculation of the parameters of the primitive unit cell of the phases, was carried out according to the positions of individual peaks. The parameters and their allowances were calculated with the help of POLYCRYSTAL software package.

3. Results and discussion.

X-ray phase method was used to analyze the composition of the ceramized glass substrate, as well as the phase composition of the substrate with the film. According to the analysis data, such phases as TiO_2 with rutile structure (SGM $P4_2/mnm$), SiO_2 with a quartz structure with various sand gravel mixes/SGM (RP_121 and $P6_222$) and with a cristobalite structure (Fd3m SGM) are present in the composition of a clean substrate without a deposited substance (see Fig. 1.). The formation of complex oxides ($CaSi_2O_5$ or $Ca_2Ti_2O_6$) was also recorded.

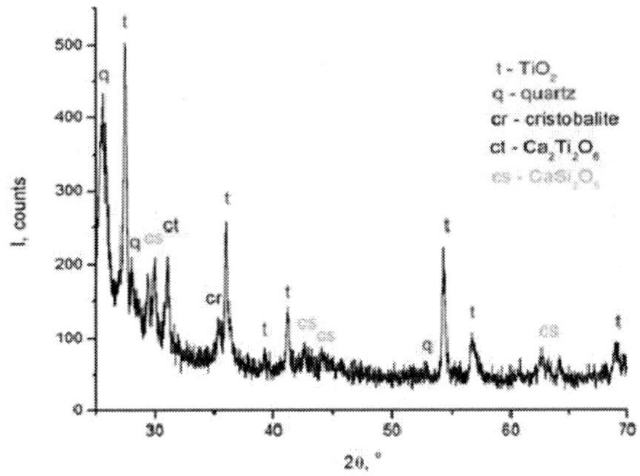

Figure 1. X-ray pattern of a clean substrate (ceramized glass).

An X-ray pattern of a film with a substrate (see Fig. 2.) detected TiO_2 phases with rutile structure, SiO_2 in quartz structure (SGM $P3_121$), $CaSi_2O_5$, $Ca_2Ti_2O_6$.

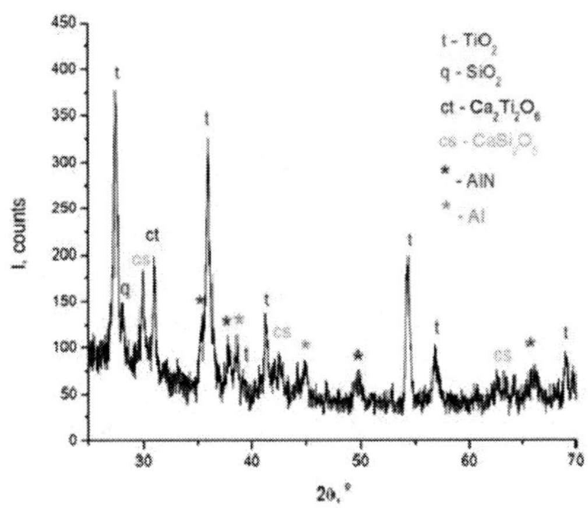

Figure 2. X-ray of the substrate with the AlN film.

However, there were no reflectances characteristic of quartz with SGM $P6_222$, as well as peaks for cristobalite. In particular, a noticeable difference in the diffraction pattern is the absence of a strong reflectance attributed to the 101 quartz reflectance with SGM $P6_222$. These differences may clearly indicate the heterogeneity of the used ceramized glass substrate, which may negatively affect the results of X-ray phase analysis. Small peaks, which can be attributed to the AlN and Al phases, are also noticeable.

When the difference X-ray pattern was obtained (see Fig. 3.), the regions with negative intensity were observed in the diffraction profile. This is explained by the effect of the part of the X-ray radiation absorbed by the film and depends on the thickness of the applied film. In order to eliminate this effect, a correction for intensity values was introduced for the final X-ray pattern along the entire profile with focus on the SEM and EDX analysis data. These data show that the ratio of the Al fraction to the N fraction was 2:1, that is, it was assumed that the sample contained 67% Al and 33% N. It is worth noting that despite the correction, there was an obvious heterogeneity in the phase composition of the conditionally pure substrate and the substrate with the film. Due to this, the range of the diffractogram under study was reduced to exclude the problem zone with missing quartz peaks (Fig. 3 shows the range under study as a rectangular region).

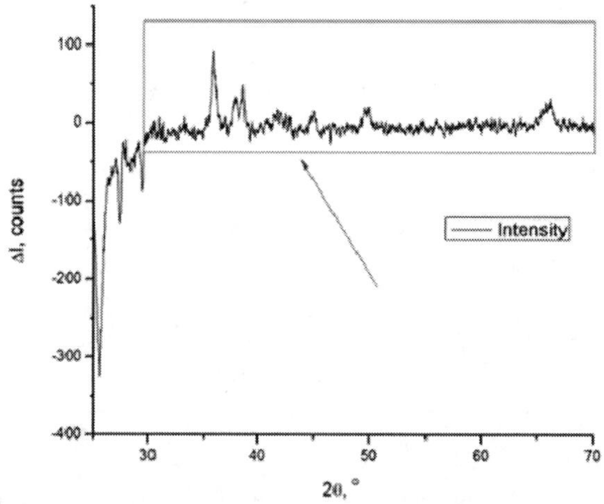

Figure 3. The difference x-ray pattern obtained by subtracting the diffractogram of the substrate from the diffractogram of the substrate with the film.

A number of clearly distinguished reflectances were also detected on the resulting diffractogram (see Fig. 4.). On the basis of the assumptions about the possible composition, a coincidence in the positions of a number of peaks for the hexagonal AlN phase was found. A mismatch of intensities for a number of this phase reflectances is noted. In particular, the peaks with hk0 indices (100 and 110) were rare, and the intensity of the 002 reflectance was significantly increased. This phenomenon may indicate the presence of texture in the sample, that is, the preferred orientation of polycrystallites of the substance. Due to the latter, the corresponding crystallographic planes are found in the reflecting position (002 in our case).

Also, in addition to the hexagonal AlN phase, a significant amount of metallic Al was found in the sample. This explains the data of the EDX analysis, which showed a deviation from the stoichiometric ratio of Al and N towards an excess of Al. In addition to the AlN and Al phases, weak reflectances, which can be attributed to V and V_2O_5, were also detected. All these data correlate quite well with the purpose of the synthesis method planned to be used for obtaining a film with V, Al and hexagonal AlN phases. The formation of vanadium oxide impurity can be due to the oxidation of metallic vanadium in the air.

The primitive unit cell parameters were calculated for the AlN and Al phases. In the case of hexagonal AlN, the parameters have the following values: a = 3.1191 ± 0.0096 Å, c = 4.977 ± 0.0098 Å, which is in good agreement with the reference data (a = 3.1114 Å, c = 4.9792 Å)[4]. For Al, the cell parameter was a lower value (a = 4.0352 ± 0.0027 Å) compared to reference data (a = 4.0494 Å) [4]. In this case, apparently, there is an error in determining the position of aluminum peaks since this phase has a conductometric calibrator with a maximum atomic packing density of 74%. As a result, the parameter cannot be much less than the standard value.

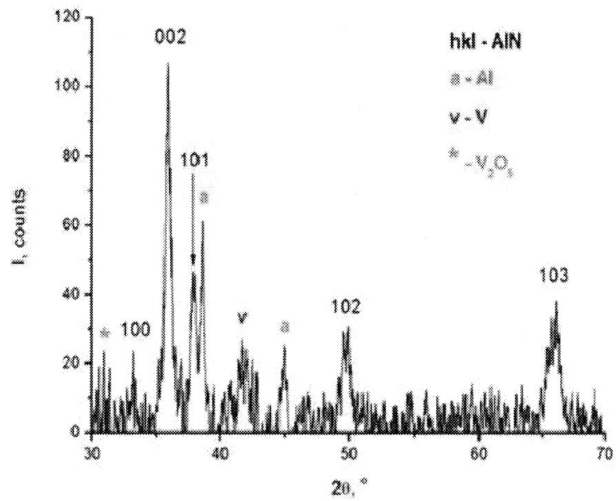

Figure 4. The difference AlN X-ray pattern normalized with the film thickness taken into account (the indices are given for the hexagonal modification).

The studies of aluminum nitride thin films through X-ray phase analysis proved the heterogeneity of the phase composition of the used substrates and the increased quantity of aluminum in the studied films. The X-ray image obtained by subtracting the curves of the "substrate" and "film+substrate" diffractograms revealed the phases AlN, Al and V deposited during the synthesis of the film. The structural parameters of the primitive unit cell of the AlN and Al phases correspond to the reference data in an acceptable way. The effect of texturing of the resulting film was detected. It manifested in a decrease in the intensity of 100 and 110 reflectations and a sharp increase in the 002 reflectation intensity.

Thus, the study shows the possibility of obtaining films of the required quality by varying the process conditions of their formation.

References:

[1] Sorokin B.P., Kvashnin G.M., Novoselov A.S., Bormashov V.S., Golovanov A.V., Burkov S.I., Blank V.D. Excitation of hypersonic acoustic waves in diamond-based piezoelectric layered structure on the microwave frequencies up to 20 GHz // Ultrasonics. 2017. V. 78. P. 162–165.

[2] Development and research of composite acoustic resonators with the"Al/(Al, Sc) /Mo/Diamond" structure with high finesse of the microwave / B.P. Sorokin, A.S. Novoselov, G.M. Kvashnin, N.V. Luparev, N.O. Asafyev, A.B. Shipilov, V.V. Aksenenkov // Physical acoustics. 2019. Vol. 65. No. 3. Pp. 325-331.

[3] Zhang Y., Wang Z., Cheeke J.D.N. Resonant spectrum method to characterize piezoelectric films in composite resonators // IEEE Trans. Ultrason. Ferroelectr. Freq. Control. 2003. V. 50. P. 321–333.

[4] Khramov A.S., Lukyanov I.V. X-ray diffraction analysis of polycrystals. Part IV. Study guide for the students of the Institute of Physics. Kazan: K(P)FU, 2010 - 76 p.

The study was carried out within the state assignment of Omsk Scientific Center of SB RAS (the project state registration number 121121700062-3)
The authors express their gratitude to the staff of Analytical and Technological Innovation Center "High Technologies and New Materials" (ATIC HTNM) of NSU for the assistance in carrying out the measurements.

Use of non-thermal plasma for decontamination of titanium implants

E. Sysolyatina[1], A. Petryakov[2], M. Abdulkadieva[3,4], E. Vasilieva[1,3], M. Chernukha[1], L. Avetisyan[1], T. Alekseeva[2], V. Barsuk[2], S. Ermolaeva[1], N. Titov[2], Yu. Akishev[2,5]

[1]The Gamaleya National Center of Epidemiology and Microbiology, Moscow, Russia
[2]State Research Center of Russian Federation TRINITI, Moscow, Troitsk, Russia
[3]Joint Institute for High Temperatures of Russian Academy of Sciences, Moscow, Russia
[4]RUDN University, Moscow, Russia
[5]National Research Nuclear University MEPhI, Moscow, Russia

e-mail: petryakov@triniti.ru

Abstract. Non-thermal plasma (NTP) has a non-specific bactericidal effect as a result of the synergistic activity of biologically active components contained in the plasma torch. As part of this work, we investigated the potential of using NTP for the antibacterial treatment of titanium implants. Applying samples made of the VT6 titanium alloy used to create implants and a strain of antibiotic-resistant staphylococcus MRSA, we showed that 1) plasma pretreatment of the sample surface led to the formation of a film containing titanium oxides which have a weak bactericidal effect reducing the colonization of the surface with staphylococcus; 2) direct plasma treatment of a polished titanium disk for 120 seconds reduced the contamination of *S.aureus* by 563 times; 3) the relief of the titanium surface itself plays an important role in the effectiveness of decontamination: the bacteria on the rough surface were less susceptible to the effects of NTP than the bacteria on the polished surface. In total, the data obtained indicate the need to continue research aimed at increasing the effectiveness of the bactericidal activity of NTP against bacteria on the surface of titanium implants.

1. Introduction

Surgical site infections (SSIs) are defined as infections that occur within 30 days after surgery or up to one year after surgery in patients receiving implants and involve either the incision or deep tissues of the surgical site [1]. Despite advances in prevention, SSIs remain a serious clinical problem, as they are associated with significant mortality and morbidity, and place considerable demands on health care resources. The incidence of SSI can be as high as 20%, depending on the surgical procedure, the follow-up criteria used, and the quality of the data collection [2], [3]. The most frequently SSI-associated microorganisms are *Staphylococcus aureus*, coagulase-negative staphylococci, *Pseudomonas aeruginosa*, *Escherichia coli* and some others [4], [5]. The risk of SSI is influenced by numerous patient- and procedure-related factors, and therefore prevention requires a comprehensive approach with systematic attention to multiple risk factors to reduce the possibility of bacterial contamination. A particular risk for the development of SSI is infection of the implant surface. The adhesion of bacterial cells to titanium implants followed by the formation of biofilms on them is still the main cause of unsuccessful osseointegration, despite the use of various coatings that prevent the colonization of the

Content from this work may be used under the terms of the Creative Commons Attribution 3.0 licence. Any further distribution of this work must maintain attribution to the author(s) and the title of the work, journal citation and DOI.
Published under licence by IOP Publishing Ltd

surface of medical devices by microorganisms and improve the adhesion of osteoblasts [6], [7]. Thus, there are good reasons for the introduction of new technologies that ensure sterility into everyday clinical practice.

Non-thermal gas plasma (NTP) is one of the bactericidal methods based on the action of physical and physical-chemical active factors. The biologically active components of non-thermal plasma include charged and metastable particles, reactive oxygen and nitrogen species (ROS and RNS), and ultraviolet light [8], [9]. NTP has a nonspecific bactericidal effect on a wide range of bacteria, including those resistant to antibiotics [10]. The targets of NTP are, first of all, the surface structures of the bacterial cell. The destruction of the lipid bilayer and the formation of pores in the membrane as a result of lipid peroxidation ultimately lead to the penetration of the active components of the plasma torch into the cell and damage to internal structures. The purpose of this work was to evaluate the potential of using low-temperature plasma for antibacterial treatment of titanium implants and to select the optimal modes of antimicrobial plasma exposure.

2. Materials and methods

2.1. Source of non-thermal plasma jet

The used NTP source was previously described in paper [11]. A non-thermal plasma jet was generated by a coaxial barrier discharge created in a narrow dielectric tube blown by argon at atmospheric pressure.

2.2. Bacterial strains and growth conditions

The Staphylococcus aureus strain SA180-F (MRSA, antibiotic resistance profile PG-OXC-E-CLI-CIP-VAN-TPM-LEV) from the collection of the Gamaleya Centre was used in the work. Bacteria were grown on BHI medium (BD, USA) at 37°C.

2.3. Samples of titanium implants

In the experiments, samples of implants made by 3D printing from titanium alloy of the VT-6 brand were used. The appearance of the samples is shown in Figure 1. As one can see, these are discs with a diameter of 15 mm and a thickness of 2.5 mm. The discs were cut by the spark method in distilled water from the rod of a long length. During the cutting process, the surface is modified by hot spark plasma. As a result, the surface of the discs after cutting by the spark method became rough. The experiments were carried out with discs whose surface was mechanically polished after their cutting off, and with discs with a rough surface, i.e. without polishing them.

a) b)

Figure 1. The appearance of samples made of titanium alloy VT6. a) disc with a mechanically polished surface; b) disc with a rough surface after its cutting off by spark method from a long rod.

2.4. Surface modification using NTP.

VT6 titanium samples with a polished and rough surface were treated with a jet of argon plasma at atmospheric pressure. The plasma jet was created by a coaxial barrier discharge with a frequency of 100 kHz and a power of 8 Watts. The jet velocity at the plasma source outlet was 30 m/s. The distance from the source outlet to the sample to be treated was 15 mm. The exposure time was varied.

2.5. Methods of examination of the implant surface before and after plasma treatment.

The study of the topography of the disk surface after it was cut off by the spark method and after mechanical polishing, as well as after prolonged treatment of the polished disk with plasma jet in the ambient air was carried out with an ADF I350 optical microscope (inverted metallographic microscope) with a maximum 1000x magnification and equipped with an ADF SPD16 digital camera with a volume of 16 megapixels (matrix size 8.7 x6.6 mm2).

Four methods of observation and registration of the surface image of the samples were used: 1) light field mode, when light falls on the sample normally from the eyepiece of a microscope; 2) dark field mode, when light falls on the sample from the side of an annular source at an angle of 45 degrees; 3) light field mode using reflected polarized light; 4) differential contrast mode, used to create contrast in unpainted transparent samples. In this case, the polarized beam from the light source is divided into two beams that pass through the sample by different optical paths. The length of these optical paths (that is, the product of the refractive index and the geometric path length) is different. Farther, these rays interfere at their intersection. This allows one to create a three-dimensional relief image corresponding to the change in the optical density of the sample, accentuating the lines and borders. However, this picture is not an accurate topographic picture.

The obtained surface topography of the samples is shown in Figures 2 and 3. It can be seen that the rough surface has numerous cavities with a characteristic depth and width of about 20-30 microns, in which bacteria can "hide" from plasma exposure. The polished surface is practically free of such cavities.

a) b) c)

Figure 2. Topography of the mechanically polished surface of VT6 titanium alloy samples at 500x magnification. The vertical and horizontal size of each frame is equal to 150 microns.
a) light field; b) differential contrast; c) polarization in reflected light: an image of a translucent speck of small dust in a form of a butterfly casually deposited on the surface of the disk is given.

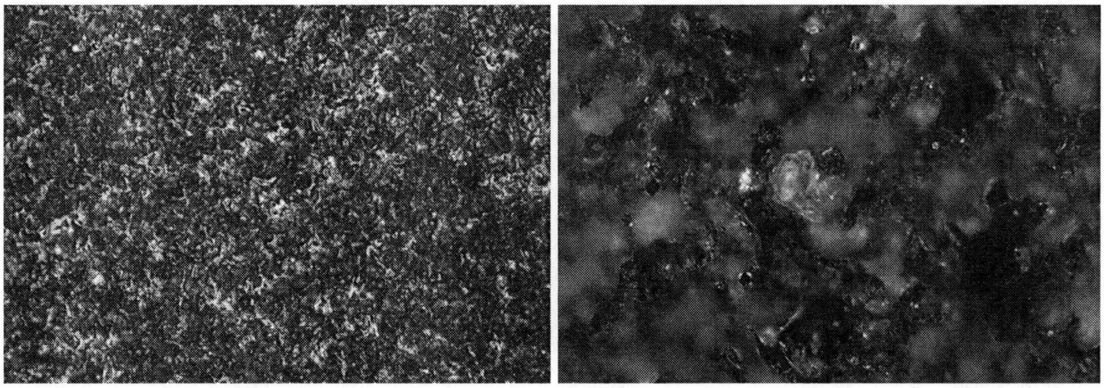

a) b)

Figure 3. Topography of the rough surface of a sample made of titanium alloy VT6 modified by hot spark plasma during cutting off the disc. a) dark field, magnification 200, vertical frame height 375 microns; b) magnification 500, vertical frame height 150 microns; polarized light, one can see an outsider orange inclusion in the center of the frame. Note the low depth of the lens focus does not allow one to see sharply the entire structure of cavities in the vertical direction. This circumstance leads to the appearance of cloudy images which correspond to the surface areas above the plane of the lens focus.

The elemental composition and topography of the initial and plasma-modified implant surfaces were studied using scanning electron microscopy (Zeiss EVO MA10 electron microscope with Bruker XFlash 5030 energy dispersive X-ray detector). This technique allows one to study the composition of the surface at a point, by area, or the distribution of elements along a line on the surface, while the depth of the information collection layer does not exceed 1 micron. The phase composition of the modified surface was studied by X-ray phase analysis using an X-ray diffractometer DRON-8 on Cu K_α radiation with a nickel filter. This method makes it possible to detect the presence of crystalline phases on an area of several square millimeters and at a depth of up to several tens of microns. Phase identification based on the results of diffractogram processing was carried out using an ICDD PDF2 X-ray file.

2.6. Titanium disc infection

5-6 bacterial colonies were seeded in BHI nutrient broth and grown for 18 hours (overnight culture). The overnight culture of bacteria was diluted with saline. Solution up to a concentration of 10^9 CFU / ml according to the McFarland standard. The resulting suspension was further diluted 10 times. Bacteria in an amount of 10^6 CFU taken in a volume of 10 μl were applied to the surface of a titanium disc and incubated at a temperature of 37°C for 1 h until complete drying. To obtain biofilms, titanium samples were immersed in the overnight culture and incubated at 37°C for 24 hours.

2.7. Antibacterial titanium disc treatment with NTP

Bacterial suspension was applied to the sample, and the discs were incubated until completely dry as described above. The infected discs were treated with NTP for 60, 120, or 300 s at a distance of 15 mm. An untreated disc was used as a control.

2.8. Bacterial load counting

Two methods were used to determine bacterial loads on the titanium disc surface: qualitative and quantitative. Fluorescent dyes included in the Live/Dead Cell Viability Assay kit (Invitrogen) were used according to the manufacturer's instructions for qualitative assessment and to determine the ratio of live and dead bacterial cells. Surface swabs were made for quantification using 100 μl of phosphate-buffered saline (PBS). The washed bacterial suspension was diluted in PBS, making a series of consecutive decimal dilutions. From each dilution, bacteria were plated on BHI agar. Colonies were counted after 24 h incubation at 37°C.

2.9. The effect of the titanium surface pre-treatment.
To evaluate the property of the titanium surface modified with NTP to affect bacterial viability, 10^6 CFU of methicillin-resistant *S. aureus* (MRSA) were applied to a disk treated with NTP 3 times for 120 s each time. The infected disk was incubated for 1 h. Alternatively, the disc was immersed in a bacterial suspension for 24 h to form biofilms. After incubation, samples were stained with Live/Dead vital dye and micrographs and washes were taken from the disc surface. An untreated disc was used as a control.

2.10. Statistics
All experiments were done in duplicate and repeated 3 times. The one-way ANOVA method was used to calculate statistical significance.

3. Results

3.1. Characteristics of the plasma jet and the barrier discharge forming the NTP jet
The dielectric barrier discharge (DBD) was created in a quartz tube with an inner and outer diameter of 2.5 and 4.5 mm, which was blown with argon at atmospheric pressure at a speed of 20-30 m/s. Inside the tube along its axis, there was a thin rod, to which a sinusoidal voltage with a frequency of 100 kHz and an amplitude of 4.5 kV was applied. A grounded electrode in the form of a thin metal foil was wrapped around the outer surface of the tube. The electric characteristics of a coaxial dielectric barrier discharge forming a non-thermal plasma jet at the outlet of the tube are shown in Figures 4 and 5. Figure 4a shows a visible image of a glowing plasma jet that blows out freely into the surrounding air. Figures 4b and 4b show shadow images of the argon jet itself both in the absence of DBD (Fig.4b) and in the presence of DBD (Fig.4b). These figures show that, firstly, the plasma jet has a smaller diameter compared to that of the gaseous jet (i.e., the plasma jet is inside the gas one) and, secondly, the argon jet exits the tube in the laminar regime. However, a comparison of Figures 4b and 4b shows that the inclusion of DBD leads to a decrease in the distance from the tube outlet, after that the laminar jet goes into turbulent mode. The occurrence of turbulent pulsations in the gas jet leads to a faster mixing of the plasma jet with the surrounding air containing oxygen and water vapor. This process is accompanied, on the one hand, by the appearance of useful OH-type radicals in the plasma jet, but, on the other hand, leads to the rapid disappearance of the primary reactive particles created by the dielectric barrier discharge.

The number and sort of reactive particles created by the dielectric barrier discharge depend not only on the composition of the plasma-forming gas but also on the electrical power inserted in the discharge. The oscillograms of current I(t) and voltage U(t) characterizing DBD are shown in Figure 5a. Sharp current splashes on the I(t) oscillogram are caused by an electrical breakdown of the gas-discharge gap happening in each half-cycle of the applied sinusoidal voltage. These breakdowns provide the formation of non-thermal plasma inside the plasma source. After every breakdown, energy is released into an argon plasma. The residence time of the gas inside the discharge zone of the plasma source is about 10^{-3} s. Thus, at the frequency of the applied voltage of 10^5 kHz, the activated gas is energized by more than 200 breakdowns, saturating the gas with numerous active particles. The growing dependence of the period-averaged electrical power W of the barrier discharge on the amplitude U of the applied voltage is shown in Fig. 5b.

Journal of Physics: Conference Series **2270** (2022) 012045 doi:10.1088/1742-6596/2270/1/012045

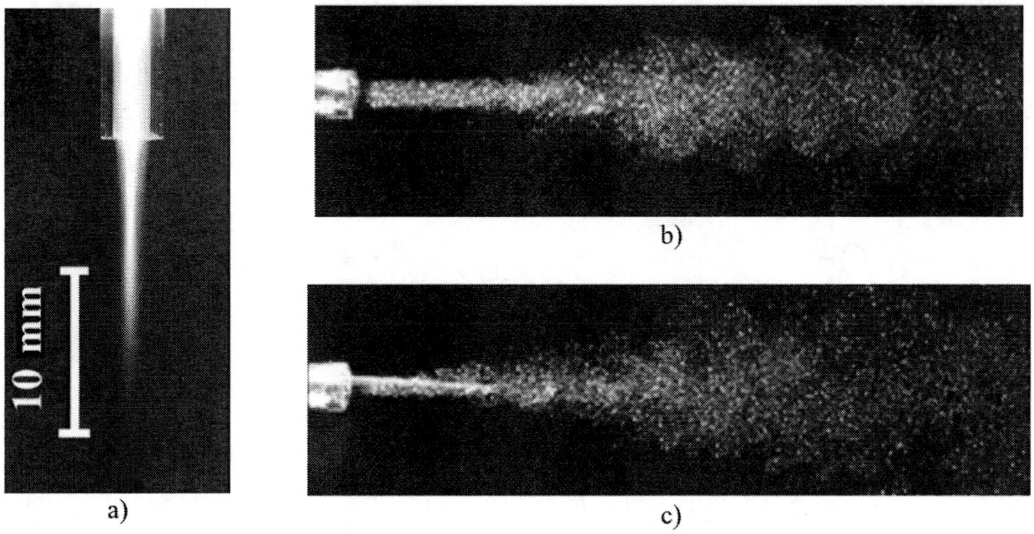

Figure 4. Visible (a) and shadow (b, c) image of an argon plasma jet flowing into free space (ambient air); the speed of the plasma jet at the discharge tube outlet is 30 m/s; shadow image of a gas jet in the absence of a barrier discharge (b) and in the presence of a barrier discharge (c). The frequency of the sinusoidal voltage exciting the barrier discharge is 100 kHz, the voltage amplitude is 4.5 kV.

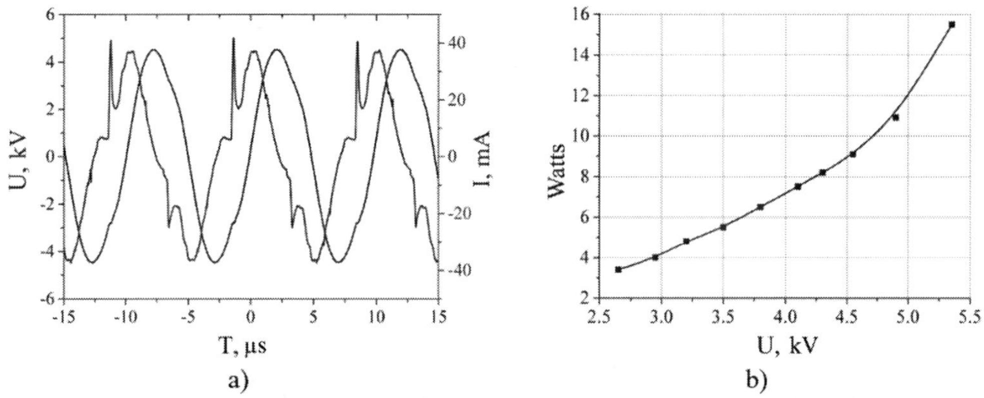

Figure 5. a) Oscillograms of the current and voltage DBD generating the plasma jet. b) The discharge electrical power dependence on the amplitude U of the applied voltage. In both cases, the velocity of the argon jet at the outlet is 30 m/s.

3.2. The results of the materials science study of the implant surface before and after plasma treatment.

The implants were made of titanium alloy of the VT6 brand, which should not contain oxygen in its composition. However, the implant surface after the spark-cutting method looked rough and gray in color, corresponding to the presence of an oxide film on its surface. Our studies of the composition of the implant surface layer with the diagnostics described above really showed a high concentration of oxygen (up to 30 or more atomic percent) due to the formation of titanium oxides during the spark-cutting. Cleaning of the implant from the gray oxide film by its mechanical removal to a state of gloss with subsequent analysis of the surface showed that the oxygen concentration on the cleaned and smooth surface of the titanium alloy sharply decreased. At the same time, the long treatment of the cleaned and smooth surface of the disc with a plasma jet in the ambient air again led to an increase in oxygen in the surface layer. The essential advantage of the oxide thin layer formation by the NTP is that the surface keeps the smoothness without the formation of any cavities. The results of the analysis are shown in Figure 6. It is clear that the presence of titanium oxides can lead to unusual results of plasma exposure to bacteria. It is important to note that the oxygen concentration on the implant surface layer depends on the exposure time of the surface by the plasma jet. This circumstance makes it possible to vary the oxygen concentration on the implant surface to the desired level if it is necessary.

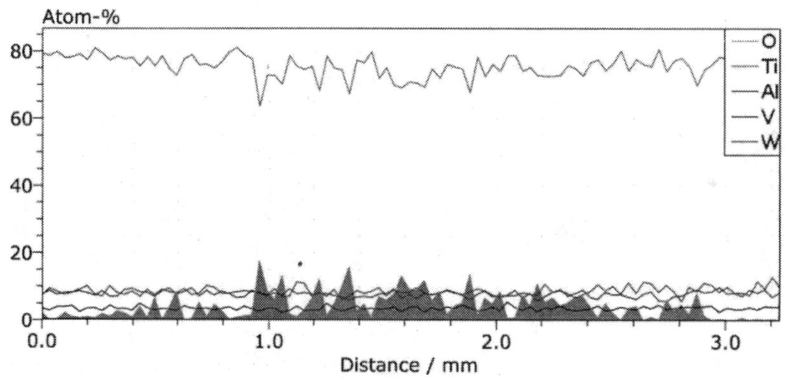

Figure 6. Radial distribution of the content of elements, including oxygen (in Atom%) on the treated and untreated implant surface by argon plasma jet. The distance of 0-0.2 mm corresponds to the untreated surface, 0.2-2.9 mm - the treated surface.

3.3. Results of NTP bactericidal activity.

3.3.1. The effect of the titanium surface pre-treatment. The rough titanium disc surface was pre-treated with NTP as described in Materials and Methods, and MRSA bacteria were applied to the surface. After 1 h incubation bacterial loads on the treated surface were lower if compared with loads on the disc that was not pre-treated with NTP. These data suggested that the titanium surface, pre-modified with NTP, had weak antibacterial properties against MRSA when the bacteria were completely dried on the sample (Fig. 7). On average, plasma pre-treatment led to a decrease in the contamination of the disc surface contaminated with S. aureus MRSA by 44.3±5%. At the same time, the plasma pretreatment of the titanium surface did not prevent the formation of biofilms during 24 h disc incubation in the bacterial liquid culture (data not shown).

a) b)

Figure 7. The effect of plasma pre-treatment of the titanium surface on the viability of *S. aureus* MRSA cells. Bacteria were visualized with the Live/Dead staining after 1 h incubation on the surface of the pre-treated or control titanium disc. a) - control, b) – disc was pre-treated with NTP. Green for live bacteria, red for dead bacteria.

3.3.2. Direct bactericidal effects of titanium treatment with NTP. Next, we applied NTP to decontaminate discs infected with MRSA bacteria. Two types of surfaces were studied, rough and polished. The efficiency of direct plasma treatment depended on the geometric properties of the surface. The Live/Dead staining showed that damaged bacteria appeared after 60 s NTP treatment when bacteria were placed on the polished surface. 300 s NTP treatment of polished discs was enough for dead bacteria to prevail in the field of view with only a few alive green bacteria (Fig. 8). On the rough surfaces treated for 300 s, Live/Dead staining revealed conglomerates formed by damaged and living bacteria in relatively similar proportions (Fig. 8).

Control Polished disc 300s Rough disc 300s

Figure 8. Bactericidal effects of infected polished and rough disc treatment with NTP. Disc surface was infected with 10^6CFU MRSA. After drying, surfaces were treated with NTP for 300s, stained with the Live/Dead marker kit, and analyzed with fluorescence microscopy.

The quantification of survived bacteria confirmed the data obtained using fluorescence microscopy. Plasma treatment of a smooth titanium disk led to a decrease in bacterial loads by 154 ± 83, 187 ± 91, and 563 ± 237 times after 60, 120, and 300 s of treatment, respectively. At the same time, the rough surface protected bacteria much better. 60 s treatment led to a decrease in the number of living bacteria by 6.7 ± 3.1 times, and an increase in exposure time to 120 and 300 s reduced the number of surviving bacteria by only 14.4 ± 6.6 and 25.1 ± 13.3 times, respectively (Fig. 9).

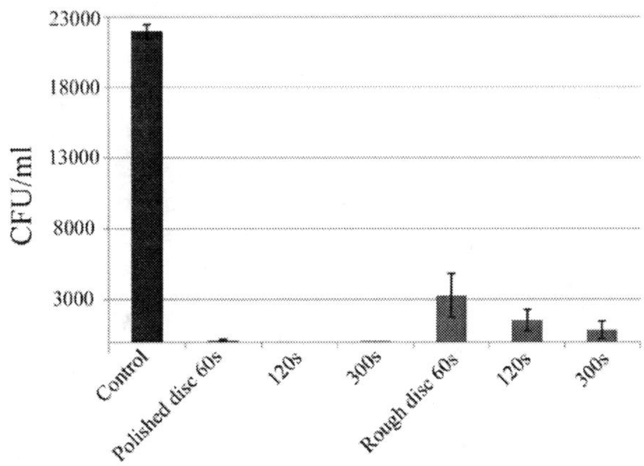

Figure 9. Bactericidal effects effect of NTP applied to rough and polished titanium surfaces contaminated with MRSA.

4. Discussion

In the framework of this work, the potential of non-thermal plasma to fight against contamination of the surface of titanium implants was studied. Investigating the effects of plasma surface pre-treatment with NTP as a means to prevent bacterial adhesion to the surface of titanium implants, and the effects of NTP treatment of titanium surfaces contaminated with MRSA, we found that 1) plasma pre-treatment of the sample leads to the formation of a film of titanium oxide, which has a weak bactericidal action; 2) direct plasma treatment of a contaminated polished titanium disk for 300 s reduced the bacterial loads by 563 times; 3) the relief of the titanium surface played a primary role in decontamination: bacteria on a rough surface are better protected from direct plasma exposure than on a polished one.

Titanium oxide and dioxide have a nonspecific bactericidal effect due to the development of free radical reactions, although the mechanisms of bacterial death have not been fully elucidated [12], [13]. Our results indicate that the pretreatment of the implant surface with NTP is able to form sufficient concentrations of surface titanium oxides for these mechanisms to come into play. It is possible that these mechanisms operate not only during surface pretreatment, but also during the action of NTP on the contaminated surface of a titanium implant. However, under the direct action of NTP on bacteria located on the surface, the bactericidal effect is much higher, which indicates a relatively minor contribution from the formed titanium oxides, and suggests the main contribution to the death of bacteria from the active components of the plasma torch. Indeed, the plasma torch contains a whole complex of antibacterial components, the effect of which is synergistically enhanced during the action of the plasma torch on the target. The main antibacterial agents of NTP include charged particles, reactive oxygen and nitrogen species formed during the interaction of a plasma torch with air, and ultraviolet radiation [8], [9].

At the same time, the obtained results indicate that the process of bacteria inactivation on the surface of implants is less efficient than, say, on the surface of an agar medium, i.e. under conditions commonly used to assess the bactericidal activity of NTP sources [10]. For the MRSA strain used in this work, a similar treatment for 300 s led to the complete destruction of living bacteria (a drop in the concentration of living cells by more than 10^6 times, data not shown) compared with a drop of 563 times on the surface of titanium. The decrease in the effectiveness of the antibacterial properties of NTP in the treatment of the surface of titanium implants can be associated with various reasons. One of these reasons is the rough surface of titanium implants used to enhance the adhesion of human cells. Roughnesses in the micrometer range, comparable to the size of the bacterial cell, protect bacteria from the impact of NTP, which was proved in this work by comparing the results of processing bacteria placed on a rough and polished surface. Other mechanisms may include the low surface moisture of the titanium compared to the same nutrient agar, which is actually a gel whose surface is wet. Additional studies on the contribution of air humidity are needed to evaluate this parameter.

5. Conclusions

Taken together, obtained data indicate that NTP has a bactericidal effect on bacteria that contaminate the surface of titanium implants, which is the sum of their total effect of the active components of the plasma torch and the activity of titanium oxides formed on the surface during processing. The relief of the titanium surface has a great impact on the outcome of the treatment. Further research is needed to increase the effectiveness of NTP impact

Acknowledgement

The work has been performed under the contract #226/3255-Д, 23.12.2020 with Sechenov University.

References

[1] European Centre for Disease Prevention and Control, *"Surveillance of surgical sire infections and prevention indicators in European hospitals – HAI-Net SSI protocol, ver. 2.2"* Stockholm ECDC, **46**, 2017

[2] Okike K and Bhattacharyya T 2006 Trends in the management of open fractures. A critical analysis *J. Bone Joint Surg. Am.* **88** 2739

[3] Lake J and Gordon O 2022 Implant-Associated spinal infections in children: How can we improve diagnosis and management? *Infect. Dis. Clin. North Am.* **36** 101

[4] Linke S, Thurmer A, Bienger K, Kleber C, Bellova P, Lutzner J and Stiehler M 2022 Microbiological pathogen analysis in native versus periprosthetic joint infections: a retrospective study *J. Orthop. Surg. Res.* **17** 9

[5] Doll K, Jongsthaphongpun K L, Stumpp N, Winkel A and Stiesch M 2016 Quantifying implant-associated biofilms: Comparison of microscopic, microbiologic and biochemical methods *J. Microbiol. Methods* **130** 61

[6] Damiati L, Eales M, Hobbs A, Su Bo, Tsimbouri P, Salmeron-Sanchez M and Dalby M 2018 Impact of surface topography and coating on osteogenesis and bacterial attachment on titanium implants *J. Tissue Eng.* **9**

[7] Wang M and Tang T 2019 Surface treatment strategies to combat implant-related infection from the beginning *J. Orthop. Transl.* **17** 42

[8] Kong M, Kroesen G, Morfill G, Nosenko T, Shimizu T, van Dijk J and Zimmermann J 2009 Plasma medicine: an introductory review *New J. Phys.* **11** 115012

[9] Sysolyatina E, Mukhachev A, Yurova M, Grushin M, Karalnik V, Petryakov A, Trushkin N, Ermolaeva S and Akishev Yu 2014 Role of the Charged Particles in Bacteria Inactivation by Plasma of a Positive and Negative Corona in Ambient Air *Plasma Process. Polym.* **11** 315

[10] Ermolaeva S, Varfolomeev A, Chernukha M, Yurov D, Vasiliev M, Kaminskaya A, Moisenovich M, Romanova J, Murashev A, Selezneva I, Shimizu T, Sysolyatina E, Shaginyan I, Petrov O, Mayevsky E, Fortov V, Morfill G, Naroditsky B and Gintsburg A 2010 Bactericidal effects of non-thermal argon plasma in vitro, in biofilms and in the animal model of infected wounds *J. Med. Microbiol.* **60** 75

[11] Akishev Yu, Zaytsev M, Kopiev V, Petryakov A and Trushkin N 2018 Dielectric barrier discharge influence on the gas-dynamic parameters of the plasma jet formed by this discharge *J. Appl. Phys.* **6** 14 [in Russian]

[12] Kim C, Lee E, Kang S, de Josselin de Jong E and Kim B 2017 Bactericidal effect of the photocatalystic reaction of titanium dioxide using visible wavelengths on Streptococcus mutans biofilm *Photodiagnosis Photodyn Ther.* **18** 279

[13] Bonyadi Z, Mirzaee M, Mehdi Ejtehadi M and Mokhtari M 2017 The bactericidal effect of simultaneous titanium oxide on common hospital bacteria *Environ. Monit. Assess.* **189** 342

CONTROL OF THE DISTRIBUTION OF THE INTERNAL CHARACTERISTICS OF THE DISCHARGE USING SUPERSONIC GAS PUMPING

B A Timerkaev[1], O A Petrova[1], R S Shamsutdinov[1], A I Sayfutdinov[1]

[1]Kazan National Research Technical University named after A. N. Tupolev-KAI, Kazan, Russian Federation

btimerkaev@gmail.com

Abstract. Theoretical studies of the distribution of glow discharge parameters in a supersonic nitrogen gas flow between the central body of the nozzle and an annular electrode around the central body are carried out. Distributions of charged particle concentrations, potential, and electric field strength are obtained. The possibility of compaction or stretching of the cathode zones has been demonstrated. In the case of the organization of a supersonic gas flow near the central body-the anode, all the near-electrode zones are pressed against the anode. In this case, the cathode is located in the area of reduced pressures. Such a way of organizing the discharge can find application in the processes of applying functional coatings.

1. Introduction

A smoldering discharge is a self-organizing discharge in the sense that after the discharge occurs in the interelectrode gap, a specific distribution of its parameters is established, such as the potential and intensity of the electric field, the concentration of charged particles, the glow pattern, which corresponds to the principle of least action. It is almost impossible to influence this distribution.

The relevance of the search for ways to control the distribution of internal discharge parameters is dictated by practical applications of glow discharge. One of the most popular applications of glow discharge is the application of protective, functional, anticorrosive coatings, in which cathode atoms knocked out by discharge gas ions are deposited on the substrate [1-3]. This process is not fast, but the quality of the coating, if the appropriate conditions are met, can be close to ideal. One of the necessary conditions for providing a high-quality coating is a strong vacuum of the working gas, in which an atom knocked out of the cathode can reach the substrate without colliding with others. On the other hand, the existence of a glow discharge requires the presence of at least five ionizing collisions of electrons in the cathode region. In this case, the condition of self-maintenance of the discharge will be fulfilled: the ions created by one primary electron in the cathode layer, when falling on the cathode, must ensure the output of at least one electron.

Thus, on the one hand, a low pressure of the working gas is required to obtain high-quality coatings, on the other hand, at such a low pressure, a glow discharge is difficult to initiate and maintain. One of the solutions to this problem is the use of a magnetic field in the cathode region. The action of the Lorentz force increases the residence time of an electron near the cathode and, accordingly, the number

Content from this work may be used under the terms of the Creative Commons Attribution 3.0 licence. Any further distribution of this work must maintain attribution to the author(s) and the title of the work, journal citation and DOI.
Published under licence by IOP Publishing Ltd

of collisions of electrons with gas atoms in the cathode region, as a result of which the main condition for maintaining a glow discharge is fulfilled.

An alternative method of solving the problem is to create regions with different concentrations of neutral particles in the interelectrode gap. As it was shown in [4-5], the organization of supersonic gas flow in a limited area of the interelectrode gap makes it possible to control the distribution of discharge parameters between the cathode and the anode. Pumping gas at supersonic speed allows you to create regions with different values of gas concentrations in the interelectrode space and in this way change the distribution of concentrations of electrons and ions, as well as the potential and intensity of the electric field. In [6], an axisymmetric supersonic flow created by an axisymmetric supersonic nozzle with a central body was used for this purpose. One of the electrodes was a central body completely immersed in the supersonic flow, and the second electrode was an axisymmetric ring with the first electrode, which was carried outside the supersonic gas flow. At the same time, an inhomogeneous gas density is created in the interelectrode space.

The aim of this work is to numerically simulate the distribution of parameters of a glow discharge with supersonic gas pumping near the central electrode at various discharge pressures and currents. As the initial distribution of the concentration of gas particles, the results of calculations performed in [6] are accepted.

2. Description of the model

The system of equations for the glow discharge plasma was written in the diffusion-drift approximation, these are k equations for the concentrations of n_k of all kinds of particles under consideration (neutral, excited particles, electrons and ions):

$$\varepsilon_0 \nabla \mathrm{E} = \rho \,, \tag{1}$$

$$\mathrm{E} = -\nabla \mathrm{V}, \tag{2}$$

$$\frac{\partial n_k}{\partial t} + \nabla \Gamma_k = R_k \,, \tag{3}$$

$$\Gamma_{e,i} = Z_{e,i}\mu_{e,i}\, n_{ke,i}\mathrm{E} - D_{e,i}\nabla n_{e,i} \,, \tag{4}$$

$$\Gamma_n = -n\nabla n_n \,, \tag{5}$$

$$\frac{\partial n_\varepsilon}{\partial t} + \Delta\, \Gamma_\varepsilon = R_\varepsilon \,, \tag{6}$$

$$\Gamma_\varepsilon = -\mu_\varepsilon\, n_\varepsilon \mathrm{E} - D_\varepsilon \nabla n_\varepsilon \,, \tag{7}$$

Here ε_0 - is the electric constant, E - is the electric field strength, which is determined by the Poisson equation (2) for the potential V, $\rho = \sum_k Z_k n_k - n_e$, Z_k - is the charge of the particle of grade k, n_k - is the concentration of particles, Γ_k - is the flow of concentrations of particles of grade k ($k = e, i, n$ for charged, excited and neutral particles), Γ_ε - the flow of density electron energies, μ_e, μ_i and D_e, D_i - mobility and diffusion coefficients of electrons and ions, $\mu_\varepsilon = \frac{5}{3}\mu_e$ - "energy" mobility and $D_\varepsilon = \frac{5}{3}D_e$ - energy diffusion coefficient, $\mu_e = \frac{e}{m}\frac{1}{\mathrm{v}_d}$, $\mathrm{v}_d = k\, n(N_2)$ - drift velocity, k - elastic collision constant, $n(N_2)$ - concentration of neutral molecules N_2, $D_e = \mu_e \frac{k_B T}{q} q$ - electron charge, k_B - Boltzmann constant, T - temperature, $n_\varepsilon = n_e \bar{\varepsilon}$ - electron energy density, $\bar{\varepsilon} = \frac{3}{2}T$ - average electron energy, $T_e = \frac{2}{3}\bar{\varepsilon}$ - electron temperature (understood as 2/3 of the average energy of the total ensemble), n - is a single vector directed towards the anode. R_e - the rate of birth and destruction of particles in reactions, R_ε - the loss of electron energy due to inelastic collisions, which are determined by plasma-chemical reactions. If M reactions occur in the plasma, as a result of which the electron density decreases or increases and P inelastic collisions of electrons with neutral particles, then the value can be calculated from the equation R_e

$$R_e = \sum_{j=1}^{M} R_j = \sum_{j=1}^{M} x_j\, k_j N_n n_e\,, \tag{8}$$

where R_j - is the reaction rate j, x_j - is the molar fraction of the target particles for reaction j, k_j - is the reaction rate constant of the corresponding inelastic process involving the electron, N_n - is the total density of neutral particles (concentration of molecules N_2). Energy losses R_ε due to inelastic collisions of electrons and heavy plasma particles are calculated by summing up the energy losses during collisions for all reactions:

$$R_\varepsilon = \sum_{j=1}^{P} x_j\, k_j N_n n_e \Delta\varepsilon_j = \sum_{j=1}^{P} \Delta\varepsilon_j R_j \tag{9}$$

where $\Delta\varepsilon_j$ - is the loss (or gain, if $\Delta\varepsilon_j < 0$) of energy in the reaction j.

Calculations were carried out for nitrogen, 73 plasma-chemical reactions from the works were used [7-9].

3. Boundary conditions

For the Poisson equation, it was believed that the anode is grounded: $Va = 0$, on the cathode $Vc = -V0$, where $V_0 = V_{00} - IR_b$, here V_{00} - is the voltage at the source, R_b - is the ballast resistance, $I = \int_S j\, ndS$ - is the current in the circuit, j the current density.

For metal electrodes, the boundary conditions for the flow of electrons Γ_e, for the energy flux density of electrons Γ_ε, for ions Γ_i, as well as excited and neutral plasma particles are:

$$n\,\Gamma_e\,\Big|_{x=0,L} = \frac{1}{4} v_{t,e} n_e - (1-\alpha)\sum_{i=1}^{N} \gamma\, \Gamma_i n\,, \tag{10}$$

$$n\Gamma_\varepsilon\,\Big|_{x=0,L} = \frac{1}{4} v_{t,e} n_e 2k_B T_e - (1-\alpha)\sum_{i=1}^{N} \gamma\, \Gamma_i n\,, \tag{11}$$

$$n\,\Gamma_i\,\Big|_{x=0,L} = \frac{1}{4} v_{t,i} n_e + \alpha\, \mu_i n_i E n\,, \tag{12}$$

$$n\,\Gamma_n\,\Big|_{x=0,L} = \frac{1}{4} v_{t,n} n_e\,, \tag{13}$$

where $v_{t,e}, v_{t,i}, v_{t,n}$ the average thermal velocities of electrons, ions, excited and neutral particles of the gas plasma $\gamma = 0.2$ are the coefficient of secondary emission of electrons from the cathode, $\alpha = 1$ on the anode, $\alpha = 0$ on the cathode.

The average energy of the emitted electron as a result of the impact of the *k-th* grade of particles:

$$\bar{\varepsilon} = \Delta\varepsilon - 2W_f, \tag{14}$$

$\Delta\varepsilon = 15.6$ eV - ionization energy of ion N_2, W_f - energy Fermi.

4. Calculation results

The system of equations (1)-(7) was solved with boundary conditions (10)-(13) in geometry 1D with axial symmetry. The cathode radius is 1.5 mm, the anode radius is 25 mm (Figure1).

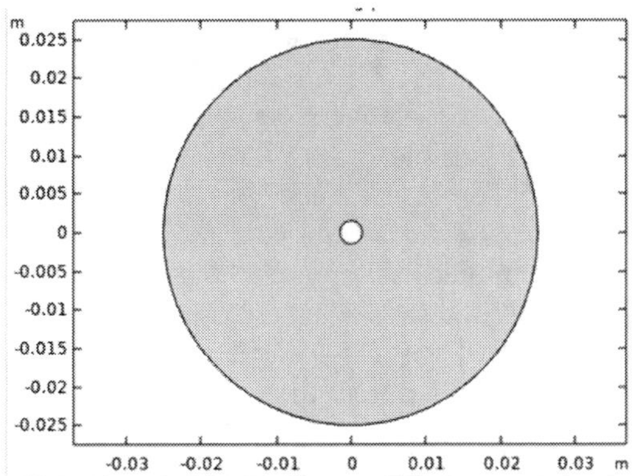

Figure 1. 1D geometry with axial symmetry.

Figure 2 shows the distribution of particle concentrations depending on the distance from the cathode center with a constant concentration of gas particles corresponding to a pressure of p0 = 5 Torr, a power supply voltage of 1000 V, a ballast resistance of Rb =100 kOhm.

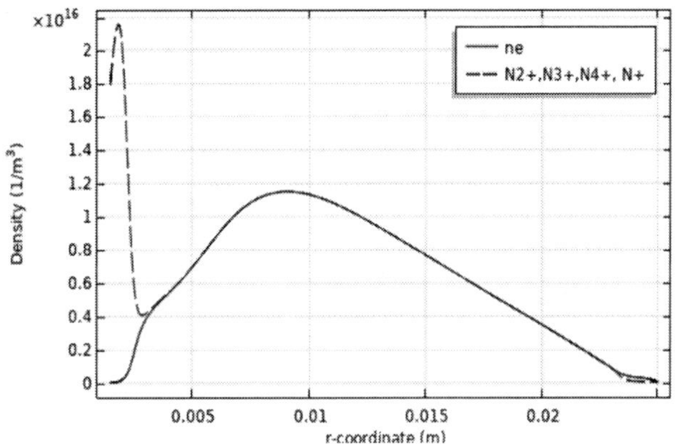

Figure 2. Distribution of electron and positive ion concentrations in the interelectrode space for a constant concentration of neutral particles.

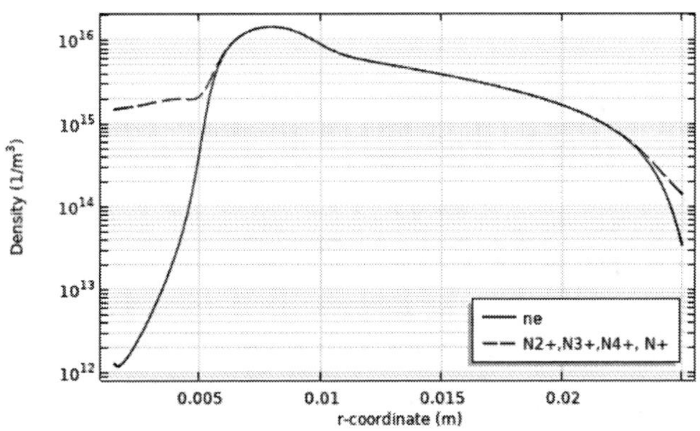

Figure 3. Distribution of electron and positive ion concentrations in the interelectrode space for the case when supersonic gas pumping is organized near the cathode-the central body, which led to a 9-fold increase in the concentration of gas particles.

This is the classical distribution of electron and ion concentrations in the interelectrode gap, corresponding to the axisymmetric geometry of the electrode arrangement. The areas of Astanov dark space, cathode glow, cathode dark space, negative glow, Faraday dark space, positive column and the area of anode potential drop are clearly visible.

When the pressure drops to 0.05 Torr at the same V_0 = 1000 V and R_b = 100 kOhm, there is no discharge. In the presence of a supersonic gas flow near the cathode-the central body, an additional concentration of neutral nitrogen molecules appears. Figure 3 shows the distributions of the concentration of ions and electrons for the case when the central electrode is a cathode and the area of increased concentration of nitrogen molecules is located near it at a distance of up to 10 mm. The concentration of neutral particles is increased by 9 times compared to the rest of the region. It follows from the figure that large concentrations of charged particles are also observed in the area of increased concentration of neutral molecules. When moving to a region with a lower concentration of neutral particles, the concentrations of charged particles also decrease. But this decrease is mainly due to the cylindrical symmetry of the electrode system. At the same time, the total current is preserved, since in this area there are high values of the electric field strength and the cross-sectional area of the discharge area.

In Figure 4a and 4b show the distributions of the concentrations of electrons, positive ions and the electric field strength in the interelectrode space for the case when the central electrode is an anode, and the area of increased concentration of nitrogen molecules is located near it also at a distance of 10 mm (n=9n0).

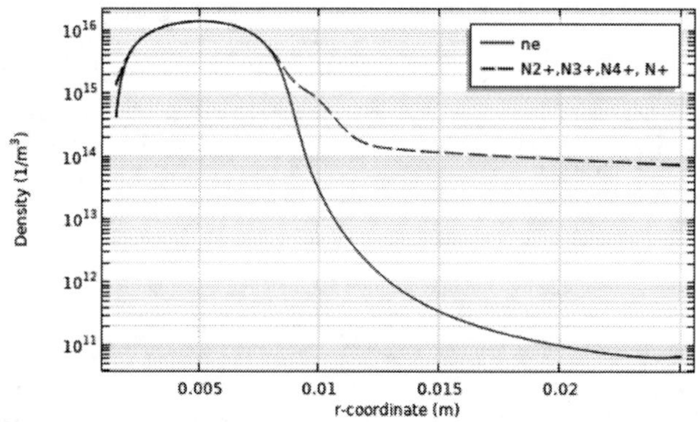

Figure 4. a) Distribution of electron and positive ion concentrations in the interelectrode space.

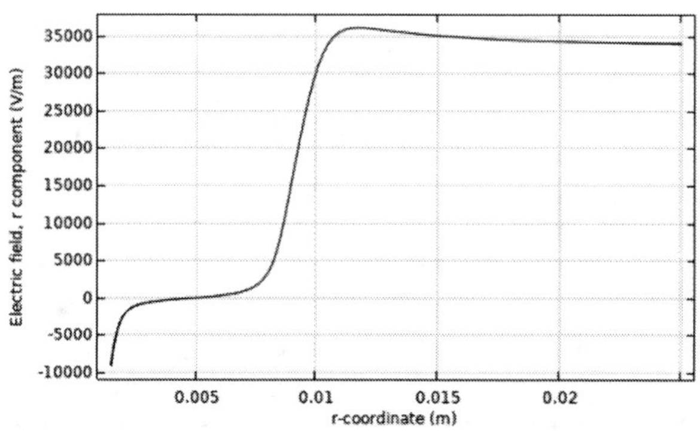

Figure 4. b) Distribution of electric field strength.

As expected, this arrangement of the electrodes and the organization of supersonic gas flow led to the displacement of all near-electrode zones to the anode. There are practically no ionization processes in the area of reduced electron concentration. In this region, the probability of collision of electrons with neutral molecules is extremely low. Therefore, the electrons released from the cathode run through this region freely, gain energy, and, falling into an area with an increased concentration of gas particles, intensively ionize gas molecules. The concentration of both electrons and ions is increasing. The ions are pulled out of this region by the applied electric field and, just like the electrons, reach the cathode without reducing their concentration. Recombination of these ions at the cathode leads to the appearance of new electrons.

5. Conclusion

As can be seen from Figure 4b, the distribution of the electric field also has a specific character. In the region of small concentrations of gas particles, the field is high, which contributes to the rapid collection of energy by electrons sufficient for repeated ionization of molecules by an electron shock.

Conclusions. As can be seen from the above analysis, an increase in the concentration of nitrogen molecules in various regions of the interelectrode space of the glow discharge significantly affects the nature of the discharge. Of particular interest is the discharge with the organization of an increased concentration of gas particles near the central body-the anode. In this case, the cathode region becomes collisionless and this type of discharge can find effective application in coating processes by cathode sputtering without magnetron effect. The proposed method allows you to control the parameters of the glow discharge.

References

[1] Saifutdinov, A. I., & Sofronitskii, A. O. (2021). Numerical Study of Breakdown and Formation Dynamics of Arc Discharge Plasma Parameters at Ultrahigh Pressures. High Energy Chemistry, 55(3), 228-232.DOI: 10.1134/S0018143921030115

[2] Shemakhin, A. Y., Zheltukhin, V. S., Shemakhin, E. Y., Terentev, T. N., & Sofronitsky, A. O. (2020, July). Experimental installation to study the RF plasma flow at low pressures with experiment data synchronization. In Journal of Physics: Conference Series (Vol. 1588, No. 1, p. 012018). IOP Publishing. DOI: 10.1088/1742-6596/1588/1/012018

[3] Fairushin, I. I., Saifutdinov, A. I., & Sofronitskiy, A. O. (2020). Numerical and Experimental Studies of the Synthesis of Copper Nanoparticles in a High-Pressure Discharge. High Energy Chemistry, 54(2), 150-153. DOI: 10.1134/S0018143920020071

[4] B.A. Timerkaev, B.R. Zalyaliev. Glow Discharge in a Transverse Supersonic Gas Flow at Low Pressures. High Temperature, 2014, Vol. 52, No. 4, pp. 471–474.

[5] B.A. Timerkaev, SaifutdinovA.I., B.R. Zalyaliev Control of the Glow Discharge Parameters at Low Pressures by Means of a Transverse Supersonic Gas Flow High Temperature, 2016, Vol. 54, No. 5, pp. 632–638. Original Russian Text © A.I2016, published in Teplofizika Vysokikh Temperatur, 2016, Vol. 54, No. 5, pp. 669–675.

[6] B.A. Timerkaev, R.S. Shamsutdinov, D.I. Israfilov, B.R. Shakirov. Smoldering discharge in a longitudinal supersonic flow. Applied Physics, 2019, №6, 34 – 41.

[7] I. Rafatov, C. Yesil. Transition from homogeneous stationary to oscillating state in planar gas discharge–semiconductor system in nitrogen: Effect of fluid modelling approach. Phys.Plasmas 25, 082107 (2018).

[8] Yu. Lebedev, A.V. Tatarinov. I.L.Epsten. Modeling of the electron microwave discharge in nitrogen. Plasma Sources Sci. Technol. 16 (2007) 726–733.

[9] Yu. Akishev, M. Grushin, V. Karalnik, A. Petryakov and N. Trushkin. On basic processes sustaining constricted glow discharge in longitudinal N2 flow at atmospheric pressure. J. Phys. D: Appl. Phys. 43 (2010) 215202 (18pp).

SPATIAL STRUCTURE OF GAS DYNAMIC CHARACTERISTICS IN A GLOW DISCHARGE WITH A SUPERSONIC AXISYMMETRIC GAS FLOW

B A Timerkaev[1], O A Petrova[1], R S Shamsutdinov[1], A I Sayfutdinov[1]

[1]Kazan National Research Technical University named after A. N. Tupolev -KAI, Kazan, Russian Federation

btimerkaev@gmail.com

Abstract. The paper presents the results of theoretical calculations of the gas-dynamic parameters of a supersonic flow in an axisymmetric Laval nozzle for the case when the central body removed from the nozzle serves as one of the glow discharge electrodes. It is shown that the presence of a supersonic flow makes it possible to establish a stationary inhomogeneous density of gas particles between the cathode and the anode of a glow discharge. The density of gas particles near the cathode and anode may differ by tens of times. The results of the study of the glow pattern of a discharge with a supersonic gas flow are presented. The dependence of the characteristics of the glow discharge on the distribution of inhomogeneous gas density in the interelectrode region is revealed. This effect opens up new possibilities for the use of discharges with supersonic gas flow for nanocoating and synthesis of nanostructures.

1. Introduction

To date, the glow discharge is widely used in gas discharge tubes as a light source, when applying protective, decorative, functional coatings, in lasers, textile, construction, military industries, medicine and other spheres of human activity. Each new application of the glow discharge is associated with new specific features. For use as a light source, it is a high efficiency and a long service life, in lasers it is high power, in medicine it is low voltage. When applying functional surfaces, a strong vacuum of the working gas in the gas discharge chamber is required. At the same time, it is necessary to provide a condition when the atom knocked out of the cathode reaches the substrate without experiencing a single collision. On the other hand, there should be a sufficient number of ionizing collisions of electrons in the cathode region. Usually this problem is solved by applying the magnetron effect, when the Lorentz force acting on the electron keeps it in the cathode region, increasing the probability of ionizing collisions of electrons in this region.

However, there are other ways to solve this problem. One of these is the organization of supersonic gas pumping in the glow discharge area. Gas-dynamic aspects of the interaction of a transverse glow discharge with a supersonic flow in a heat source model based on the Navier-Stokes equations were numerically studied in [1]. Axisymmetric flows are calculated for a uniform oncoming flow. In [2–5], studies of a glow discharge in miniature flows at atmospheric pressure were carried out. Nozzles with a diameter of about 100 μm serve as electrodes in such discharges. Such discharges are called plasma microjets. In [2], a method was proposed for creating microplasma in the atmosphere by controlling a glow discharge with a gas flow emanating from a nozzle. Such an organization of the discharge makes

Content from this work may be used under the terms of the Creative Commons Attribution 3.0 licence. Any further distribution of this work must maintain attribution to the author(s) and the title of the work, journal citation and DOI.
Published under licence by IOP Publishing Ltd

it possible to effectively use the entire volume of the glow discharge plasma for surface treatment. The authors of [3] studied the glow discharge plasma formed as a result of the intersection of two flows of glow discharge plasma in helium. The authors of [4] carried out two-dimensional numerical simulation of such discharges. The characteristics of an atmospheric glow discharge plasma in a complex gas flow consisting of helium surrounded by a flow of nitrogen or a mixture of nitrogen and oxygen were studied in [5]. The external flow in this case shielded the discharge plasma from the atmosphere. The method of influencing the characteristics of a glow discharge using a supersonic flow was also investigated and developed in [6–9].

The organization of supersonic gas pumping made it possible to form regions with different concentrations of neutral particles in the gas discharge chamber, which in turn led to a different spatial distribution of the electrophysical characteristics of the glow discharge.

In the same work, the study of a glow discharge in a supersonic gas flow has been further developed. The main difficulty in organizing the interaction of a glow discharge with a supersonic gas flow is that the discharge tends to bend around the supersonic jet. In contrast to the previously considered methods of organizing gas pumping, in this paper it is proposed to immerse one of the electrodes in a supersonic flow so that it is completely surrounded by it. And the other electrode should be made in the form of a ring around the first one. At the same time, the supersonic gas flow will create an inhomogeneous density on the gap of the glow discharge from the cathode to the anode without the possibility of bending the gas flow by the discharge.

2. Theoretical representation

Before conducting an experiment to study the effect of supersonic gas flow on the spatial distribution of glow discharge zones, a numerical simulation of gas flow through a Laval nozzle with a central body was carried out in the Ansys software package. The simulation allowed us to obtain a theoretical representation of the glow pattern, as well as the distribution of gas density, gas flow velocity and temperature, as well as the Mach number in the discharge chamber.

An ideal gas model was used in the calculations. The system of Navier-Stokes equations was solved together with the energy balance equation. The parameters of the Laval nozzle with a central body were taken as parameters of the nozzle, which was supposed to be used in experiments.

Figure 1 shows the results of calculations of gas flow velocity distributions (Figure 1a) and Mach numbers (Figure 1b) in an infinite axisymmetric space at an inlet pressure of 760 Torus and an initial outlet pressure of 10 Torus. In the core of the supersonic flow, the outflow velocity reaches up to 2 M. The boundaries of the supersonic gas flow are also clearly visible. There is a border region near the central body where there is no supersonic flow.

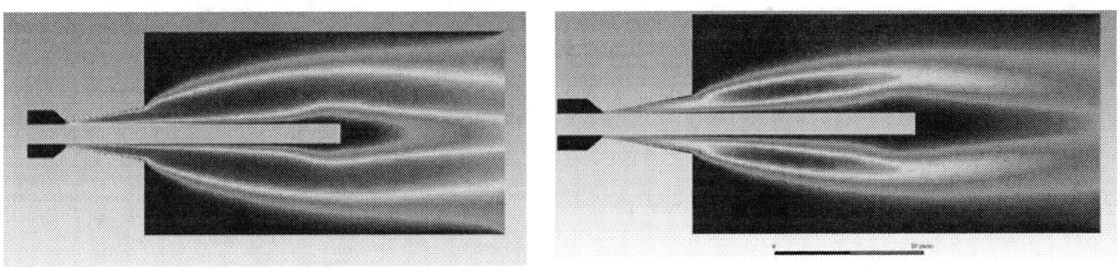

a) distribution of gas flow velocity. b) distribution of Mach numbers.

Figure 1. Results of simulation of gas flow outflow in the Ansys program.

Figure 2 shows graphs of the dependence of the gas flow rate and temperature on the distance from the surface of the cylindrical central body in the plane passing through the anode plane at a distance of 10 mm from the nozzle outlet section. The surface of the cylindrical central cathode body corresponds to the coordinate of 1.5 mm.

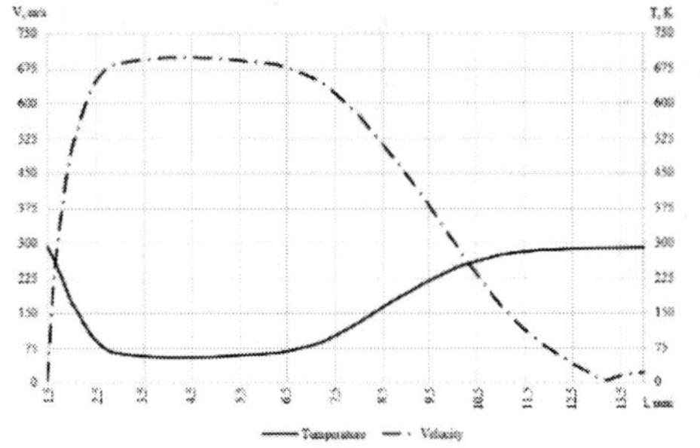

Figure 2. Distribution of gas flow velocity and temperature in the radial direction in the plane of the electrodes.

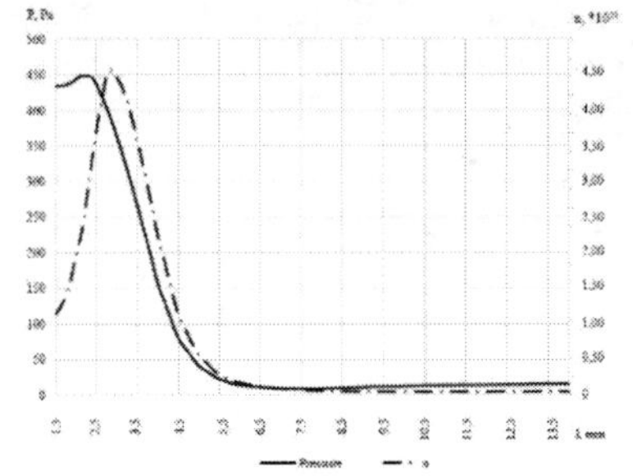

Figure 3. Distribution of the concentration of flow particles and pressure on the slice in the anode plane.

A boundary condition of adhesion was set on the cathode surface, so the flow velocity near the cathode surface drops to zero. The sound velocity is achieved at a distance from the cathode equal to ~ 0.25 mm. The core of the supersonic flow falls at distances from 1 mm to 6 mm from the cathode surface. The flow velocity in this area reaches values of more than Mach 2.

Figure 3 shows the distributions of particle concentration and gas pressure in the same section.

According to the calculations carried out in the Ansys program, it can be seen that the gas flow formed by the Laval nozzle makes it possible to create an inhomogeneous distribution of gas particles in the interelectrode space. In particular, the concentration of particles in the core of the gas stream is about fifty times greater than in the rest of the flow. Such a distribution of particle concentration should lead to compression of the cathode regions. In this regard, it is easier to place the ring electrode in a sparser area of the discharge chamber. If there is a cathode in the rarefied zone, then all the near-electrode regions will have to shift to a zone with a supersonic gas flow near the anode.

3. Experimental setup

The experimental setup (Figure 4) included a vacuum system, a gas supply system, an electric power supply system, as well as measuring instruments and a gas discharge chamber.

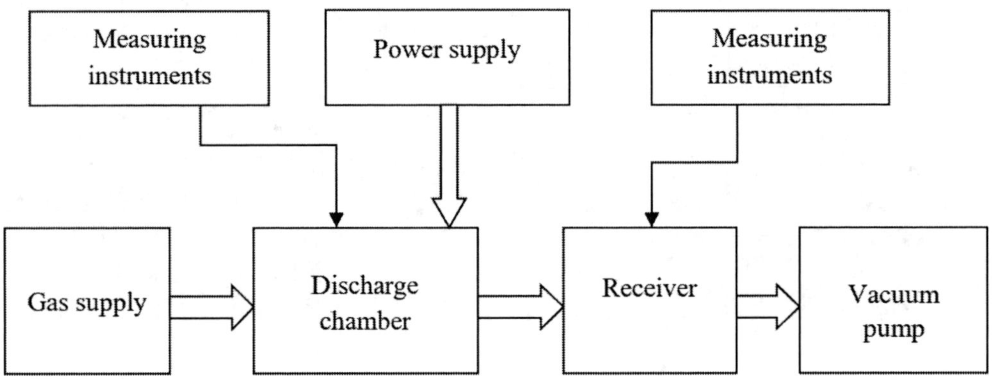

Figure 4. Experimental setup diagram.

The gas discharge chamber is a vacuum dome, into which hoses for pumping and supplying air are connected, as well as two electrical wires for connecting the cathode and anode. Under the dome, the Laval nozzle with a central body is axisymmetrically arranged, which serves as one of the electrodes around which the second ring electrode is located, and the confuser (Figure 5).

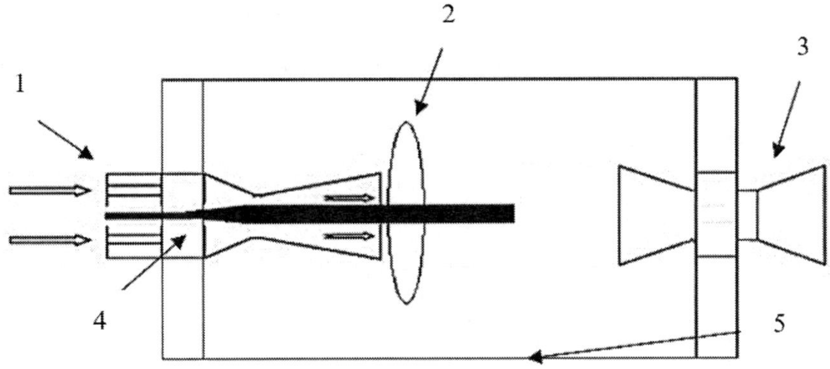

Figure 5. Discharge device with supersonic gas flow. 1 - supersonic nozzle with a central body; 2 - ring electrode-the second electrode; 3 – confuser; 4 - The central body is the first electrode; 5 - Attachment of the nozzle - confuser system.

The diameter of the nozzle in the critical section d1 = 3.5 mm, and in the output section D1 = 8 mm, the diameter of the cylindrical central body throughout is d2 = 3 mm, the area of the critical section S1 = 2.6 mm2, the area of the output section S2 = 50.3 mm2. The length of the nozzle diffuser is l1 = 12 mm, while the central body protrudes beyond the nozzle at a distance of l2 = 30 mm. At a distance of l3 = 10 mm from the outlet section of the nozzle, an annular electrode with a diameter of d3 = 14 mm is located axisymmetrically to the central body.

4. Experiment

First, a discharge was ignited in the chamber, while in the absence of a gas flow, near-electrode zones were observed, as well as a Faraday dark space (Figure 6a, the central body is the cathode). Further, when the gas flow was supplied and a supersonic flow was formed, the near-electrode regions were pressed against the central body (cathode). Along with this, there was an increase in the Faraday dark space, and the positive table was practically absent (Figure 6b).

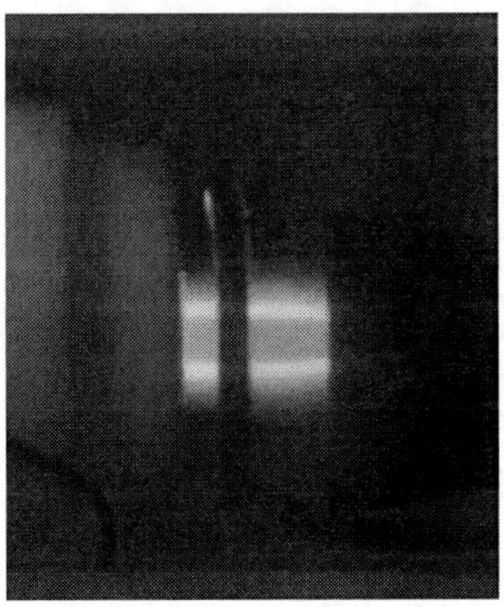

a) discharge in a stationary gas.

b) discharge in a supersonic gas flow. The flow velocity is directed from left to right.

Figure 6. The picture of the discharge glow. U=500 V, I=50 mA

Such a glow is possible provided that a region with a high content of particles injected by a Laval nozzle with a supersonic gas flow is formed near the cathode. On the slice passing through both electrodes, the area with an increased concentration of gas particles occupies about 3 mm. At the same time, the region with supersonic flow velocity extends up to 7 mm from the surface from the central electrode. As a result of the experiment, it was found that the experimental studies are in good agreement with the results of the calculation of the glow discharge and with the discharge glow patterns for the case when the central body is a cathode.

5. Conclusions

Theoretical and experimental studies of the possibility of controlling the structure and parameters of a glow discharge in a supersonic gas flow between the central body of the nozzle and an annular electrode around the central body are carried out. This experiment differs from previous studies in that such a relative position of the electrodes in relation to each other excludes the possibility of the discharge of the gas flow. Due to supersonic gas pumping through the glow discharge region, it was possible to narrow the near-electrode discharge regions. As the discharge current increases, the positive column increases outside the region with supersonic gas pumping. Such a way of organizing the discharge can be used in the processes of applying functional coatings.
Low-temperature plasma can be used for the synthesis of various nanostructures and is well suited for the modification of various surfaces. This is shown in many works [10-12].

References

[1] Gromov V. G., Levin V. A., Ershov A. P., Shibkov V. M. // High Temperature. 2006. Vol. 44. No. 2. P. 178.
[2] Yokoyama T., Hamada Sh., Ibuka Sh., Yasuoka K., Ishii Sh. // Journal of Physics D: Applied Physics. 2005. Vol. 38. P. 1684.
[3] Shirai N., Ibuka Sh., Ishii Sh. // IEEE transactions on plasma science. 2008. Vol. 36. No. 4. P. 960.
[4] Tochikubo F., Shirai N., Uchida S. // Applied Physics Express 4. 2011. Vol. 4. No. 5. P. 056001-1-056001-3.
[5] Tomita K., Urabe K., Shirai N.i, Sato Y., Hassaballa S., Bolouki N., Yoneda M., Shimizu T., Uchino K. // Japanese Journal of Journal of Applied Physics. 2016. Vol. 55. P. 066101-1-066101-5.
[6] B.A. Timerkaev, B.R. Zalyaliev. Glow Discharge in a Transverse Supersonic Gas Flow at Low Pressure. High Temperature, 2014, Vol. 52, No. 4, pp. 471-474.
[7] B.A. Timerkaev, SaifutdinovA.I., B.R. Zalyaliev Control of the Glow Discharge Parameters at Low Pressures by Means of a Transverse Supersonic Gas Flow High Temperature, 2016, Vol. 54, No. 5, pp. 632–638. Original Russian Text © A.I2016, published in Teplofizika Vysokikh Temperatur, 2016, Vol. 54, No. 5, pp. 669–675.
[8] B.A. Timerkaev, R.S. Shamsutdinov, D.I. Israfilov, B.R. Shakirov, Glow discharge in a longitudinal supersonic flow. Applied Physics, 2019, (6), стр. 34–41.
[9] R.S. Shamsutdinov, B.A. Timerkaev. The influence of a supersonic flow of gas at glow discharge. Journal of Physics: Conference Series 1870 (2021) 012019 doi:10.1088/1742-6596/1870/1/012019.
[10] Saifutdinov, A. I., & Sofronitskii, A. O. (2021). Numerical Study of Breakdown and Formation Dynamics of Arc Discharge Plasma Parameters at Ultrahigh Pressures. High Energy Chemistry, 55(3), 228-232.DOI: 10.1134/S0018143921030115
[11] Fairushin, I. I., Saifutdinov, A. I., & Sofronitskiy, A. O. (2020). Numerical and Experimental Studies of the Synthesis of Copper Nanoparticles in a High-Pressure Discharge. High Energy Chemistry, 54(2), 150-153. DOI: 10.1134/S0018143920020071
[12] Sadikov, K. G., Sofronitskiy, A. O., & Larionov, V. M. (2017, November). The effect of electrically conductive additives on the plasma pyrolysis of heavy hydrocarbons. In Journal of Physics: Conference Series (Vol. 927, No. 1, p. 012046). IOP Publishing. DOI: 10.1088/1742-6596/927/1/012046

Journal of Physics: Conference Series **2270** (2022) 012048 doi:10.1088/1742-6596/2270/1/012048

ELECTRIC ARC SYNTHESIS OF SILICON NANOSTRUCTURES

B A Timerkaev[1], V Y Turutin[1], V S Feltsinger[1], A A Zalyalieva[1], R R Vakhitov[1], R S Shamsutdinov[1]

[1]Kazan National Research Technical University named after A. N. Tupolev-KAI, Kazan, Russian Federation

btimerkaev@gmail.com

Abstract. The paper proposes a method for the synthesis of silicon carbide of microscopic sizes and silicon Nano formations in the form of nanotubes and nanofilaments in an argon arc. Graphite electrodes with silicon inclusions were used as arc discharge electrodes. The electrodes were placed in a plasma reactor in an upright position at an argon pressure of 485 Torr. During the discharge, carbon and silicon atoms evaporated from the anode surface. As a result of the experiment, deposits of silicon carbide were formed on the surface of the graphite cathode, and carbon-silicon nanostructures were formed on the surface of the anode. The synthesized samples were analyzed on an electron microscope.

1. Introduction

Silicon nanostructures, such as silicon carbide (SiC), silicon nanofilaments, due to their exceptional hardness, wear resistance and chemical inertia, are widely used in the manufacture of turbine blades, diesel engines, light-emitting devices, field emission displays, nanosensors, as well as abrasive materials. Silicon carbide is mainly produced by the well-known Acheson method. Chemical vapor deposition (CVD) and pulsed laser ablation are also used for the synthesis of silicon nanostructures. These processes have some advantages, such as, for example, simplicity and the possibility of obtaining products in large volumes. But at the same time, there are some limitations, especially in terms of cost, processing time and purity of products. Plasma methods of synthesis of nanomaterials from silicon are proposed as alternative methods in [1-6]. Silicon carbides have already been successfully synthesized in various plasma sources, such as a DC arc embedded in a dielectric liquid [1,2].

In the article [1], for example, the synthesis of silicon carbide was carried out in an arc plasma immersed in fuel oil. Crystalline silicon was placed in a small depression in a graphite anode. In the arc discharge, carbon atoms were released from hydrocarbons and, simultaneously, silicon boiled in the graphite anode hole. Silicon carbide was deposited on the cathode of an electric arc sunk into fuel oil.

In [2], a similar synthesis method is presented with electrodes immersed in a dielectric liquid, but hydrocarbon liquids are taken as the medium. The authors carried out three types of experiments. In the first case, a prolonged spark discharge was ignited in liquid cyclohexane between a silicon anode and a graphite cathode, in the second case - in liquid cyclohexane between two silicon electrodes, in the third case - in liquid tetramethylsilane between two graphite electrodes or two tungsten electrodes. The authors of this work also studied the possibilities of synthesis of nanoscale silicon carbides in pulsed discharges. They managed to find modes in which nanoscale silicon carbides were synthesized. The

Content from this work may be used under the terms of the Creative Commons Attribution 3.0 licence. Any further distribution of this work must maintain attribution to the author(s) and the title of the work, journal citation and DOI.
Published under licence by IOP Publishing Ltd

results obtained showed that pulsed discharges between Si - C and Si – Si electrodes in cyclohexane produce Si nanoparticles with a C shell, but not SIC nanoparticles. However, prolonged spark discharges in tetramethylsilane do produce SIC nanoparticles.

Coleman and co-authors [3] investigated the process of carbonation of silicon nanoparticles to SiC. Depending on the experimental conditions, either hollow or solid morphologies of SIC nanoparticles can be obtained. The authors found that the plasma first heats Si nanoparticles (with a C-shell) and initiates the diffusion of C into Si to generate SiC. Then energy is released (exothermic reaction), which additionally heats the particles, which leads to a complete structural transformation into SiC. This reaction takes about hundreds of milliseconds.

High-frequency induction and microwave plasmas are also worthy of attention for the synthesis of nanoscale silicon carbides.

Experiments with the production of silicon carbides in an argon arc with graphite electrodes were demonstrated in the work of V.I. Podgorny et al. [[4]. The experiment was carried out under the following conditions: the discharge current is 60-75 A, the interelectrode gap is about 0.1-1 mm, the argon pressure is 100-300 Torus, burning time 20–30 minutes

The samples from the work [5] synthesized under conditions similar to those of the argon-arc discharge in the work of V.I. Podgorny et al., are carbon nanotubes and nanodiamonds. The elemental composition of the samples showed that they consist entirely of carbon. The authors suggest that Si acted as a catalyst for the formation of carbon structures, and silicon nanotubes and silicon carbide were transferred higher due to the high temperature and energy released during the synthesis of C-nanostructures.

In [6], the synthesis of ultrafine-grained nanoparticles of amorphous silicon carbide (a-SiC) was carried out by decomposition of hexamethyldisilane in an arc plasma based on a magnetically stabilized sliding arc discharge. The results showed that the synthesized products are a mixture of a-SiC nanoparticles and excess free carbon. The average size of the nanoparticles was about 9 nm, and there were many functional groups and oxidized coatings on the surface of the nanoparticles. Photoluminescent properties showed that a-SiC nanoparticles had strong blue-green radiation. when excited at a wavelength of 340-420 nm, which may be due to surface functional groups, oxidized upper layers and the effect of quantum limitation due to ultrathin dimensions. The correlation between the arc current and the crystal structure of the product showed that the low input energy contributes to the synthesis of a-SiC nanoparticles, while the high input energy contributed to the transition from an amorphous structure to b-SiC.

The electric arc method has its advantages over other methods of obtaining nano- and micro-sized structures of silicon and silicon carbide, due to the coexistence of regions with large temperature gradients. Near the arc column, the temperature can reach ten thousand degrees, and already at a distance of literally a few millimeters, the temperature will be about a thousand degrees. Effective synthesis of silicon nanostructures in an arc plasma does not require the presence of a catalyst, as for carbon nanostructures, and heating to a temperature at which the growth of formations begins occurs in a short period of time (about 100 ms). In this paper, an electric arc method is proposed for the synthesis of silicon nanostructures and the conditions under which some types of carbon-silicon and silicon structures differing in size by several orders of magnitude are created are described.

2. Experimental setup

The experimental setup for the synthesis of silicon nanostructures consisted of a vacuum chamber, power supply systems, vacuuming, and gas supply. Graphite rods with a thickness of about 14 mm were chosen as electrodes. The cathode was pointed at the end, and inclusions of crystalline silicon were placed on the anode. The cathode and anode had a vertical arrangement. During the experiment, it was possible to adjust the interelectrode distance. Before the experiment, the vacuum chamber was pumped out by a vacuum pump, and then purged and filled with argon to a pressure of 485 Torr. To initiate the arc, contact was briefly established between the electrodes and it was broken. At the same time, an arc was ignited

Journal of Physics: Conference Series **2270** (2022) 012048 doi:10.1088/1742-6596/2270/1/012048

between the electrodes. During the experiment, the voltage between the electrodes was maintained at about 26 V at a current of 45 A. The experiment lasted about 1 minute.

3. Experimental results

The samples obtained as a result of the experiment were studied using an electron microscope. Figure 1 shows images of silicon carbide. Their size is about 20-40 microns. Smaller formations of carbides are visible on the faces of some particles. The crystals have clear edges and smooth flat faces.

Figure 1. Silicon carbide crystals. An increase of 300 times.

Figure 2 shows images of nanoscale silicon carbide crystals with a size of about 200-500 nm. In the same figure, silicon nanowires are visible, the length of which reaches 10 microns.

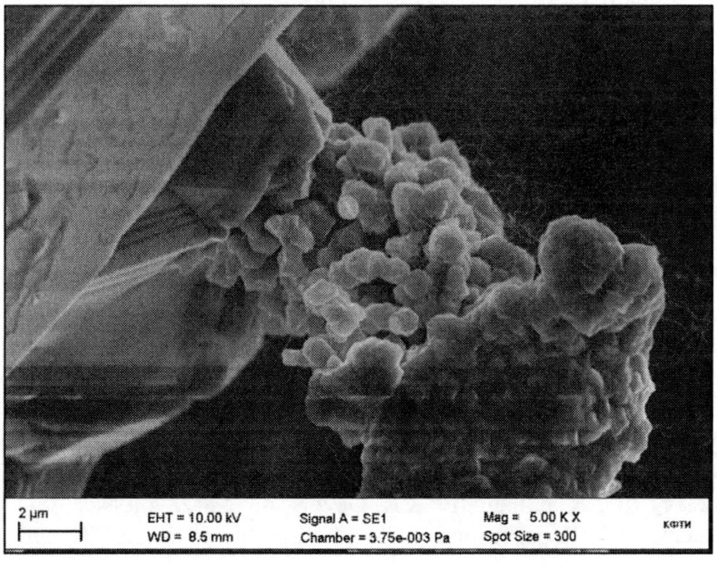

Figure 2. Nanoscale silicon carbide crystals. An increase of 5000 times.

Figure 3 shows silicon nanowires synthesized during the same experiment at a magnification of 5000 times. The diameters of these filaments are about 50-100 nm, and the length is about 10 nm. There are also growths on the surface of the formed filaments, most likely from silicon oxide. The average length of surface neoplasms is approximately 10 nm.

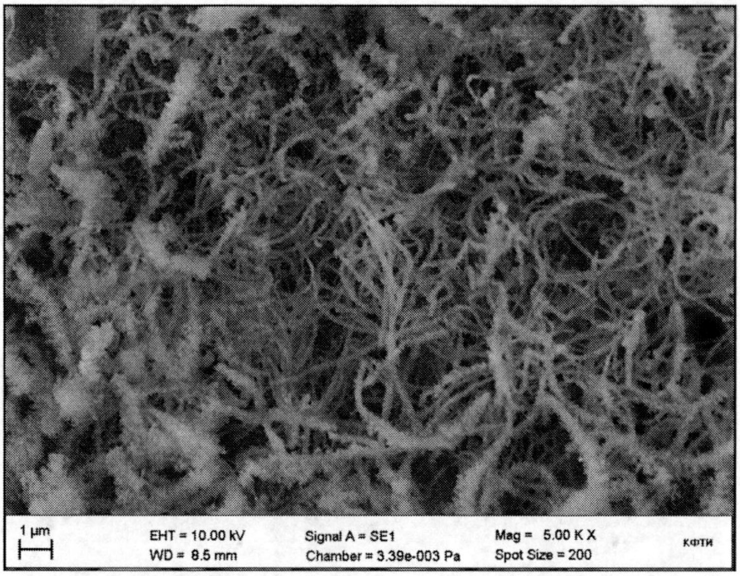

Figure 3. Silicon nanowires. An increase of 5000 times.

The nanowires are indirect, arranged randomly, while filling the entire volume as much as possible. There are many oxidized coatings on the surface of nanowires. The elemental composition of this sample showed, in addition to silicon, the presence of oxygen.

4. Conclusion

In an argon arc at a pressure of 485 Torr, it was possible to synthesize silicon carbide of microscopic dimensions and silicon nanofilaments coated with multiple oxidized nanostructures.
Low-temperature plasma can be used for the synthesis of various nanostructures and is well suited for the modification of various surfaces. This is shown in many works [7-10].

References

[1] B.A. Timerkaev, V. S. Felzinger, A. Akhmetvaleeva et al. Synthesis of silicon carbide in arc discharge in fuel oil. 2021. Journal of Physics: Conference Series. J. Phys.: Conf. Ser. 1870:012005.

[2] Ahmad Hamdan, Douaa El Abiad, Min Suk Cha. Synthesis of Silicon and Silicon Carbide Nanoparticles by Pulsed Electrical Discharges in Dielectric Liquids. 2021. Plasma Chem Plasma Process 41(6). P. 1647-1660

[3] D. Coleman, T. Lopez, O. Yasar-Inceoglu et al. Hollow silicon carbide nanoparticles from a non-thermal plasma process. J. Appl. Phys. 117(19):193301.

[4] V. I. Ppodgorny, B.Z. Belashev, R.N. Osaulenko, A.N. Ternovoy. Obtaining samples of carbides in an arc discharge plasma. // ZhTF. 2013. Vol. 83. Issue 7. pp. 77-81.

[5] B.A. Timerkaev , R.R. Shaikhattarov, R.K. Gevorgyan et al. Synthesis of nanodiamonds and carbon nanotubes in siliconargon arc.2021 J. Phys.: Conf. Ser. 1870:012015.

[6] 6. Jiawen Zhou, Cheng Wang, Ming Song, Xianhui Chen, Weidong Xi. Simple synthesis of ultrafine amorphous silicon carbide nanoparticles by atmospheric plasmas. 2021. Materials Letters. 299:130072.

[7] Saifutdinov, A. I., & Sofronitskii, A. O. (2021). Numerical Study of Breakdown and Formation Dynamics of Arc Discharge Plasma Parameters at Ultrahigh Pressures. High Energy Chemistry, 55(3), 228-232.DOI: 10.1134/S0018143921030115

[8] Shemakhin, A. Y., Zheltukhin, V. S., Shemakhin, E. Y., Terentev, T. N., & Sofronitsky, A. O. (2020, July). Experimental installation to study the RF plasma flow at low pressures with experiment data synchronization. In Journal of Physics: Conference Series (Vol. 1588, No. 1, p. 012018). IOP Publishing. DOI: 10.1088/1742-6596/1588/1/012018

[9] Fairushin, I. I., Saifutdinov, A. I., & Sofronitskiy, A. O. (2020). Numerical and Experimental Studies of the Synthesis of Copper Nanoparticles in a High-Pressure Discharge. High Energy Chemistry, 54(2), 150-153. DOI: 10.1134/S0018143920020071

[10] Sadikov, K. G., Sofronitskiy, A. O., & Larionov, V. M. (2017, November). The effect of electrically conductive additives on the plasma pyrolysis of heavy hydrocarbons. In Journal of Physics: Conference Series (Vol. 927, No. 1, p. 012046). IOP Publishing. DOI: 10.1088/1742-6596/927/1/012046

PRODUCTION OF HYDROGEN FROM HEAVY HYDROCARBONS

B A Timerkaev[1], G R Ganieva[2], B R Shakirov[1], A A Zalyalieva[1], R S Shamsutdinov[1]

[1]Kazan National Research Technical University named after A. N. Tupolev-KAI, Kazan, Russian Federation
[2]Kazan Federal University, Kazan, Russian Federation

btimerkaev@gmail.com

Abstract. The mechanism of the effect of an arc discharge on heavy hydrocarbon raw materials is studied in the work. The mechanism of formation of charged particles in micro-discharges with recessed electrodes is shock ionization. As a result of the interaction of hydrocarbon raw materials with an arc discharge, long hydrocarbon molecules are broken down, light hydrocarbons such as ethylene, hydrogen, and methane are formed. In the case of the organization of a micro-arc discharge, a gentle thermal regime also provides a gentle splitting of long hydrocarbon molecules, and in this case gasoline fractions of oil are formed in greater quantities compared to the case of exposure to raw materials by a conventional arc discharge.

Currently, search and implementation work on the transition to alternative energy sources is underway all over the world. The main goal of these works is the transition to green energy, the transition to renewable energy sources and the reduction of carbon dioxide emissions into the atmosphere. Green energy is mainly focused on solar, wind, and geothermal energy. This can also include the energy of hydroelectric power plants and, to a lesser extent, nuclear power. One of the main directions of green energy is hydrogen energy. Hydrogen energy is technologically quite difficult to implement. But at the same time, hydrogen is an environmentally friendly energy carrier. The most attractive is the production of hydrogen from water, using solar energy at southern latitudes as energy for the separation of water molecules into oxygen and hydrogen.

At the transition stage, the main source of hydrogen production is the steam conversion of methane and natural gas. Scientists have been dealing with this problem for many years. Various methods of conversion of hydrocarbon gas have already been developed. One of these methods is the plasma chemical method. The plasma chemical method of methane conversion to produce hydrogen and the synthesis of carbon nanotubes can be considered absolutely "green" in terms of carbon dioxide emissions into the atmosphere. Moreover, in terms of energy efficiency, the hydrogen obtained in this way, taking into account the higher combustion temperature, is only slightly inferior to the energy of burning methane directly. Carbon nanotubes in this process also belong to the target product. Solar energy can also be used as an energy source for this process. The advantages and problems of hydrogen energy are discussed in detail in [1, 2].

Recently, discharges in gas bubbles have been used to decompose hydrocarbons in order to obtain various target fractions (see, for example, [3-6]). In [4], a method of decomposition of ethanol in a

Content from this work may be used under the terms of the Creative Commons Attribution 3.0 licence. Any further distribution of this work must maintain attribution to the author(s) and the title of the work, journal citation and DOI.
Published under licence by IOP Publishing Ltd

microwave discharge plasma and in an electric arc was applied. The main gas products were hydrogen, acetylene, methane and ethylene, of which up to 80 vol. % was hydrogen. In [5,6], an arc discharge was ignited in a gas bubble in fuel oil. In such a discharge, the authors of the work managed to synthesize carbon nanotubes. The generation of hydrogen and hydrogen-containing components can take place both at the stage of electrolysis and during discharge lit.

In this work, an arc sunk into the fuel oil was used to produce hydrogen from fuel oil and synthesize nanotubes. Experiments were carried out with copper cooled and graphite electrodes. In the course of experimental studies, the current values ranged from 30 to 80A, and the voltage from 20 to 50V, depending on the interelectrode distance. When an electric arc is immersed in fuel oil, the arc burns in a gas bubble. The high temperature of the plasma displaces the liquid from the arc location area and the discharge burns in the vapors of hydrocarbons that have evaporated into this gas bubble from its external boundaries. The hydrocarbon vapors contained in the gas bubble are exposed to high temperatures and are also attacked by plasma electrons. As a result, one hundred percent conversion of hydrocarbons can occur. As gases accumulate in the gas bubble, the bubble expands and rises upward under the influence of the Archimedes force. If the electrodes are immersed in the fuel oil deep enough, the gas bubble breaks away from the main bubble and floats up.

Let's analyze the phenomena that occur during the decomposition of hydrocarbon raw materials by microarc discharge. During the experiment, the heat removal from the arc is carried out due to the natural convection of fuel oil, which, as it were, washes the electrode. In an electric arc embedded in hydrocarbon raw materials, a pressure is set that corresponds to the depth of immersion of the micro-arc discharge and the surface tension force at the boundary of the gas-vapor bubble. Since the depth of the fuel oil is only a few centimeters, then the pressure will be on the order of atmospheric.

The part of the fuel oil that is in direct contact with the arc will be in a boiling state, supplying various oil fractions to the discharge area. Surface boiling and relatively low thermal conductivity of fuel oil prevent overheating of the bulk of fuel oil and its coking. During the experiment, the gas pressure in the chamber (above the liquid surface) is maintained approximately equal to atmospheric pressure.

The features of discharges with recessed electrodes are that they are realized in volumes with boiling liquid walls, and the characteristics and conditions of their existence are determined by the electrical power and parameters of the liquid, both in the liquid state and in the vapor state.

The composition of the gas obtained during the experiment was studied chromatographically on a gas-analyzing chromatograph "Crystal 5000.2". It was found that in the released gases hydrogen is 36%, ethylene is more than 47% and methane is about 7%. Thus, the main gas products of decomposition of fuel oil by microarc discharge are ethylene, hydrogen and methane.

Also, during the experiment, carbon deposits formed on the electrodes, which were analyzed on an electron microscope. It turned out that the deposits contain a large number of carbon nanotubes (fibers) of multilayer and single-layer structure with a characteristic diameter of 9 -15 nm and different lengths of the order of 1 micron (Figure 1).

Thus, the mechanism of the effect of an arc discharge on heavy hydrocarbon raw materials has been studied. The mechanism of formation of charged particles in micro-discharges with recessed electrodes is shock ionization.

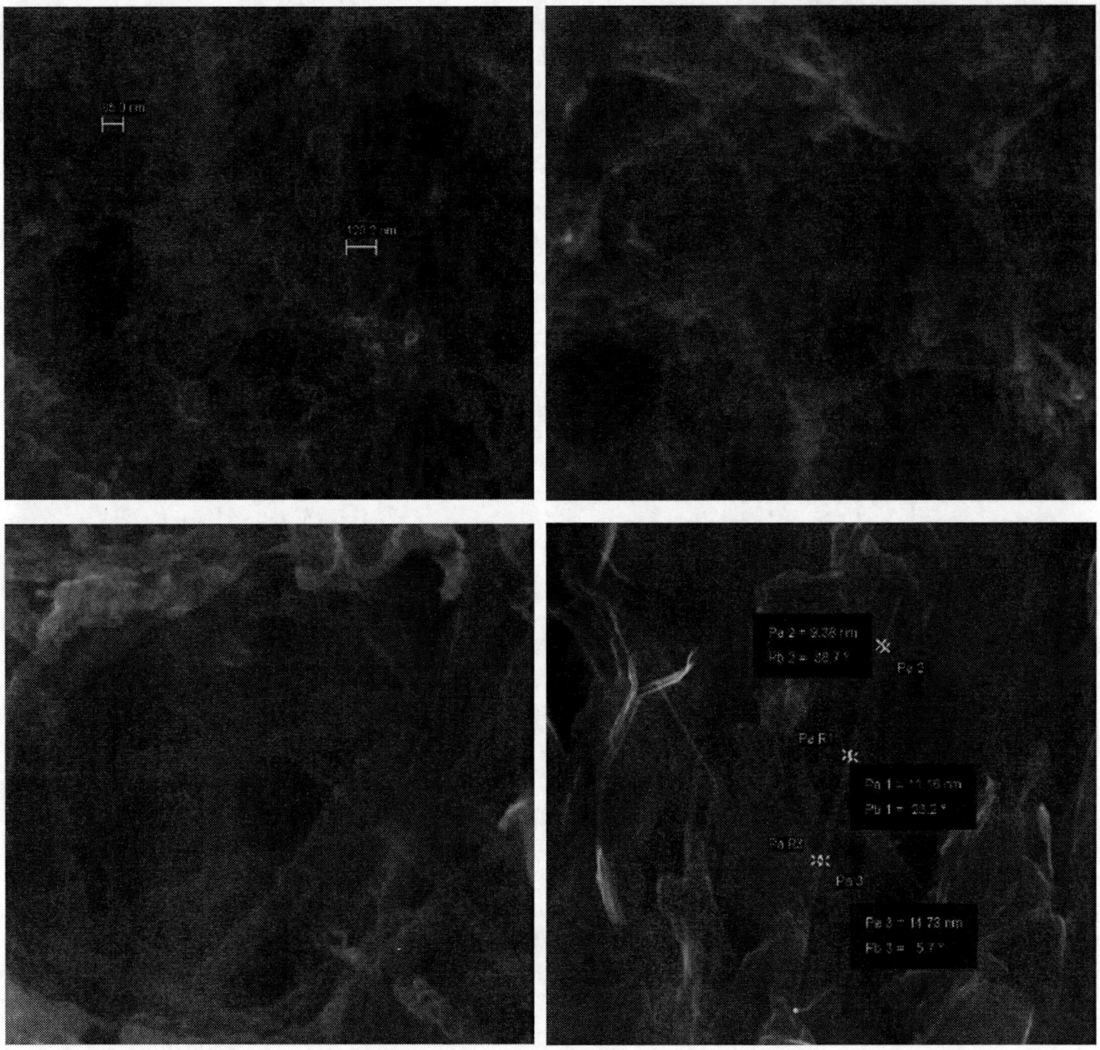

Figure 1. Electron microscopic image of carbon deposits on arc discharge electrodes.

As a result of the interaction of hydrocarbon raw materials with an arc discharge, long hydrocarbon molecules are broken down, light hydrocarbons such as ethylene, hydrogen, and methane are formed. In the case of the organization of a micro-arc discharge, a gentle thermal regime also provides a gentle splitting of long hydrocarbon molecules, and in this case gasoline fractions of oil are formed in greater quantities compared to the case of exposure to raw materials by a conventional arc discharge.

Low-temperature plasma can be used for the synthesis of various nanostructures and is well suited for the modification of various surfaces. This is shown in many works [7-10].

References

[1] V.N. Fateev, V.I.Porembsky, S.A.Grigoriev, et al. Development and research of hydrogen energy systems at the National Research Center "Kurchatov Institute". News of higher educational institutions. ENERGY PROBLEMS. 2021;23(2):128-148.

[2] R.S. Yarullin, I.Z. Salikhov, D.Z. Cherezov, A.R. Nurislamova. Prospects of hydrogen technologies in power engineering and in the chemical industry. News of higher educational institutions. ENERGY PROBLEMS. 2021;23(2):70-83.

[3] S. Samukawa, M.Hori, S. Rauf, K. Tachibana, P. Bruggeman, G. Kroesen, J.C. Whitehead, A.B. Murphy, A.F. Gutsol, S. Starikovskaia, U. Kortshagen, J.C. Boeuf, T.J. Sommerer, M.J. Kushner, U. Czarnetzki, N. Mason // J. Phys. D: Appl. Phys. 2012. V.45. 253001.

[4] K.A. Averina, R.S. Borisov a, Yu.A. Lebedev. Microwave discharges in liquid hydrocarbons: current state. High Energy Chemistry, 2020, T. 54, No. 3, pp. 228-234.

[5] G.R. Ganieva, B.A. Timerkaev. Decomposition of heavy hydrocarbons in argon arc with the sunken electrodes // Journal of Physics: Conference Series 669 (2016) 012061166.

[6] G.R. Ganieva, B.A. Timerkaev, D.I. Ziganchin and M.M. Auhadeev, Electrical microdischarges in liquids and prospects of their application in plasma chemistry/ Journal of Engineering Physics and Thermophysics, 2014. Vol 87, No3, pp 677-681.

[7] Saifutdinov, A. I., & Sofronitskii, A. O. (2021). Numerical Study of Breakdown and Formation Dynamics of Arc Discharge Plasma Parameters at Ultrahigh Pressures. High Energy Chemistry, 55(3), 228-232.DOI: 10.1134/S0018143921030115

[8] Shemakhin, A. Y., Zheltukhin, V. S., Shemakhin, E. Y., Terentev, T. N., & Sofronitsky, A. O. (2020, July). Experimental installation to study the RF plasma flow at low pressures with experiment data synchronization. In Journal of Physics: Conference Series (Vol. 1588, No. 1, p. 012018). IOP Publishing. DOI: 10.1088/1742-6596/1588/1/012018

[9] Fairushin, I. I., Saifutdinov, A. I., & Sofronitskiy, A. O. (2020). Numerical and Experimental Studies of the Synthesis of Copper Nanoparticles in a High-Pressure Discharge. High Energy Chemistry, 54(2), 150-153. DOI: 10.1134/S0018143920020071

[10] Sadikov, K. G., Sofronitskiy, A. O., & Larionov, V. M. (2017, November). The effect of electrically conductive additives on the plasma pyrolysis of heavy hydrocarbons. In Journal of Physics: Conference Series (Vol. 927, No. 1, p. 012046). IOP Publishing. DOI: 10.1088/1742-6596/927/1/012046

Hydrophilization of the surface of a natural hydrophobic material by non-equilibrium low-temperature plazma

G R Rakhmatullina[1], V P Tikhonova[1], D K Nizamova[1], R F Akhverdiyev[1], K F Bigeyeva[1]

[1]Federal State Budgetary Educational Institution of Higher Education "Kazan National Research Technological University", Kazan, Russia

Gulnaz-f@yandex.ru

Abstract. We considered the effect of non-equilibrium low-temperature plasma for hydrophilization the surface of fish skins. Fish skin is a hydrophobic material in its nature. The relevance of this issue is justified by fact that the dressing of animal skin is a series of consecutive liquid processes. Natural structure of fish skins promotes the poor diffusion of working solutions, so it hinders a chemical modification of materials. The plasma processing of fish skin provides the partition of supramolecular structure of collagen (basic protein of dermis) due to disruption of weak intermolecular bonds. This leads to better diffusion of chemical materials and as a result, liquid processes intensify and time production cycle reduces.

1. Introduction

Increasing demand for genuine leather products makes the producers to expand raw material base. Now you can often see on the market the products from exotic kinds of skin (crocodile, ostrich).The uniqueness of genuine leather is indisputable. The scientists have not still been able to create a worthy alternative to this material [1].

To create leather we explored fish skins, which were waste of the food industry. Fish skin is a hydrophobic material in its nature. This significantly impedes the passing of liquid technological processes of leather production. We tried to increase the hydrophilic properties of skin surface by the flux of non-equilibrium low-temperature plasma. Such kind of plasma has been widely considered by many authors as a tool for modification the structure of light industry materials [2].

Materials and methods

For determining effective mode of plasma modification we examined the samples of salmon skins by plant tanning. This method was selected as the most ecofriendly one. As a criterion for plasma exposure we chose the time indicator for soaking water droplet by surface of material.

Modification was carried out at the following parametres: plasma-forming gas – argon, gas consumption -0,04 g/s, pressure-26,6 Pa, power discharge – from 500 to 2000W, processing time – from 3 to 9 minutes.

2. Results

Most of the fish have two layers of dermis: the upper layer is formed by a thin inter-layer of loose connective tissue – it surrounds the scales; the lower layer consists of a dense connective tissue. We did the research on changes in absorbive capacity of the upper and lower layers.

The results of the change of water drop soak time from the upper layer of dermis can be seen in figure1.

Content from this work may be used under the terms of the Creative Commons Attribution 3.0 licence. Any further distribution of this work must maintain attribution to the author(s) and the title of the work, journal citation and DOI.
Published under licence by IOP Publishing Ltd

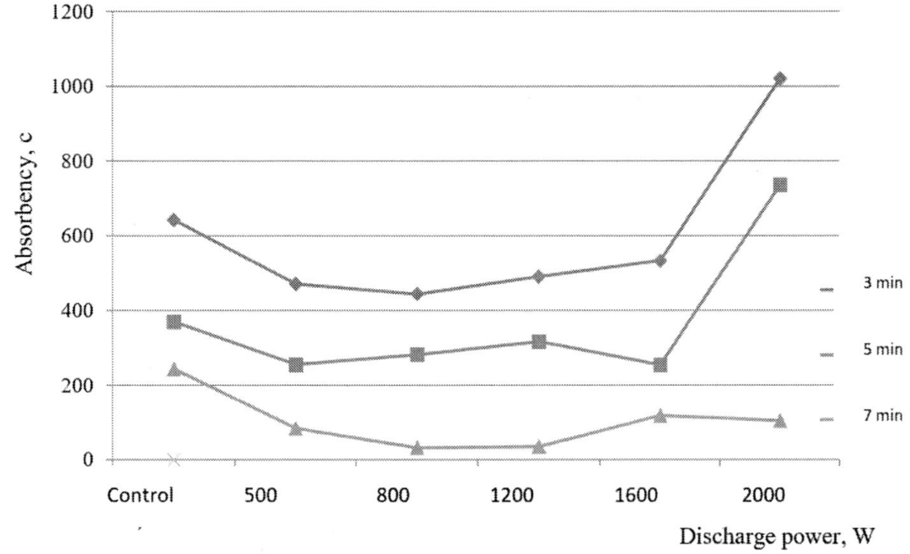

Figure 1 – The change in absorbency of the upper layer of the dermis depending on plasma discharge power and modification time

Figure 1 shows that curves have a similar character. As discharge power Increases to 800 W the soaking time of water droplet decreases and then increases. The best result is fixed on the sample modified by non-equilibrium low-temperature plasma flow at 800W of power discharge for 7 minutes.

The results in change of time absorption of a drop of water from lower layer of dermis are shown in figure 2.

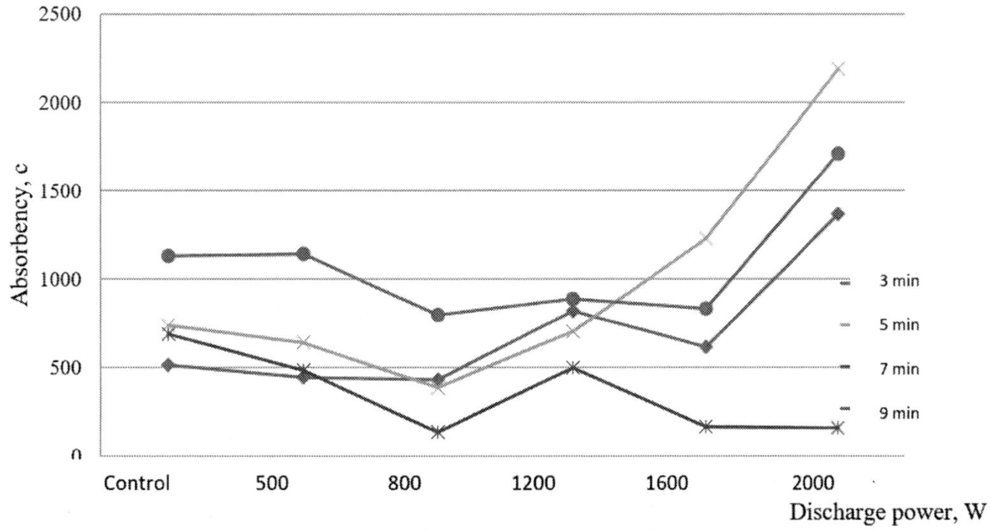

Figure 2 – The change of absorbency from the lower layer of dermis depending on power discharge and time of modification

The curves presented in figure 2 show that the time of water absorption reduces with increasing discharge power and reaches its peak at 800W. The increasing of power leads to deterioration of absorbency. Further increasing of power leads to deterioration of absorbency. Thus the shortest time of water absorbency is observed at a discharge power of 800 W and the time processing for 7 minutes. We would also like to note that the time of absorbency by the upper layer of dermis of the test samples is in the range of 230 – 640 sec. (it depends on skin topography) and the lowest layer of dermis is in the range of 500 – 1100 sec. So the lower layer is more hydrophobic in its nature.

3. Conclusion.
The change of dermis absorbency as a result of plasma modification is connected with separation of supramolecular structure of collagen (basic protein of dermis) due to the violation of weak intermolecular bonds. The increase of dermis hydro-helicity promotes better diffusion of working solutions. As a consequence, liquid processes intensify and production time is reduced.

References
[1] Timoshina YA et al 2020 Application of preliminary plasma modification the surface of synthetic materials in the process of application of functional metal coatings. Journal of Physics: Conference Series. – 2020. – V. 1588 – P. 012053
[2] Rakhmatullina G R 2010 Scientific and technological bases of covering dyeing of leather using non-equilibrium low-temperature plasma Sciences. *Ed. KNRTU*, Kazan 300

Journal of Physics: Conference Series **2270** (2022) 012051 doi:10.1088/1742-6596/2270/1/012051

Application of plasma methods for the synthesis of nanostructures

A A Masyanov[1]

[1]Kazan National Research Technical University named after A.N. Tupolev–KAI, Kazan, Russia

sasha.masyanov@mail.ru

Abstract. There are many plasma methods for the synthesis of nanostructures, for example, the plasma-chemical vapor deposition method, the cathode-arc deposition method, the plasma jet method, the ion sputtering method, etc. Each of the listed methods can be organized in various ways, and each method is suitable for creating certain nanostructure.

1. Introduction

Gas-discharge plasma is widely used in various applications [1–29], including those used for the synthesis of nanostructures. Nanostructures can be synthesized by various methods and methods, however, plasma methods for the synthesis of nanostructures play a special role. Different methods for synthesizing nanostructures use different approaches, some methods use a top-down approach, while others use a bottom-up approach. The top-down approach involves the production of nanostructures by grinding large bodies to nanoscale values. And the "bottom-up" approach involves the assembly of nanostructures from atoms and molecules, i.e. there is an enlargement of objects to nanoscale values. Plasma methods involve the use of gas-discharge plasma to transform a substance into an atomic state. Further, the substance in the atomic form, under certain conditions, participates in the formation of nanostructures.

2. Plasma methods for the synthesis of nanostructures

There are many plasma methods for the synthesis of nanostructures, for example, the plasma-chemical vapor deposition method, the cathode-arc deposition method, the plasma jet method, the ion sputtering method, etc. Each of the listed methods can be organized in various ways, and each method is suitable for creating certain nanostructures.

2.1. Plasma chemical vapor deposition

Plasma chemical vapor deposition can be used to synthesize thin nanosized films. In this case, films are formed from a material in a gaseous state, which passes into a solid state on a substrate. As a rule, these processes take place in sealed chambers at low pressures. The electric discharges used in this process can be direct current, alternating current, and also the HF range. In addition to electrical discharges, microwave plasma can be used. The process can be supplemented with laser radiation. The substrate on which the films are deposited may have a heating element to maintain a certain temperature. There are computerized systems that are capable of applying films with a given composition, respectively, and properties in an automated mode by plasma-chemical vapor deposition,

Content from this work may be used under the terms of the Creative Commons Attribution 3.0 licence. Any further distribution of this work must maintain attribution to the author(s) and the title of the work, journal citation and DOI.
Published under licence by IOP Publishing Ltd

as well as performing experiments and saving operating modes, all this ensures high repeatability of results.

2.2. Cathode-arc deposition
The basis of cathode-arc deposition is the vapor deposition deposition process, in which the evaporation of the cathode material occurs as a result of the action of an electric arc. The process proceeds as follows: first, electrodes and a substrate are installed in a sealed chamber, on which it is planned to deposit the cathode material. Further, air is pumped out of the chamber, the volume of the chamber is ventilated with inert gases, then an inert gas, for example, helium, argon, etc., is pumped into the chamber to the required pressure and an arc discharge is initiated. The sprayed material is deposited in the form of a powder, which, depending on the material and conditions, can be nanoparticles, nanotubes, fullerenes.

2.3. Plasma jet method
As in the case of cathode-arc deposition, the plasma jet method is based on the process of deposition by condensation from the vapor phase. However, depending on the material being deposited, chemical vapor deposition processes can also be involved in the deposition of coatings by this method. This method is also called plasma spraying, its essence lies in the fact that a sprayed material is fed into the plasma jet, which melts when it enters the area of the plasma jet and is directed to the substrate in the form of drops, nanoclusters, individual atoms and molecules.

2.4. Ion sputtering
The method of method of ion sputtering is based on the sputtering of a material as a result of the collision of heavy particles such as ions with it. The target can be bombarded not only with ions, but also with atoms and molecules. Not all incident heavy particles are able to cause the emission of atoms and molecules from the target surface, for this they must have sufficient (threshold) kinetic energy equal to 20-50 eV.

Before the start of the process, air is pumped out of the chamber for the purpose, then an inert gas is pumped into the chamber space to the required pressure. As a rule, with a two-electrode system at a pressure below 0.5 Pa, the glow discharge cannot be ignited, therefore, in this case, this is the lower limit of the pressure value in this process. A voltage of the order of several kilovolts is applied to the electrodes, which exceeds the ionization potential and the breakdown voltage, this leads to the fact that electrons begin to be emitted from the cathode, that is, secondary electron emission occurs.
These electrons escaping from the cathode are directed towards the anode, while on their way they experience inelastic collisions with gas atoms and thereby ionize them. The inert gas ions formed in this way hit the target surface (which is also the cathode) and knock out atoms from its surface.
The use of a more complex four-electrode (triode) system makes it possible to carry out the ion sputtering process at much lower pressures compared to the two-electrode system.

The highest productivity can be achieved using a magnetron sputtering system. A magnetic system consisting of permanent magnets is installed on the reverse side of the cathode. Thus, a closed magnetic field is formed perpendicular to the electric field created between the electrodes on the front side of the cathode. In crossed electric and magnetic fields, electrons are efficiently captured near the surface of the sputtered target. Thus, an electron that has fallen into such a "trap" can only get out of it by colliding with another particle, but even then, it will only switch to a new trajectory, located a little further. Such transitions will continue until the fields are weakened. Those. each electron will make more collisions. Thus, to increase the efficiency of ionization, which will lead to an increase in the density of the ion current, and, consequently, to an increase in the rate of sputtering of the target material. In this case, the pressure at which the spraying process will take place can be less than 0.1 Pa. The ion sputtering process can also be supplemented with laser radiation. The ion sputtering method mainly produces films from a few nanometers to hundreds of micrometers. In addition to films, it is also possible to obtain nanoparticles using a transverse gas flow.

Journal of Physics: Conference Series **2270** (2022) 012051 doi:10.1088/1742-6596/2270/1/012051

3. Conclusion

Various plasma methods were considered in the work. It can be concluded that for the creation of certain nanostructures, it is worth choosing the appropriate methods.

I would like to express my gratitude to A. O. Sofronitskiy for his consultation while writing this paper.

References

[1] Saifutdinov, A. I., & Sofronitskii, A. O. (2021). Numerical Study of Breakdown and Formation Dynamics of Arc Discharge Plasma Parameters at Ultrahigh Pressures. High Energy Chemistry, 55(3), 228-232.DOI: 10.1134/S0018143921030115

[2] Shemakhin, A. Y., Zheltukhin, V. S., Shemakhin, E. Y., Terentev, T. N., & Sofronitsky, A. O. (2020, July). Experimental installation to study the RF plasma flow at low pressures with experiment data synchronization. In Journal of Physics: Conference Series (Vol. 1588, No. 1, p. 012018). IOP Publishing. DOI: 10.1088/1742-6596/1588/1/012018

[3] Saifutdinova, A. A., Sofronitskiy, A. O., Timerkaev, B. A., & Saifutdinov, A. I. (2020). Plasma-Chemical Decomposition of Hydrocarbons on the Basis of the Micro-Arc Discharge with Disc Electrodes Rotating in the Bulk of Raw Materials. Russian Physics Journal, 62(11), 2132-2136.DOI: 10.1007/s11182-020-01957-0

[4] Fairushin, I. I., Saifutdinov, A. I., & Sofronitskiy, A. O. (2020). Numerical and Experimental Studies of the Synthesis of Copper Nanoparticles in a High-Pressure Discharge. High Energy Chemistry, 54(2), 150-153. DOI: 10.1134/S0018143920020071

[5] Asadullin, T. Y., Galeev, I. G., Sofronitskiy, A. O., & Gizeev, M. M. (2019, November). Acoustic impact on electric discharge parameters during sterilization of freeze-dried products. In Journal of Physics: Conference Series (Vol. 1370, No. 1, p. 012016). IOP Publishing. DOI: 10.1088/1742-6596/1370/1/012016

[6] Fairushin, I. I., Saifutdinov, A. I., Sofronitskiy, A. O., Timerkaev, B. A., & Dautov, G. Y. (2019, October). Development of plasma reactor design for synthesis of copper nanoparticles using multi-scale simulation. In Journal of Physics: Conference Series (Vol. 1328, No. 1, p. 012088). IOP Publishing. DOI: 10.1088/1742-6596/1328/1/012088

[7] Asadullin, T. Y., Galeev, I. G., & Sofronitskiy, A. O. (2019, October). The application of pulsed discharge for sterilization of freeze-dried product. In Journal of Physics: Conference Series (Vol. 1328, No. 1, p. 012071). IOP Publishing. DOI: 10.1088/1742-6596/1328/1/012071

[8] Timerkaev, B. A., Ganieva, G. R., Kaleeva, A. A., Israfilov, Z. K., & Sofronitskii, A. O. (2019). Growing of carbon nanotubes from hydrocarbons in an arc plasma. Journal of Engineering Physics and Thermophysics, 92(5), 1248-1252. DOI: 10.1007/s10891-019-02040-3

[9] Dautov, G. Y., Kashapov, N. F., Dautov, I. G., & Sofronitskiy, A. O. (2018, July). Research of the influence of the geometry of the discharge chamber on the characteristics of the arc plasmatron. In Journal of Physics: Conference Series (Vol. 1058, No. 1). IOP Publishing. DOI: 10.1088/1742-6596/1058/1/012035

[10] Timerkaev, B. A., Andreeva, A. A., & Sofronitskiy, A. O. (2017, November). Discharge creeping along the surface in the process for producing nanomaterials. In Journal of Physics: Conference Series (Vol. 927, No. 1, p. 012068). IOP Publishing. DOI: 10.1088/1742-6596/927/1/012068

[11] Sadikov, K. G., Sofronitskiy, A. O., & Larionov, V. M. (2017, November). The effect of electrically conductive additives on the plasma pyrolysis of heavy hydrocarbons. In Journal of Physics: Conference Series (Vol. 927, No. 1, p. 012046). IOP Publishing. DOI: 10.1088/1742-6596/927/1/012046

[12] Timerkaev, B. A., Andreeva, A. A., & Sofronitskiy, A. O. (2017). Discharge creeping along the surface in the process of cleaning and strengthening of the materials surface. In Journal of

Physics: Conference Series (Vol. 789, No. 1). IOP Publishing. DOI: 10.1088/1742-6596/789/1/012063

[13] Sadikov, K. G., Sofronitskiy, A. O., & Dautov, I. G. (2017). Functional plasma sprayed coatings on magnesium ceramic substrates. In Journal of Physics: Conference Series (Vol. 789, No. 1, p. 012043). IOP Publishing. DOI: 10.1088/1742-6596/789/1/012043

[14] Timerkaev, B. A., Sofronitskiy, A. O., & Andreeva, A. A. (2016). Carbon nanotubes formation in the decomposition of heavy hydrocarbons creeping along the surface of the glow discharge. In Journal of Physics: Conference Series (Vol. 669, No. 1, p. 012062). IOP Publishing. DOI: 10.1088/1742-6596/669/1/012062

[15] Timerkaev, B. A., & Shamsutdinov, R. S. (2020, July). The Influence of Supersonic Gas Stream on Spatial Structure of Glow Discharge. In Journal of Physics: Conference Series (Vol. 1588, No. 1, p. 012061). IOP Publishing. DOI 10.1088/1742-6596/1588/1/012061

[16] Galeev I G, Asadullin T Y 2017 Improving of stability of the volumetric glow discharge in the gas flow J. of Phys. Conf. Ser. 789(1) 012012

[17] Asadullin T Y, Galeev I G 2017 Combined method of stabilization of the glow discharge in gas flow J. of Phys. Conf. Ser. 927(1) 012003

[18] Galeev I G, Asadullin T Ya 2016 Obtaining fullerene-containing soot during combustion of gaseous hydrocarbons in an external electric field J. of Phys. Conf. Ser. 669 012016

[19] Galeev I G, Asadullin T Y 2016 Enhancing of the glow discharge stability in chamber with cathode sections coated by a discontinuous dielectric coating J. of Phys. Conf. Ser. 669(1) 012015

[20] Shakirov A S, Galeev I G and Asadullin T Ya 2020 Features of formation of fullerene-containing soot during combustion gaseous and liquid hydrocarbons J. of Phys. Conf. Ser. 1588(1) 012013

[21] Shamsutdinov, R. S., & Timerkaev, B. A. (2021, April). The influence of a supersonic flow of gas at glow discharge. In Journal of Physics: Conference Series (Vol. 1870, No. 1, p. 012019). IOP Publishing. DOI 10.1088/1742-6596/1870/1/012019.

[22] Cheklaukova E.P., Glebov A.N., Shavaleeva S.M., Aleev I.N. Nanomagnetic liquids in energy-saving technologies. Zh. "Bulletin of the Tupolev KSTU". No. 3, 2008, pp. 62-65.

[23] Glebov A.N., Cheklaukova E.P., Ogurtsova V.R., Shavaleeva S.M. Magnetic fluids in energy-saving technologies / Bulletin of the Tatarstan Branch of the Russian Ecological Academy. 2007. No. 2. p. 81.

[24] Shavaleeva S.M., Domracheva Yu.K. Obtaining dichroic coatings on glass by magnetron sputtering. / In the collection: Chemistry and Engineering Ecology - XX. Proceedings of the international scientific conference (school of young scientists) dedicated to the 100th anniversary of the formation of the Tatar ASSR. 2020. pp. 239-241.

[25] Napalkov, O. G., Saifutdinov, A. I., Saifutdinova, A. A., & Timerkaev, B. A. (2021). Simulation of the Carbon Synthesis Process in Atmospheric-Pressure Microwave Discharge in an Argon–Ethanol Gas Mixture. High Energy Chemistry, 55(6), 525-530.

[26] Timerkaev, B. A., Shakirov, B. R., Kaleeva, A. A., & Saifutdinov, A. I. (2021). Arc-Assisted Synthesis of Germanium Nanocrystals in Argon. High Energy Chemistry, 55(5), 402-406.

[27] Timerkaev, B. A., Kaleeva, A. A., Petrova, O. A., Suleymanov, R. M., Sorokina, A. R., Ibragimov, I. S., & Saifutdinov, A. I. (2021, April). Synthesis of semiconductor nanostructures in an argon arc. In Journal of Physics: Conference Series (Vol. 1870, No. 1, p. 012013). IOP Publishing.

[28] Saifutdinova, A., Timerkaev, B., & Saifutdinov, A. (2020, September). Synthesis of nanodiamonds from fuel oil processing products using an arc discharge. In 2020 7th International Congress on Energy Fluxes and Radiation Effects (EFRE) (pp. 867-869). IEEE.

[29] Timerkaev, B. A., Kaleeva, A. A., Timerkaeva, D. B., & Saifutdinov, A. I. (2020). Synthesizing germanium nanotubes in an electric Arc plasma. Russian Journal of Physical Chemistry A, 94(3), 613-617.

First-Principle Molecular Dynamics Study of Methane Hydrate

M B Yunusov*, R M Khusnutdinov

Institute of Physics, Kazan (Volga region) Federal University, Kazan, Russia

mukhammadbek@mail.ru

Abstract. Currently, much attention of the scientific community and the gas industry is focused on the structural, physical and thermodynamic properties of gas hydrates. This interest is explained by the fact that there is a prospect of using natural gas hydrates as a new fuel source. This article presents the results obtained during the first-principle molecular dynamic study of the thermal and electronic properties of hydrates. For hydrates of methane with cubic sI and hexagonal sH configurations, the average heat capacity C_V was computed. The densities of electronic states are studied for filled and unfilled configurations of sI and sH hydrates. The spectra of electron energy were calculated for sI hydrate, which has unfilled molecular cages. Also, for methane and ethane hydrates, the binding energies between the framework and the gas molecule are calculated.

1. Introduction

Great interest to the energy industry are crystalline compounds called hydrates. These are substances of variable composition, the clathrate lattice of which is formed by water molecules. Molecules of low-molecular gases can be enclosed in lattice cavities. Gas hydrates are compounds of variable composition and are described by the formula M*nH2O (M is a hydrate-forming molecule, n is a hydrate number, the average number of H_2O molecules per guest molecule). The hydrate framework is formed by water molecules that are connected by a hydrogen bond. The crystal lattice of water has cavities in which low molecular weight compounds, including methane and ethane, can be enclosed. Hydrates containing natural gases are mainly formed in permafrost zones and on oceanic shelves. Van-der-Waals interactions between the framework and the guest molecule ensure the energy stability of hydrates [1-3]. Gas hydrate stability is insured by repelling the water frame from the internal guest molecules. The configuration stability and electronic features of various gas hydrates were investigated in Ref. [4] by quantum mechanical modeling in the framework of a generalized gradient approximation. The authors have shown that the inclusion of gases (CO_2, CO, CH_4, H_2) in molecular cells stabilizes the structure of hydrate. Also, the authors in Ref. [4] calculated the binding energies between the gas molecule and surrounding molecular cavity. For carbon dioxide, the binding energy in the hydrate was −2.36 eV, for carbon monoxide hydrate −1.57 eV, for methane hydrate −0.58 eV, for hydrogen hydrate −0.36 eV.

The type of hydrate structure depends on several factors: temperature, pressure, type of gas molecule, etc. The cubic configuration sI is the most common structure in nature, next in popularity is sII (cubic) and sH (hexagonal) [2]. The most common molecule of gas in hydrates is methane. It predominantly crystallizes into a cubic sI configuration, which has both large molecular water cells

Content from this work may be used under the terms of the Creative Commons Attribution 3.0 licence. Any further distribution of this work must maintain attribution to the author(s) and the title of the work, journal citation and DOI.
Published under licence by IOP Publishing Ltd

(d=58.6 *nm*) and small water cells (d=51 *nm*). These clusters can contain CH_4 molecules with a diameter of 43.6 *nm* [1, 2]. Such a system usually forms when methane and water come into contact at pressure p~6.0 *atm* and temperature T=300 K [3].

At present, much attention of the scientific community and the gas industry is focused on the structural, physical and thermal properties of gas hydrates. This interest is explained, firstly, by the fact there is a prospect of using natural gas hydrates as a new fuel source. Presumably, there is ~ 10^{16} m^3 of gas in the form of gas hydrates, and it is much larger than the reserves of usual natural gas [1-3]. Secondly, this interest is due to the fact that the hydrate formation problem is acute in the gas industry. Passing through wellbores and gas pipelines, natural gas is deposited on the walls in the form of hydrate and reduces the throughput of pipes.

2. Simulation details

To study hydrates, quantum-mechanical molecular dynamics simulations were performed using density functional theory. All simulation was performed using the VASP software package [5, 6]. Using the data obtained by X-ray diffraction, the algorithm for optimizing the positions of hydrogen atoms, and considering that total dipole moment is equal to zero, the sI and sH crystal structures were reconstructed [7-10]. Models of methane and ethane hydrates were obtained by including CH_4 and C_2H_6 molecules in free molecular cages.

An NVT-ensemble was chosen as the simulated system. The temperature was varied in the range T= [180; 260] K in 20K increments. The Nose–Hoover thermostat was used to establish the state of thermodynamic equilibrium. As mentioned above, the density functional theory was used for calculations. This method uses the electron density as a key quantity instead of the wave function. Interelectronic interaction is calculated using the generalized gradient approximation GGA [5,6] . The PBE functional [7] is used to take into account the exchange-correlation corrections. The interaction of electrons with ions is described using smoothed pseudopotentials (PAW potential). The basis set consisted of plane waves, and a cutoff energy of 245 eV was used. Since the initial configuration of the system was optimized for energy and dipole moment, a small number of ion steps were required for complete relaxation. The duration of the ion step was 1 fs. Within one ion step, at least 5 electronic optimization steps are calculated, with a convergence energy of 10^{-4} eV.

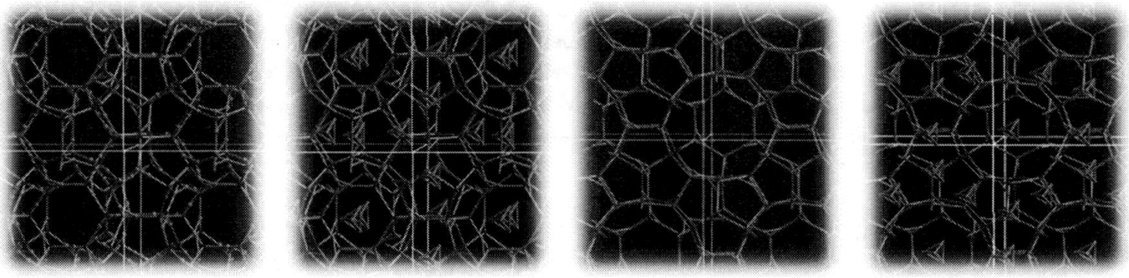

sI sI+CH_4 sH sH+CH_4

Figure 1: Empty and methane-filled hydrate molecular cavities with sI and sH configurations.

3. Results and discussion

First, the average heat capacity of empty clathrate frameworks and methane hydrates was calculated at a constant volume in the temperature range T = [180; 260] K (see table 1). For this, we used an sI cubic cell with a = 12 Å with 8 methane molecules in the cavities (178 ions in total) and an sI cubic cell with empty molecular cavities (138 ions in total). For the hexagonal configuration sH, a rectangular cell was created with parameters a = 12 Å, b = 21 Å, c = 10 Å. For sH, two cases were also studied: the case with the inclusion of 12 methane molecules in the cavity (264 ions in total) and the case of empty molecular cavities (204 ions in total). Also, based on the equilibrium lattice parameters, the methane hydrate densities sI (912 kg/m^3) and sH (898 kg/m^3) were estimated. A good agreement was reached with the known data [8-10].

Table 1: Calculated and experimental values of heat capacity.

	C_V (J/kg·K)		$C_{V,\,EXP}$ (J/kg·K)
	sI	sH	
empty cages	3007±152	3093±12	3300 [21]
CH_4	2362	2569±77	2160±100 [18], 2306 [19]

Second, for the empty configuration sI, the DOS was obtained (see Fig. 2). The obtained DOS spectra agree well with the data given in Ref. [4]. Also at present work, the band structure was obtained (see Fig. 3). The energy band structure gives information about how the energy of an electron will change when moving along a certain direction in the space of k-vectors. As a rule, lines connecting points of high symmetry are considered as distinguished directions of the k-vector. The sI configuration has a cubic lattice, and the cubic Brillouin zone was chosen for it. As can be seen from Fig. 3, electrons with energy –20 ÷ 0 eV (lower energy levels and valence band) have a direct form relationship $E(\mathbf{k})$. The conduction band spectra (energy range 0 ÷ 5 eV) has extremums at the different points. The upper bound of the band gap has a maximum at point R and a minimum at point W. The calculated band gap in the sI crystal is about 5.0 eV. This suggests that hydrates have strong dielectric properties.

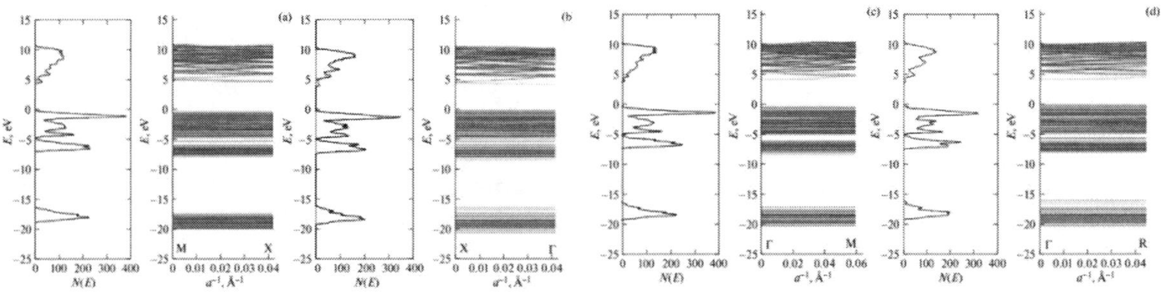

Figure 2: Density of states $N(E)$ and the band structure $E(\mathbf{k})$ of sI unfilled configuration.

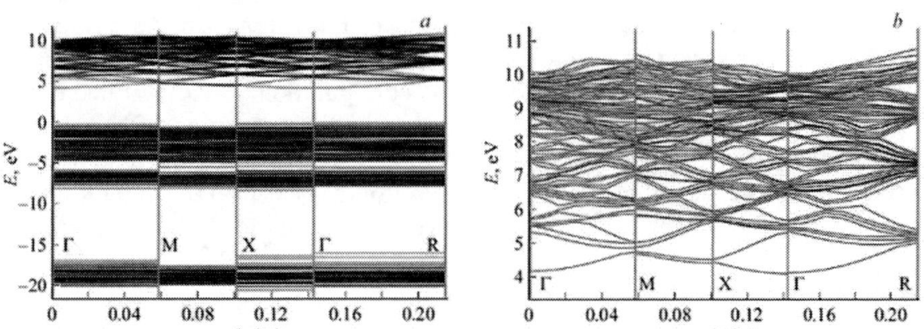

Figure 3: Complete energy band structure of sI crystal (a) and band structure in the energy range 4÷11eV (conduction band) in different directions.

Third, for hydrates with the sI configuration containing methane and ethane, the bond energies between the guest molecule and the surrounding molecular cavity were calculated. The methane molecule fills all 8 cavities in the sI structure, and the binding energy calculated for it was –0.25 eV. This value differs from the value obtained in Ref. [4], which is equal to –0.58 eV. The reason for the differences may be that different potentials and cutoff energies are used. And also in that different lattice parameters are used, in this work the value a = 12.03 Å is used, while in Ref. [4] the value a = 11.62 Å is used. The ethane molecule fills only 6 large cavities in the sI configuration, and the calculated binding energy for it was –0.27 eV. As mentioned earlier, the presence of gas molecules in the cavities stabilizes the frame, reducing the total energy of the cell. This is confirmed by the negative value of the binding energy of the gas molecule in the hydrate. When comparing the DOS $N(E)$ for the unfilled and filled hydrate frameworks (see Fig. 4), in the second case, an electron density peak is detected at an energy of -10 eV. The observed shift of the electron density to the low–energy region when gas molecules are included in the cavity indicates a decrease in the cell energy, which leads to the stabilization of the gas hydrate.

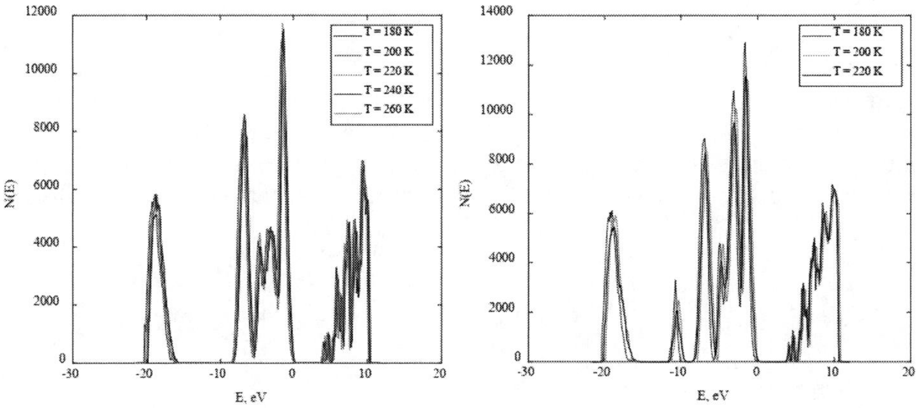

Figure 4: Density of electronic states *N(E)* of hydrates with empty cavities (left column) and cavities filled with methane (right column).

4. Conclusions

This paper presents the quantum-mechanical simulations results of the hydrates for a wide temperature range T=[180;260] K. This temperature range corresponds to the phase of the stable state of gas hydrates in nature. We have carried out detailed studies of the electronic and thermal properties of various crystalline forms of methane and ethane hydrates. First, for methane hydrates and empty configurations sI, sH, good agreement was found between the computed data for heat capacity C_V and density and experimental data [4, 8-10]. Second, for the configuration sI with unfilled cages, the DOS and the energy band structure are calculated. Third, the binding energies of methane and ethane molecules with the hydrate lattice sI were calculated; the values were –0.25 eV and –0.27 eV, respectively. Based on a comparison of the density of electronic states of filled and unfilled hydrates of CH_4, it is confirmed that the inclusion of these molecules in the cavity leads to energy gain.

Acknowledgments
The computational part of the paper was supported by the Russian Science Foundation (project no. 22-22-00508). This article is also partially supported by the Kazan Federal University Strategic Academic Leadership Program (PRIORITY-2030).

References

[1] Sloan E D, Koh C A 2007 *Clathrate Hydrates of Natural Gases* (CRC Press, Taylor&Francis USA)

[2] Yunusov M B, Khusnutdinoff R M, Mokshin A V 2021 *Phys. Solid State* **63** 372
[3] Khusnutdinoff R M 2017 *Colloid J.* **79** 152
[4] Casco M E, Silvestre-Albero J, Ramirez-Cuesta A J 2015 *Nature Commun.* **6** 6432
[5] Kresse G, Hafner J 1994 *J. Phys.: Condens. Matt.* **6** 8245
[6] Kresse G, Joubert D 1999 *Phys. Rev.* **59** , 1758
[7] Perdew J P, Burke K, Ernzerhof M 1996 *Phys. Rev. Lett.* **77** 3865
[8] Makogon Yu F 2010 *Geology And Mineral Resources Of World Ocean* №2 5
[9] Istomin V I, Yakushev V S 1992 *Natural gas hydrates* (Nedra. M.)
[10] Gabitto J F, Tsouris C 2010 *Journal of Thermodynamics* **2010** 12

Experimental and computational study of arc discharges in powerful alternate arc heater

V V Mironov[1], A S Kozaev[1], A S Negodyaev[1*]

[1]Joint stock company state scientific center of the Russian Federation «Keldysh Research Center», Moscow, Russia

*negoandr@gmail.com

Abstract. The paper deals with the results of computational and experimental research on the formation process of electric arc discharges in a three-phase Zvezda-type alternating current arc heater. A flow modelling in the arc heater, as well as an experiment on providing a visual control for the location and the shape of arc discharges were conducted. A device of a visual control system is presented. The analysis of the results obtained from experimental research is carried out.

1. Introduction

Alternate current arc heaters are a fairly versatile tool for conducting various high-temperature tests. For example, the oil and gas sector are highly interested in the field of plasma application for recycling, waste disposal and hydrocarbon treatment in order to obtain non-hazardous substances in the output [1]. At the same time, certain types of unstable operation of arc heaters are quite often manifested in experiments, which is largely due to the behavior of electric arcs and the effect of shunting.

For the effective use of plasma torches, one of the key factors is service life (the overhaul period). A shunting process, the interaction between an electric arc and elements of a gas path not provided by design or by operating mode, has a great influence on service life. The shunting process is a rather negative factor because current and voltage characteristics of the electric arc heater and a gas heat rate change due to shunting occurrence. Apart from that, purity of an outgoing flow of a working fluid can be disrupted, and the erosion of metallic electrodes rises significantly.

The objective of this paper is a study of specific features on the formation of arc discharges and their shunting in the plasmatron of unique design Zvezda developed and used in the Keldysh Research Center [2].

2. Shunting process

The most distinctive process, namely a shunting of an arc, cannot remain unnoticed when studying electric and physical processes occurring in a gas path of the plasmatron. One can say that a shunting process is a specific case of electric breakdown. It was investigated in works of many authors, for example, in the reference [3]. There are several types of shunting. In fact, they can be divided into a large-scale shunting process and a small-scale one. The scheme of shunting processes is shown in Figure 1.

Content from this work may be used under the terms of the Creative Commons Attribution 3.0 licence. Any further distribution of this work must maintain attribution to the author(s) and the title of the work, journal citation and DOI.
Published under licence by IOP Publishing Ltd

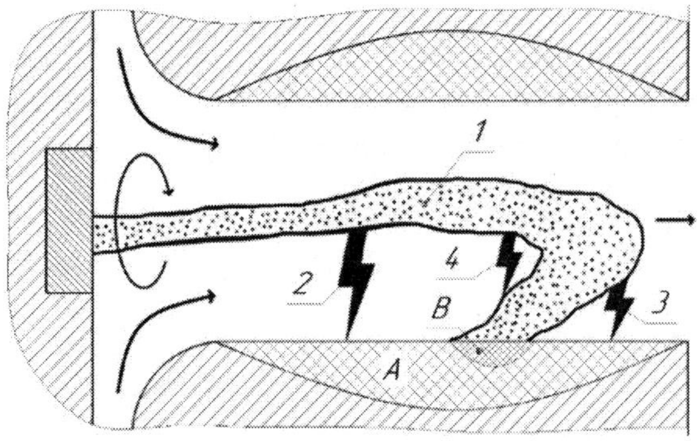

Fig. 1. Scheme of a large-scale shunting process and a small-scale one.

If a large-scale shunting process, electric breakdown (2) occurs between the arc (1) located on the electrode axis and its wall. A new arc section appears and it is moved by a working fluid flow towards the output section of the heater. The shunted section of a discharge gradually fades. The described process is periodic. A large-scale shunting process regulates an average length of arc column and an average drop of the potential on it, a destroyed zone of a wall surface of the electrode-chamber (zone A), pulse or other characteristics of the arc and of the flow.

A small-scale shunting process (3) between arc-electrode occurring in gas layers that are close to chamber walls mainly determines a specific erosion of the material of an electrode. Electric breakdown (4) between arc –arc happens if a small-scale shunting process. It occurs in a hinge of the arc and it indirectly affects the rate of erosion of an electrode. The reason for this is that while shunting between arc-arc, a spot of the arc does not move, warming under it the surface of an electrode-chamber (zone B).

This paper deals with interactions between arc discharge and design elements such as a mixing chamber and confusers, which is a small-scale shunting process within the frames of the accepted classification.

3. Special features of the Zvezda-type arc heater

A scheme of the Zvezda-type alternating current electric arc heater [2, 4] is shown in Figure 2.

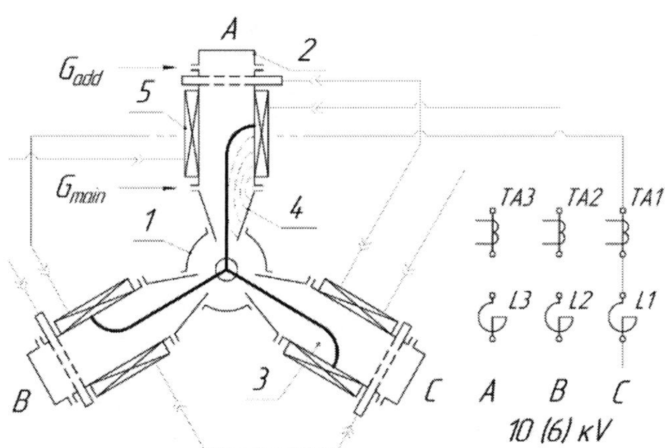

Fig. 2. Power supply system of the Zvezda-type arc heater.

The arc heater consists of 3 identical arc chambers arranged under the $2\pi/3$ angle to each other and a common mixing (damper) chamber (1). An arc chamber consists of a backplate (2), an electrode-chamber (3) and a compression channel (confuser) (4) in which the intensification of gas heating is performed by means of a gas dynamic compression of the arc due to the radial velocity component. To increase erosion resistance of electrodes, the rotation of arc spots interacting with the surface of the electric arc chamber is used. For this reason, each electrode is equipped with magnetic coils (5) each of which is included into the circuit of one of phases. These coils provide coincidence by current phase in an arc and by magnetic field strength. The phases of supply network (*A, B and C*) are connected to electrodes. The induction coil (*L1-L3*) is included in each phase in sequence with the arc; moreover, it is possible to change a value of induction with the help of switching in order to ensure stable burning of each arc and to have opportunity in changing a value of the arc current. Figure 2 illustrates measuring current transformers (*TA1-TA3*) with the help of which a control for arc currents is performed. Electrode-chambers are separated from a conical compression channel by an electric insulator (a vortex generator) each of which is equipped with small nozzles through which the tangential feed of the working gas is realized. The main working gas flow rate G_{main} is supplied between a confuser and an electrode-chamber via insulators. The remaining part of gas flow rate G_{add} (10-30%) is supplied from a backplate side. A mixing chamber (1) has an output nozzle whose axis is perpendicular to the plane of the figure.

Arcs` ignition process in the investigated electric arc heater is initiated by electric potential difference between an electric arc chamber and a compression channel. At that moment, the arcs interact according to the Zvezda (star) scheme with a zero point on the metal because compression channels are electrically connected between each other. Consequently, both electric arc chambers and confusers are eroded. After passing through compression channels, the arcs are pulled to the mixing chamber and they converge in a zero point in the plasma.

4. Calculations using engineering methods

In accordance with the objective set before the experiment, an operating mode of the plasmatron was calculated. The determination of operating parameters was conducted using engineering methods based on empirical dependence that showed its reliability and satisfactory accuracy on the Zvezda-type arc heater operating on air as a working fluid. The following parameters were determined:

1. Working fluid mass flow rate and pressure were calculated by the formula (in accordance with the reference [5]:

$$G = \dot{m} = \frac{p_{st} F_{cr} A(k)}{\sqrt{RT}} \tag{1}$$

where $A(k) = \left(k\left(\frac{2}{k+1}\right)^{\frac{k+1}{k-1}}\right)^{1/2}$, $\dot{m}$ is air mass flow rate, p_{st} is stagnation pressure at the nozzle, F_{cr} is a nozzle throat area, k is a ratio of specific heat, R and T are gas constant and gas temperature.

2. Gas temperature in output section (in accordance with the reference [2]:

$$T = 2.6 \cdot 10^3 \cdot \frac{I^2 \cdot p \cdot d_{av}}{G_1^2}^{0.095} \tag{2}$$

where G_1 is a mass flow rate via an electrode $G_1 = G/3$, d_{av} is an average value between the diameter of an electrode and the diameter of a confuser`s output section, I is arc current. The value $d_{av} = 4.13$ cm for the Zvezda-type arc heater considered in this paper.

Table 1 illustrates the calculated gas dynamic parameters of the chosen mode in accordance with the above formulas.

Table 1. Calculated gas dynamic parameters.

G [kg/s]	p [MPa]	I [A]	T [K]	k	$A(k)$	ν [g/mol]	R [J/K·kg]
0.21	1.14	280	3919.74	1.21	0.65	27.03	307.56

It is necessary to determine the shape and the location of arc discharges relative to a gas path of the arc heater in order to implement a numerical modelling of the plasmatron. Some assumptions were taken to simplify calculations, namely arc discharges were presented as a static one-piece construction in the shape of needles, which are converged at one point laying in the central plane of electrodes and in the geometric center of the mixing chamber. The tip of a needle in these assumptions determines the location of an anchor spot of arc discharge. Therefore, it is necessary to determine physical parameters of needles for setting up a problem in a Flow Simulation Solver of Solid Works Software. The diameter of an arc channel was determined by a method described in the reference [2]:

$$d = 0{,}08\frac{I^{0,6}}{H^{0,2}p^{0,2}} \quad (3)$$

where I is current force, H is magnetic field strength and p is pressure, $H = \frac{I*w}{L}$, where w is a number of turns in a coil, L is a coil length. The diameter of an arc channel was so equal to 0.28 cm.

It is necessary to determine the location of anchor spots of an arc in order to identify the geometry of arc discharges. The length of arc discharges was estimated on the base of reports for the detection of defects of the plasmatron that present data about an average location of the anchor spot of the arc while working with operating modes similar to the calculation mode. Therefore, the final length of arc discharge was 461.9 mm.

Thermal power of AC arc was estimated by the following formula:

$$P = I \cdot U \cdot \cos\varphi \quad (4)$$

where U is voltage in arc discharges (according to the reference [2]): $U = 1.84 \cdot 10^3 \cdot \left(\frac{G_1 \cdot p \cdot d_{\text{cp}}}{I}\right)^{1/3}$, φ is an angle of phases shift. The values of radiative and convective heat flux were deducted from the obtained value of a thermal power to a cooling jacket of the arc heater, which were separately calculated under the conditions of a thermal balance.

5. Experimental research

The achievement of the task set in the part of experimental research is related to the necessity of visual observation of special features of the formation of electric arc discharges in a chamber, as well as the analysis of compliance of the special features to classical conceptions. To ensure opportunity for observation, the Zvezda-type plasmatron was modified in terms of construction from the backplate side of the mixing chamber. A pressure measuring node has undergone changes, which was redesigned to a compartment for a video camera. The scheme of the node is shown in Figure 3.

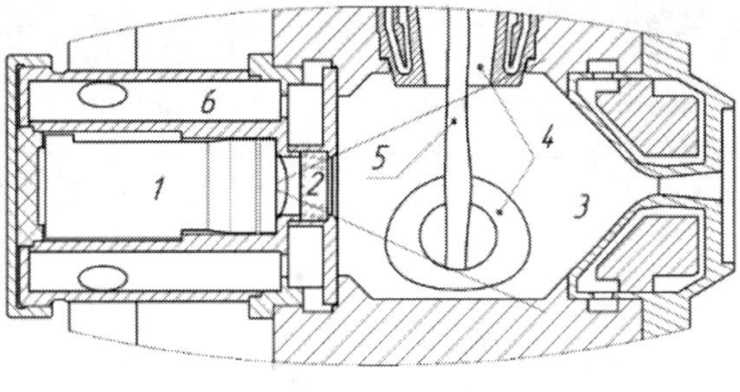

Fig. 3. Scheme of the visual observation node.

The Air Pro 3 HD Sports video camera (1) is a basis core of a visual observation system. It has the following characteristics: 12 MP resolution matrix, 120-Hz frame frequency at the resolution of 1280x720, aperture is f/2.8, observation angle is 160°. The video camera is located in such a way that its axis coincides with the axis of the mixing chamber of the arc heater (3) and that its axis traverses a geometrical point of connection of axes of three electrodes. Quartz glass (2) protects the Air Pro camera against pressure and high temperatures. An optical light filter protects the camera against electromagnetic radiation from electric arc discharges. Arc discharges (5) passing through compression channels (4) are connected in the mixing chamber. Cooling jackets (6) ensure the operational performance of the system. The video camera (1) showed a good stability at all operating modes of the plasmatron under the conditions of strong electromagnetic fields.

6. Results obtained and their analysis

The processing and the analysis of experimentally obtained data showed that a theoretical calculation with the use of engineering methods was correctly fulfilled. The difference in calculated and measured values is in the orders of error. The analysis of a video record showed that this mode does not ensure a stable nominal operation of the plasmatron because the pictures of convergence of arcs to each other are periodically replaced by the pictures of divergence of arc discharges in the mixing chamber by the angles from 10,75 ° up to 45,17 ° (Fig. 4). The angle of an average deviation of core flows with arc discharges was equal to 33,85 °. The developed method showed with a high accuracy (within the range of error) the coincidence of calculated values of gas dynamic processes that occur in the Zvezda-type electric arc heater with experimental parameters. Consequently, one can say that the developed method and calculations performed with it describes processes occurring in the arc heater with sufficient accuracy. Additionally, one can make a conclusion about the correct approach to a computer modelling of the flow.

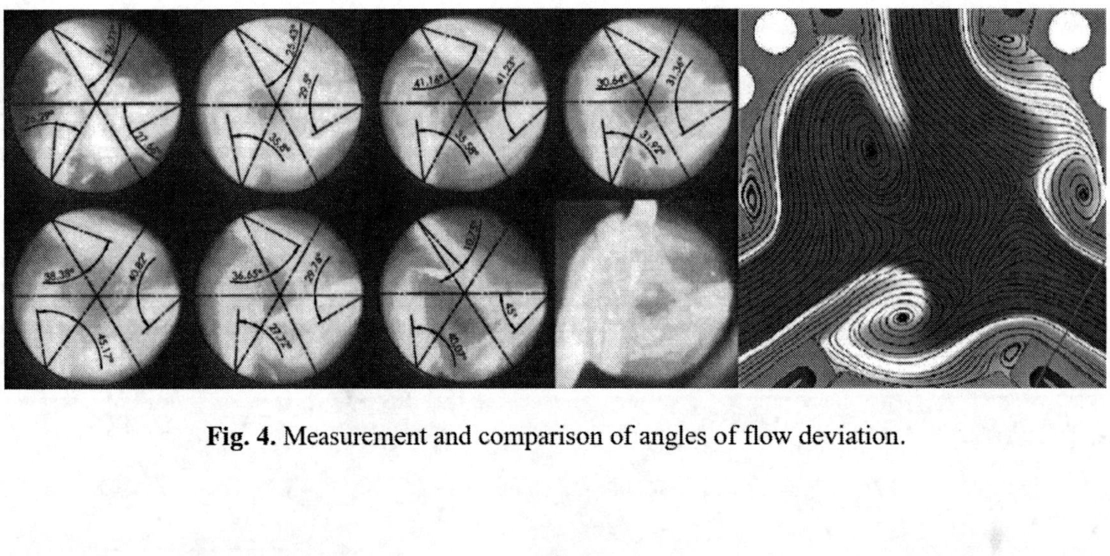

Fig. 4. Measurement and comparison of angles of flow deviation.

7. Conclusion

In conclusion, it is worth organizing the results obtained within the framework of conducted research, which concerns one of the most important characteristics (in terms of industrial use) of the electric arc heater in question, namely service life (the overhaul period). Service life in its turn depends on peculiarities of the interaction between an arc discharge and elements of the plasmatron design. The presence and intensity of such interaction in the mixing chamber zone and the exit of compression channels (confusers) were investigated in the present paper. A successful attempt to organize a visual observation system over a process of the formation and of the existence of arc discharges was made. Based on the results of the work done, the following conclusions were made:

1. A method for calculating main gas dynamic parameters of a gas path of the arc heater was developed on the base of a commercial software complex.

2. The view and the location of arc discharges differ from the classical conception of the presence of straight-lined cores that are converged at one point of the geometric center of intersection of the electrodes and the mixing chamber axes.

3. The picture illustrating shapes and the location of arc discharges is not stationary, but periodic. The period time is about 12-30 Hz. This fact seems to be related to gas dynamic special features of a flow formation in the gas path.

4. The shapes of arc discharges coming out of the compression channels are straight linear segments, nevertheless they are not converged at one point in the geometric center, and they deviate by an average value of 33,85° at the exit of confusers, in the width range from 10,75° up to 45,17°. This fact is due to gas dynamic special features of a flow formation in gas path. The deviation of arc discharges causes their interaction with side faces of the compression channels of an adjacent phase and their intensive destruction because of high temperatures.

5. The calculations made in the Solid Works FlowSimulation software complex indicate that the direction of trajectories of a gas velocity flowing out from a compression channel has a deviation from the axis of an electrode. This fact is well consistent with the observed deviation of arc discharges from the axis of an electrode in the experiment, it indicates the influence of gas-dynamic effects on the shape, and the location of the arc discharges.

Thus, increasing service life of the Zvezda plasmatron considered scheme is closely related to the organization of the flow in the gas path.

References

[1] Kashapov N.F., Nefedev E.S., Timerkaev B.A., Fakhrutdinov B.A. The Decomposition of Heavy Hydrocarbons by Electric Arc Plasma // 36th International (Zvenigorod City) Conference on Plasma Physics and CTF, 9-13 February 2009.

[2] Koroteev A.S., Mironov V.M., Svirchuk Yu.S. Plasmatrons: designs, characteristics, calculation // Moscow: Mashinostroenie Publ., 1993, 296p.

[3] Zhukov M.F. The Fundamentals for Calculations of Plasmatrons of Linear Circuits // Novosibirsk: Academy of Sciences of the USSR, Siberian Branch of the Institute of Thermal Physics, MINNEFTEKHIMMASH USSR, Design Bureau ENERGOKHIMMASH, 1979, 149 p.

[4] Svirchuk Y.S., Golikov A.N. Three-Phase Zvezda-Type Plasmatrons.-IEEE Transaction on Plasma Science, December 2016, volume 44, number 12, p.3042-3047.

[5] Dorofeev A.A., The Fundamentals of the Theory of Thermal Rocket Engines // Moscow: Publ. of the Bauman MSTU, 1999, 415p.

Journal of Physics: Conference Series **2270** (2022) 012054 doi:10.1088/1742-6596/2270/1/012054

A new approach to estimating kinetic rate factors of phase transformations

Dinar T. Yarullin

Kazan Federal University, Kazan, Russia

E-mail:YarullinDT@gmail.com

Abstract. An original approach is proposed to estimate of the main kinetic factors that determine the rate of phase transformations in condensed matter. These factors define the frequency of atom transitions from different phases during phase transformation induced by undercooling, high pressures or mechanical impact. On the example of crystallization of a model amorphous system, it is shown that this approach gives a correct estimation of the crystal growth and crystal decay rates both at the initial stage of the phase transition and at the stage of coalescence of crystalline structures.

1. Introduction

The key factors determining the rate of phase transitions in condensed systems are the rate of particles transition g^+ from the parent phase to a new phase and the rate of particles transition g^- from the daughter phase back to the parent phase [1-3]. For example, in the case of crystallization of an amorphous system, g^+ will determine the rate of atom transitions from the amorphous phase to the crystalline phase through their attachment to the surface of crystalline nuclei. On the other hand, g^- will characterize the decay of crystalline nuclei and the transition of atoms from the ordered to the initial amorphous phase. The kinetic factors g^+ and g^- are fundamental quantities, since they are used to determine the growth rate and nucleation rate of new phase nuclei [4–7]. Moreover, the values of these parameters strongly depend on the size of nuclei, which complicates their estimation at the stage of intensive growth of the new phase [8, 9].

The values of the kinetic factors g^+ and g^- are difficult to estimate by experimental and simulation methods. Experimental methods can give approximate values based on indirect calculations, for example, if the surface self-diffusion of atoms is quantitatively known [10]. Estimation of kinetic factors can be performed on the basis of simulation data by analyzing the average times of the first appearance of nuclei of a certain size, also called MFPT analysis [11-13]. However, this method can be applied to estimate g^+ near nuclei of critical size [14]. In the case of larger nuclei, the accuracy of the g^+ estimate suffers greatly. Moreover, both experimental and simulation methods do not allow one to correctly estimate the value of the kinetic factor g^-. Therefore, there is a need to develop algorithms that make it possible to determine the values of g^+ and g^- with high accuracy for nuclei of different sizes.

In this work, we present an original numerical approach to estimating g^+ and g^- by the results of molecular dynamics simulations of the crystallization process in an amorphous system. The advantage

Content from this work may be used under the terms of the Creative Commons Attribution 3.0 licence. Any further distribution of this work must maintain attribution to the author(s) and the title of the work, journal citation and DOI.
Published under licence by IOP Publishing Ltd

of this approach is in the possibility to estimating g^+ and g^- for crystalline nuclei of completely different sizes and shapes [15]. The developed approach can be applied to quantitative characterization of atomic rearrangements occurring at ultrahigh temperatures under the ionized gas effect, for example, during plasma or laser heating of materials [16, 17].

2. Estimation of the kinetic rate factors

According to the basic definition, the quantities g^+ and g^- characterize the frequency of particles attachment to the surface of a nucleus and the frequency of particles detachment from the surface of a nucleus, respectively. Hence it follows that these quantities depend on the number of attached particles k^+ and detached particles k^- per unit time Δt [15, 20]. Therefore, the dependences of g^+ and g^- on the nucleus size N can be determined as follows [14]:

$$g^+(N) = \frac{k^+(N)}{\Delta t}, \quad (1)$$

$$g^-(N) = \frac{k^-(N)}{\Delta t}. \quad (2)$$

In Eqs. (1) and (2), the time interval Δt must be comparable with the time step of integrating the equations of motion in the molecular dynamics method.

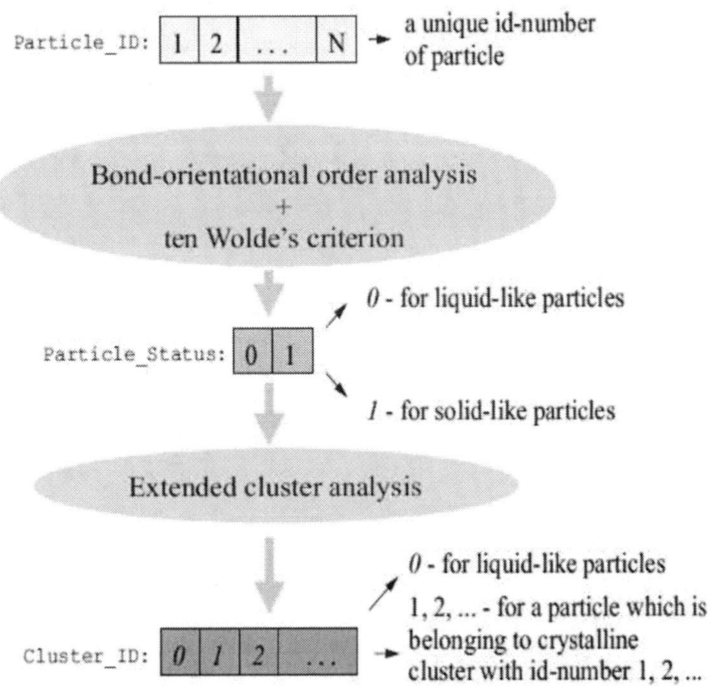

Figure 1. Scheme of the developed procedure for estimating the parameters k^+ and k^-.

We have developed an original numerical procedure that allows one to estimate the N-dependences of the parameters k^+ and k^- with high accuracy [see Figure 1]. High accuracy is achieved due to the fact that all calculations are performed without using of fitting parameters and any approximations. At the initial stage, all particles of the system are assigned unique ID numbers as it shown in Figure 1. The belonging of particles to any phase is determined by the method of bond order analysis, which makes it possible to assign identification number 0 to atoms of the amorphous parent phase and number 1 to particles of the crystal phase. Cluster analysis is carried out through the calculation of the local orientational order parameter q_6, based on the correlation condition [18,19]:

$$\left| \sum_{m=-6}^{6} q_{6m,i}(t) \cdot q^*_{6m,j}(t) \right| > 0.5, \;\; i,j = 1,2,\dots,N, \tag{3}$$

where

$$q^*_{6m,j}(t) = \frac{q_{6m,j}(t)}{\sqrt{\sum_{m=-6}^{6} \left| q_{6m,j}(t) \right|^2}}, \tag{4}$$

$$q_{6m,i}(t) = \frac{1}{n_i} \sum_{j=1}^{n_i} Y_{6m}(\theta_{ij}, \phi_{ij}). \tag{5}$$

Here, $Y_{6m}(\theta_{ij}, \phi_{ij})$ are the spherical harmonics; the quantities θ_{ij} and ϕ_{ij} determine the polar and azimuthal angles, respectively; n_i is the number of nearest neighbors of i-th atom. The realization of the condition (3) for at least seven particles j from the nearest environment of a particle with index i is sufficient for inclusion of the i-th particle into the crystalline phase.

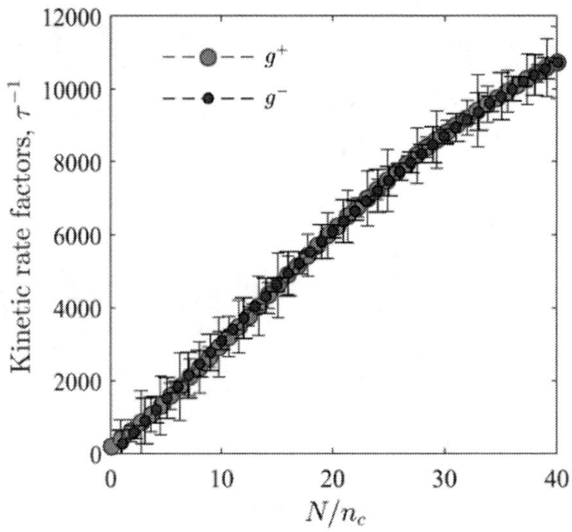

Figure 2. Calculated kinetic factors g^+ and g^- as functions of the reduced nucleus size N/n_c, where $n_c = 50$ is the critical nucleus size.

Based on the cluster analysis results, the particles receive the numerical characteristic "Cluster ID", which characterizes their belonging to a crystalline nucleus. Based on the change of the particles' "Cluster ID" at different times, we determine the number of attached and detached particles for a nucleus of a certain size *N*. Figure 2 shows the dependences of the kinetic rate factors g^+ and g^- as functions of the nucleus size, which are estimated using Eqs. (1) and (2) for a crystallizing amorphous Lennard-Jones system [15]. The shape of these dependencies is consistent with the results of theoretical calculations, which confirms the correctness of the proposed approach [20].

3. Conclusions

Thus, this work presents an original approach to estimating the kinetic rate factors of crystallization. We have shown that this approach makes it possible to estimate the dependence of these factors on the size of crystalline nuclei and this approach is applicable to estimating the nucleus growth rate near the critical size. The developed approach and the obtained results are the basis for the development of a complex method aimed at the description of the kinetics of structural-phase transformations in condensed systems of completely different types [21].

Acknowledgments

This study is supported by the Kazan Federal University Strategic Academic Leadership Program. The author acknowledges the Foundation for the Development of Theoretical Physics and Mathematics "Basis" (project No. 20-1-2-38-1).

References

[1] Shneidman V A 2001 *J. Chem. Phys.* **115** 8141
[2] Weinberg M C et al. 2002 *C. R. Chimie* **5** 765
[3] Galimzyanov B N, Yarullin D T, Mokshin A V 2018 *JETP Letters* **107** 629
[4] Galimzyanov B N, Mokshin A V 2021 *J. Non-Cryst. Solids* **570** 121009
[5] Mokshin A V, Khusnutdinoff R M, Galimzyanov B N 2020 *Phys. Chem. Chem. Phys.* **22** 4122
[6] Mura F, Zaccone A 2016 *Phys. Rev. E* **93** 042803
[7] Baidakov V G, Tipeev A O 2019 *J. Non-Cryst. Solids* **503-504**, 302
[8] Mokshin A V, Galimzyanov B N 2015 *J. Chem. Phys.* **142** 104502
[9] Galimzyanov B N, Ladyanov V I, Mokshin A V 2019 *J. Cryst. Growth* **526**, 125214
[10] Mokshin A V, Galimzyanov B N, Yarullin D T 2019 *JETP Letters* **110** 511
[11] Wedekind J, Reguera D 2008 *J. Phys. Chem. B* **112** 11060
[12] Mokshin A V, Galimzyanov B N 2017 *Russian Journal of Physical Chemistry B* **11** 473
[13] Mokshin A V, Galimzyanov B N 2018 *Journal of Rheology* **62** 265
[14] Song H, Sun Y, Zhang F, Wang C Z, Ho K M, Mendelev M I 2018 *Phys. Rev. Materials* **2** 023401
[15] Yarullin D T, Galimzyanov B N, Mokshin A V 2020 *J. Chem. Phys.* **152** 224501
[16] Fairushin I I, Vasiliev M M, Petrov O F 2021 *Molecules* **26** 6974
[17] Fairushin I I, Khrapak S A, Mokshin A V 2020 *Results in Physics* **19** 103359
[18] ten Wolde P, Ruiz-Montero M, Frenkel D 1995 *Phys. Rev. Lett.* **75** 2714
[19] Steinhardt P, Nelson D, Ronchetti M 1983 *Phys. Rev. B* **28** 784
[20] Kashchiev D 2000 *Nucleation: Basic Theory with Applications* (Butterworth-Heinemann, Oxford)
[21] Khustutdinoff R M, Galimzyanov B N, Mokshin A V 2018 *Journal of Experimental and Theoretical Physics* **126** 83

Journal of Physics: Conference Series **2270** (2022) 012055 doi:10.1088/1742-6596/2270/1/012055

Investigation of a high-frequency magnetron sputtering system operation modes during copper thin films deposition

N A Tumanov[1], D V Kirillov[1], E V Vorob`ev[1]

[1]Bauman Moscow State Technical University, Moscow, Russia

tna148@yandex.ru

Abstract. The results of studying the radio frequency (RF) magnetron sputtering (MS) operation parameters during copper sputtering are presented. A comparison of the growth rate dependence on power during the operation of MS is made in RF and DC modes. The pressure dependences of the RFMS growth rate and bias target voltage are obtained. The prospect of using RFMS for deposition of smooth coatings is shown.

1. Introduction

The problem of creating high-quality thin-film coatings is one of the most urgent among the modern directions in the development of ion-plasma technology. Currently, protective, wear-resistant, reinforcing, decorative coatings and technologies for metallization of rolled materials, manufacturing elements of electronic and microelectronic products, as well as optical systems are in demand. One of the ways to form thin films is the magnetron sputtering (MS) method, which makes it possible to obtain coatings from almost any materials (metals and their alloys, dielectrics, semiconductors) with high performance and good adhesion.

Recently, in many areas, there has been a tendency to increase the requirements for the quality of coatings, the production technologies of which are well known and debugged. In particular, for modern high-quality optical mirrors [1], neutron supermirrors [2] and substrates for graphene synthesis [3], multilayer coatings for erosion diagnostics [4], the requirements for coating roughness are continuously growing. The most common method for achieving the set goals is the method of molecular beam epitaxy [5]. This method allows obtaining the smoothest coatings, but it has low productivity and requires expensive equipment. A good alternative to molecular beam epitaxy can be radio frequency magnetron sputtering (RFMS) [1–3, 6]. Although the roughness of the films obtained by RFMS is somewhat higher than that of epitaxial films, this disadvantage is fully compensated by the performance of the method when using the technology on an industrial scale. At the same time, much less attention is paid to RFMS in the scientific literature compared to other types of MS, which indicates the prospects of research on this topic.

2. Experiment details

In the work, an industrial HF MRS with a copper target was studied. The deposition scheme and a photo of the stand are shown in Figure 1. The vacuum chamber was pumped down to a base pressure of $5 \cdot 10^{-6}$ Torr using a vacuum system based on a spiral and turbomolecular pumps. A capacitive manometer MKS Instruments 627D was used for vacuum measurement. Deposition process was performed in the pressure range $5 \cdot 10^{-3} \dots 5 \cdot 10^{-4}$ Torr and powers 100…1000 W (reflected power is less than 1% in

Content from this work may be used under the terms of the Creative Commons Attribution 3.0 licence. Any further distribution of this work must maintain attribution to the author(s) and the title of the work, journal citation and DOI.
Published under licence by IOP Publishing Ltd

RF mode at 13.56 MHz) with preliminary substrate cleaning by an ion beam. Power was supplied by an Applied electronics APEL-M-5PDC-650-2 DC power supply and an Advanced Energy CESAR RF power generator with a Navio matching device. Power supplies switched manually. The coating growth rate was measured using an Inficon STM-2 thin film monitor. The deposition was performed on single-crystal (110) silicon wafers without additional heating at a power of 1 kW, a pressure of $7{,}5 \cdot 10^{-4}$ Torr and a distance of 130 mm between the substrate and the MS. The coating thickness was 500 nm at a growth rate of 2.26 nm/s in RF mode and 3.54 nm/s in DC mode. The surface was studied using a scanning electron microscope (SEM) Thermo Scientific Quattro S.

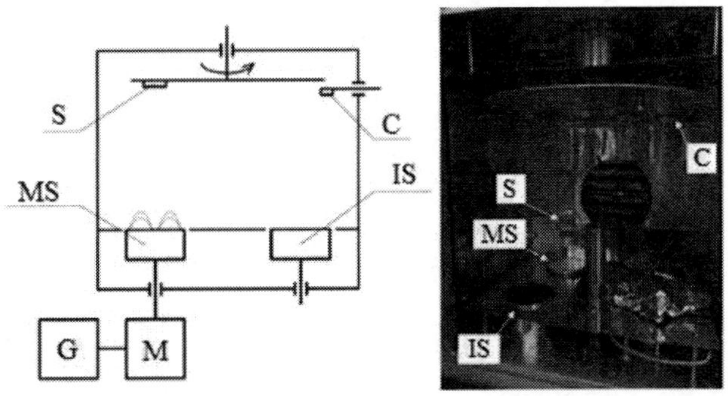

Deposition scheme stand view

G – RF generator, M – matching device, MS – magnetron sputtering system, S – substrate, IS – ion source, C – crystal growth rate sensor

Figure 1. Deposition scheme and general stand view.

3. Results and discussion

Figure 2 shows the pressure dependences of the growth rate and bias voltage of the RFMS target at a constant power (1 kW).

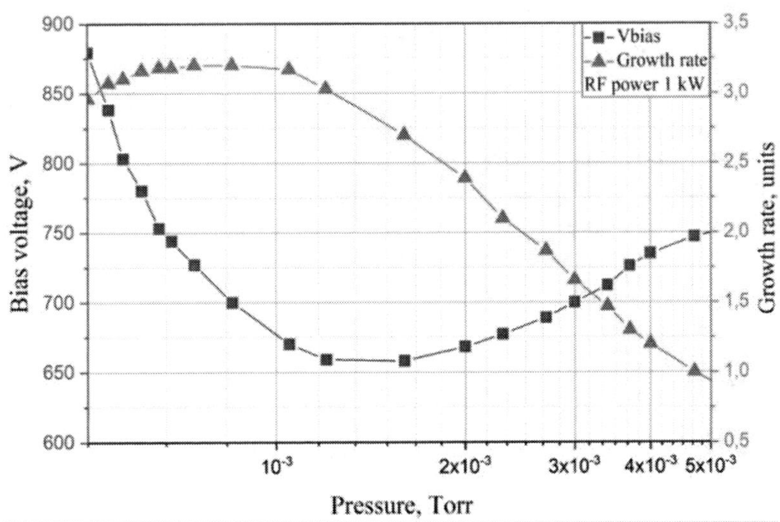

Figure 2. Dependences of the growth rate and RFMS target bias voltage on pressure.

It has been established that there is an optimal pressure at which the maximum growth rate is achieved. With a power of 1 kW, this pressure is $7{,}5 \cdot 10^{-4}$ Torr. In this case, no explicit dependence of the growth rate on the self-bias voltage of the MPC target is observed. At the optimum pressure, the dependence of the coating growth rate on power was measured (Figure 3). In the entire power range, the coating growth rates in the RFMS mode were approximately 1.5 times lower than in the direct current mode under the same other conditions.

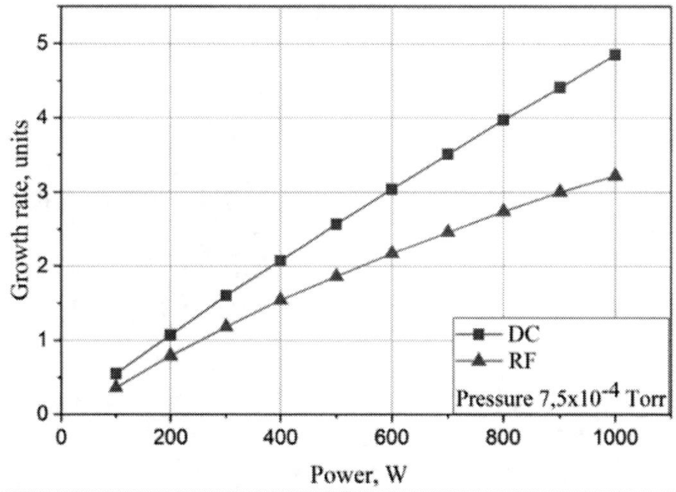

Figure 3. Dependence of the growth rate on the RFMS and DCMS power.

Figure 3 shows SEM images of two coatings 500 nm thick, obtained using RFMS and DCMS under the same conditions.

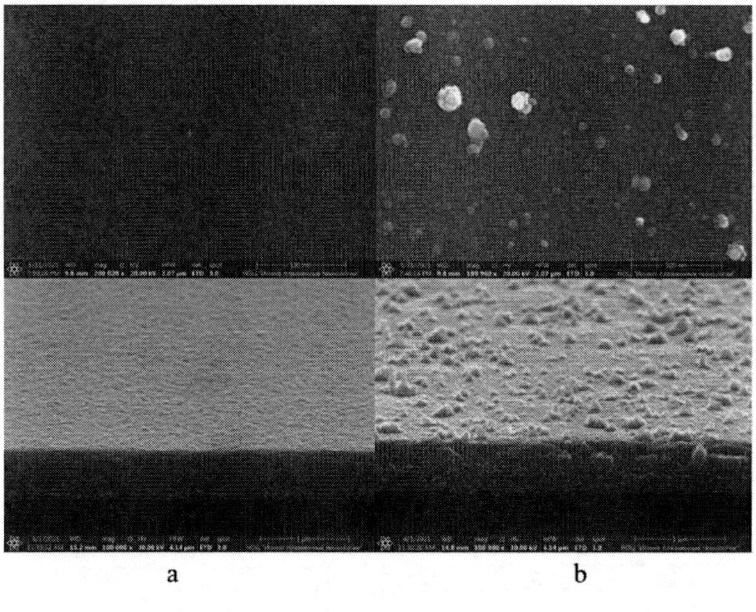

Figure 4. Copper coatings obtained by the RFMS (a) and DCMS (b).

The coating obtained with RFMS is much smoother than the coating applied with DCMS. It can be concluded that the use of high-frequency magnetron sputtering systems is promising for obtaining coatings with the lower roughness.

4. Conclusion

In conclusion, operation modes of RFMS during sputtering of copper films were considered. The most optimal pressure at which the maximum growth rate is achieved is determined. In this mode, coatings were obtained using RFMS and DCMS. Further study of the coatings using a scanning electron microscope showed that the films obtained by RFMS are smoother. This indicates that the use of RFMS is promising for obtaining coatings with the lower roughness.

References

[1] Magnetron sputtering technique for analyzing the infuence of RF sputtering power on microstructural surface morphology of aluminum thin flms deposited on SiO2/Si substrates / Somayeh Asgary, Elnaz Vaghri, Masoumeh Daemi, Parisa Esmaili, Amir H. Ramezani, Saim Memon, Siamak Hoseinzadeh // Applied Physics A. 2021. No. 127, 752.

[2] Influence of sputtering power on properties of titanium thin films deposited by RF magnetron sputtering / Noormariah Muslim, Y.W. Soon, C.M. Lim, N.Y. Voo // ARPN Journal of Engineering and Applied Sciences. 2015. No. 16. P. 7184 – 7189.

[3] Fabrication of high-quality single-crystal Cu thin films using radio-frequency sputtering / Seunghun Lee, Ji Young Kim, Tae-Woo Lee, Won-Kyung Kim, Bum-Su Kim, Ji Hun Park, Jong-Seong Bae, Yong Chan Cho, Jungdae Kim, Min-Wook Oh, Cheol Seong Hwang, Se-Young Jeong // Sci. Rep. 2014. No. 4, 6230.

[4] Procedure of Predicting the Lifetime of the Accelerating Electrode of an Ion-Extraction System of a Gas-Discharge Ion Source / R. V. Akhmetzhanova, Ye. V. Vorob'evb, D. V. Dukhopel'nikovb, S. G. Ivakhnenkob,V. A. Obukhova, G. A. Popova, V. V. Svotinaa, and M. V. Cherkasovaa // Journal of Surface Investigation: X-ray, Synchrotron and Neutron Techniques. 2019, Vol. 13, No. 6, P. 1217–1224.

[5] High quality, transferrable graphene grown on single crystal Cu (111) thin films on basal-plane sapphire / Reddy, K. M. [et al] // Appl. Phys. 2011. No. 98, 113117.

[6] Effects of RF magnetron sputtering deposition process parameters on the properties of molybdenum thin films / P. Chelvanathan, S.A. Shahahmadi, F. Arith, K. Sobayel, M. Aktharuzzaman, K. Sopian, F.H. Alharbi, N. Tabet, N. Amin // Thin Solid Films. 2017. No. 638. P. 213 – 219.

Modification of the materials based on protein fibers in a low-pressure high-frequency plasma flow

G R Rakhmatullina[1], E A Pankova[1], O V Fukina[1], M Khayytov[1], L V Chapaeva[1]

[1]Federal State Budgetary Educational Institution of Higher Education "Kazan National Research Technological University", Kazan, Russia

Gulnaz-f@yandex.ru

Abstract. The effect of low-pressure high-frequency (HF) plasma treatment on the structure of the materials formed by bundles of natural protein (collagen) fibers is considered. The outer surface of the material is being affected by a constant flow of ions and a pulse-periodic flow of electrons. During this procedure the electrons flow at opposite sides of the sample in antiphase. This leads to that an alternating electric field is created, which intensity is estimated approximately at 105 V/m, inside the material. As a result, in the capillaries and pores, a breakdown occurs and charged particles are forming during this process. The recombination of these particles releases an energy of 15.76 eV on the surface of pores and capillaries, which is transferring to the surface molecules of the protein, which leads to the bulk modification of the material based on protein fibers. Such volumetric processing provides an effective modification of structural elements, allowing you to significantly improve the quality of liquid treatments in the production of collagen-containing materials and their properties in general.

1. Introduction

Modification of materials of various nature with the help of low-pressure RF plasma is a current area of research. Materials based on natural protein fibers, which have a complex structural organization, so as the dermis (the main layer of the animal skin) has a multi-level system formed as a result of interweaving bundles of collagen fibers, the fibers themselves and the fibrils and subfibrils that form them [2].

The most complete picture of the changes taking place as a result of plasma treatment can be obtained on the basis of studies of the structure of the leather material.

Materials and methods

As objects of our research, samples of the dermis of sheepskin skins were considered in the work. An experimental high-frequency low-pressure capacitive discharge plasma unit was used to modify materials based on protein fibers. The evaluation of the results of modification of materials based on protein fibers by high-frequency low-pressure plasma was carried out in terms of the welding temperature and the degree of change in the dimensions of the structural elements of the dermis.

2. Results

Traditionally, technological processes, during the obtaining semi-finished leather products from animal skins, can be divided into two categories. The first category includes processes aimed at separating the collagen structure (soaking, ashing, pickling, etc.). The second category is aimed at fixing the resulting structural organization (tanning, retanning, etc.). It should be noted that the first group of processes is characterized by a decrease in the welding temperature, and the second, on the contrary, by an increase in this indicator. (figure 1)

Content from this work may be used under the terms of the Creative Commons Attribution 3.0 licence. Any further distribution of this work must maintain attribution to the author(s) and the title of the work, journal citation and DOI.
Published under licence by IOP Publishing Ltd

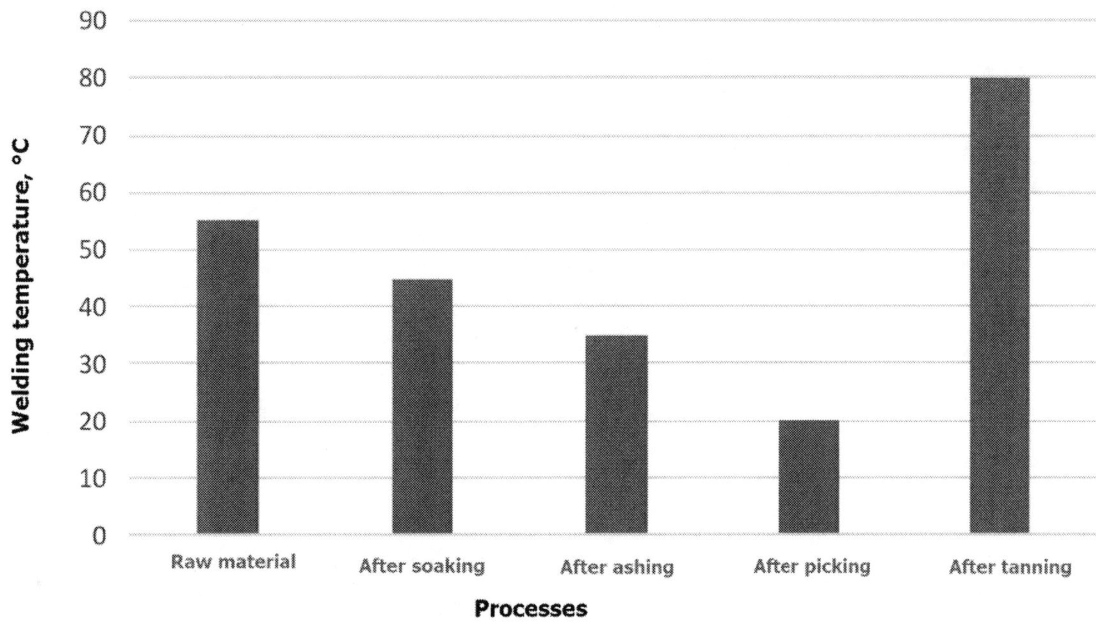

Figure 1 – The change in the temperature of derma welding during the technological cycle of semi-finished leather production.

It was found that HF plasma treatment of a material based on natural protein fibers (collagen) occurs due to the following factors. Ions with the energy of 70-100 eV are bombarding the outer surface of a material. And a burning non-self-sustaining discharge in the supramolecular interstructural space causes recombination of ions on the surface of capillaries and pores which leads to volumetric processing of materials. The transfer of energy to the atoms of the near-surface layer of the material leads to the removal of contaminants, the rupture of some of the non-valent bonds, which makes it possible to change the conformation of protein molecules and leads to the separation of the supramolecular structure of collagen [1].

With the help of electronic microscopy, a comparative assessment of the structural elements of the dermis of control (untreated with plasma) and experimental (treated with plasma) samples was carried out. It was found that the fibers of unmodified samples are characterized by a denser arrangement of fibrils, with an average fibril diameter of 103 nm and an interfibrillar distance of 5.5 nm. In the case of HF plasma modification of the samples, the average fibril diameter is 124 nm, and the interfibrillar distance is 39.1 nm, which indicates an increase in the fibril diameter by 20%, and the interfibril distance - by 7 times. These changes provide a deeper diffusion of working solutions, which improves the efficiency of technological processes.

To confirm this assumption, measurements were made of the temperature of dermal welding, which is a characteristic of the degree of structuring of the material. A sample of material based on protein fibers treated with HF plasma was characterized by a lower welding temperature compared to the control sample from the soaking to the pickling included (fugure 2), which indicates a better flow of these processes as a result of deep and uniform diffusion of working solutions deep into the dermis. As a result, the welding temperature rises after the tanning process, which is a prerequisite for obtaining a semi-finished product with higher strength properties.

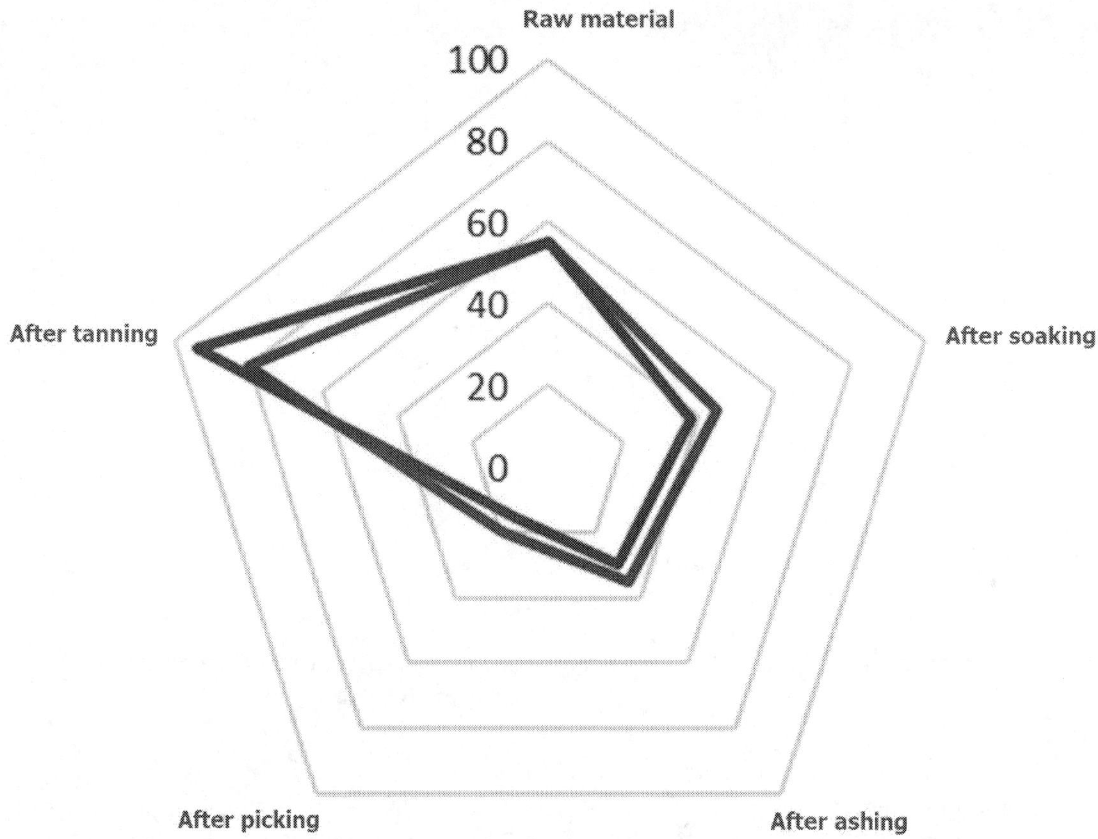

Figure 2 – Changes in the derma welding temperature during the technological cycle of semi-finished leather production (blue color - without plasma processing of raw materials, red color - with plasma processing of raw materials).

3. Conclusion.

Thus, the modification of materials based on natural protein fibers with the use of low-pressure RF plasma allows for a better separation of the supramolecular structure of the material. Such changes in the structure increase the efficiency of subsequent treatments and, as a result, obtain a semi-finished leather product with higher mechanical properties.

References

[1] Abdullin I Sh 2000 High-frequency plasma-jet processing of materials at reduced pressures. *Theory and practice of application Ed. Kazan*. Kazan 348

[2] Rakhmatullina G R 2010 Scientific and technological bases of covering dyeing of leather using non-equilibrium low-temperature plasma Sciences. *Ed. KNRTU*, Kazan 300

Confidence of nano-measurements of Hf-Ti-N multilayer PVD condensates from the plasma phase

M M Mironov[1], M M Grebenshchikova[1]

[1]Kazan National Research Technological University, Kazan, Russia

mironov.medunstrument@yandex.ru

Abstract. The microhardness and thickness of PVD coating layers made of titanium and hafnium nitrides were measured experimentally. The standards of TC 441 were analyzed. The sequence (protocol) of the analysis of nano-measurements and obtaining reliable measurement results was given. To obtain a quantitative value of the reliability of measurements for superhard nanocoats is proposed to use for educational purposes the calculation of the confidence interval using the Student criterion.

1. Formation of nitride coating and its evaluation

The coating was condensed from the vapor-plasma phase on a device with three electric arc evaporators in a nitrogen atmosphere with a pressure of 0.2 Pa. Two evaporators with magnetic stabilization of the arc produced a phase of metallic hafnium, and one cathode evaporated titanium.

The discharge current of the evaporators was 60-75 A. A base was placed in the center of the chamber, which rotated and a reference voltage of 200 V was applied to it. The rotation frequency of the base was 2.5 min^{-1}. The coating was condensed in layers. The layout of electric arc evaporators is shown in Figure 1.

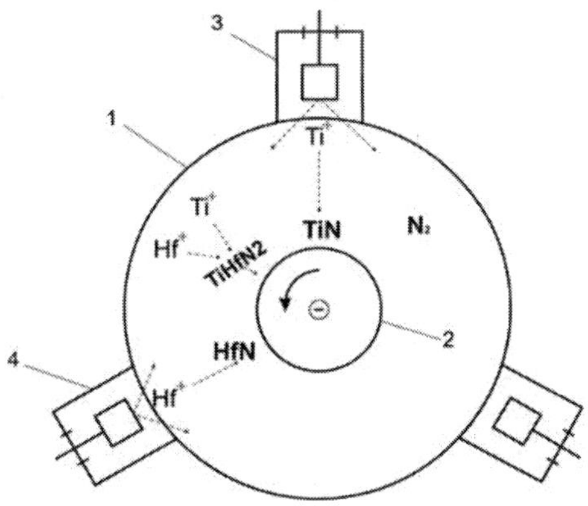

Figure 1. The scheme of the combined coating formation (top view). Description: 1 - vacuum chamber, 2 - rotating cylindrical sample holder, 3, 4 - arc evaporators.

Content from this work may be used under the terms of the Creative Commons Attribution 3.0 licence. Any further distribution of this work must maintain attribution to the author(s) and the title of the work, journal citation and DOI.
Published under licence by IOP Publishing Ltd

Journal of Physics: Conference Series **2270** (2022) 012057 doi:10.1088/1742-6596/2270/1/012057

The formation of the nitride layer occurs when the sample passes through the spray zone of the arc evaporator. Two layers of hafnium nitride and a layer of titanium nitride are condensed sequentially. The density of hafnium is greater than the density of titanium. Therefore, the thickness of the two layers of hafnium nitride is approximately equal to the thickness of the titanium nitride layer. As the rotation speed of the product increases, the thickness of the layers decreases, and their number increases. At the same time, the total thickness of the coating formed during the same time remains the same. The condensation conditions were described in detail in the materials of previous studies [1].

The formation of such a multilayer coating structure contributes to the overlap of the intergranular boundaries of steel. This prevents the formation of through pores even in a thin coating. The coating surface contains a phase of microdrops.

Figure 2 shows a section of the entire thickness of the coating with layers of nitrides. It was recorded by a high-resolution spectroscope on the Auriga station platform.

Multiple overlapping of the substrate phase separation zones by nanoscale layers leads to the formation of bridges, their coalescence and disruption of the ordered crystalline columnar structure of the coating. Another type of nanostructure is formed. The formation of a new phase of $HfTiN_2$ was revealed. It is not marked in Bruker catalogs.

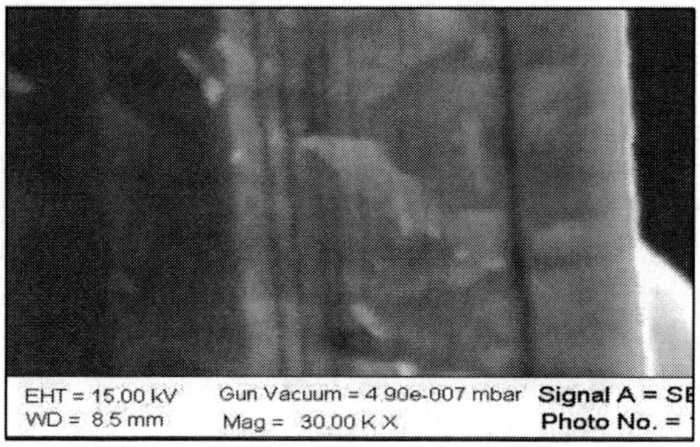

Figure 2. Section of a multilayer coating with a total thickness of 2500 nm.

The results of measuring the main characteristics of the coating and layers are shown in Table 1. Microhardness was measured on a PMT-3 microhardness meter in manual mode at loads of 20 and 50 grs.

Table 1 - Characteristics of Hf-Ti-N nitride coatings

Compound	Layer thickness, nm		Hardness, GPa	Roughness parameter, Ra, microns	Phase	Number of layers	Confidence
	All	One					
TiN	2400	2400	3.2-14.5	0.05	TiN_2, TiN	1	0.8
HfN	2000	2000	4.8-7.6	0.2	HfN_2, HfN	1	0.8
(Ti+Hf)N	2500	10-20	4.8-12.0	0.1	HfTiN, $HfTiN_2$	70 - 90	0.8

The thickness of the layers of titanium and hafnium nitrides was determined by calculation using the growth rate of the coating on a fixed substrate. The actual thickness of the layers was measured experimentally by the coating section based on the total thickness and the number of turns of the substrate during coating application. The confidence probability was set.

2. A reliability of the measurement results of the characteristics of nanoobjects

The Technical Committee of TC 441 of the State Standard "NANOTECHNOLOGY" has practically completed regulatory documentation in the field of nanotechnology over the past 10 years. The standards carry out technical regulation according to methods, measuring systems and requirements for nanodispersions, fibers, coatings and composites. The main standards and terms in the field of nanotechnology are given in the sources [2, 3].

An important component is the concept of the reliability of the measurement results of the characteristics of nanoobjects. The reliability of the results should be formed taking into account GOST (ISO) 17025-2019 General requirements for the competence of research and calibration laboratories. Mandatory fulfillment of these requirements is associated with the inherent properties of nanosystems. These are properties such as:

- the presence of a linear size of objects outside the visible light range (1-100 nm);
- lack of clear phase boundaries;
- comparability of the dimensions of these boundaries with the linear dimensions of nanoobjects;
- uncertainty of the characteristics of nanoobjects.

The uncertainty of measurements is physically related to the variety of conditions and measurement methods and the variability of nanosystems over time, as well as the low reliability of measurement results. The problems of nanometrology are directly related to measuring equipment, calibration, standards and standardized methods.

The obtained multilayer nanoscale coatings based on hafnium and titanium nitrides and their properties should be considered within the framework of TC 441 standards.

To confirm the reliability of the results of research and measurements, should be used the content of GOST (ISO) 17025-2019 General requirements for the competence of research and calibration laboratories.

The measurement results should contain information about deviations of values or ranges of the confidence interval with a given confidence probability. In our opinion, the most acceptable calculation method is the calculation of the interval based on the Student's criterion and the mean square deviation for a given probability. Determining the confidence probability of measurement results, i.e. the inverse problem, has a complicated mathematical apparatus and requires knowledge and skills in the field of mathematical analysis. The measurement results must be obtained on verified, calibrated measuring equipment using standardized methods and in standardized (conditioned) conditions. Using the results of calculating the confidence interval, it is possible to interpret these results with a certain confidence probability. The description of research results should be carried out in terms and definitions recommended by regulatory documents.

During the assessment of the reliability of the results of nano-measurements according to GOST (ISO) 17025-2019 for superhard nanocoats according to GOST R 57408-2017, it is advisable to be guided by the provisions of the standards in the order (protocol): terms, nanoobject, characteristics of nanoobjects, GOST R ISO 22309-2015. GSI. Electron probe microanalysis. Quantitative analysis using energy dispersion spectrometry for elements from 11 and above and GOST R ISO 16243-2016. GSI. Chemical analysis of the surface. X-ray photoelectron spectroscopy. Registration and presentation of data.

And in conclusion, in our opinion, it is useful to pay attention to the terms "nanocoating" and "nanolayer" for composite materials, which are sometimes interpreted equivalently. It is also necessary to observe the difference in the equivalent terms "film" and "coating", which may not be taken into account when translating.

References

[1] M M Mironov, L T Bayazitova, M M Grebenshchikova. PVD Barrier coatings with antimicrobial function for medical implants. In the collection of low-temperature plasma in the processes of applying functional coatings. Kazan 2020, pp.158-161.

Journal of Physics: Conference Series **2270** (2022) 012057 doi:10.1088/1742-6596/2270/1/012057

[2] Register of developed nanoindustry standards. [Electronic resource] https://www.gostifo/fiop.site > .527
[3] List of national and interstate standards assigned to TC 441. [Electronic resource] http://www.gostifo/tk_441_perechen ...

Three-color pyrometry study of radiation and gas dynamic processes in a heterogeneous plasma-vortex ring

L Y Volodin[1]

[1]Bauman Moscow State Technical University, Moscow, Russian Federation

volodinlu@yandex.ru

Abstract. The paper presents the results of a study of a heterogeneous plasma vortex ring using the three-color image pyrometry technique. The vortex ring is formed during deceleration of a supersonic pulsed aluminum plasma jet in atmospheric air. A technique for calibrating a digital camera and a post-processing process for images are proposed that makes it possible to visualize the internal structure of the vortex. As a result of the study, the effect of vortex separation of condensed and gas flow components was experimentally demonstrated.

1. Introduction

It was reported on the design of a new type of pulsed sources of high-power broadband radiation with a high efficiency of converting stored energy into radiation [1-3]. The sources are based on the pulsed supersonic plasma jet generation and its transformation into a large-scale plasma-vortex formation with an afterglow time that is orders of magnitude greater than the energy deposition time. The use of active metals as plasma-forming material leads to combustion reactions intensification in the vortex volume with atmospheric air [1]. Despite a number of studies, the plasma vortex formation is a relatively poorly understood form of a highly turbulent vortex flow [4-6]. To optimize the characteristics of the radiation sources, it is necessary to conduct a number of fundamental research aimed at studying the physical processes occurring in the plasma formations.

In previous studies [7, 8], the radiation spectrum of a plasma vortex was recorded and the effect of partial optical transparency of the radiation emerging from its volume was discovered. This work is a continuation of research and is aimed at studying the radiation-gas dynamic processes occurring in a plasma-vortex formation in various spectral ranges using the three-color image pyrometry method.

Three-color pyrometry is widely used to study the temperature fields of various heated objects, including high-speed radiating flows [9-12]. The traditional technique makes it possible to determine the color temperature using various spectral channels. As a rule, closely spaced and free from the molecular bands and atomic lines spectral ranges are used. In this study we deliberately register both molecular and gray condensed phase radiation emitted from the optically transparent volume to visualize transport processes in a plasma vortex ring.

2. Experimental

2.1. Digital camera calibration

The image pyrometry system was build using Canon 550D digital camera with CMOS sensor. The main camera features that make it possible to use it for scientific research are 14-bit ADC and possibility to

Content from this work may be used under the terms of the Creative Commons Attribution 3.0 licence. Any further distribution of this work must maintain attribution to the author(s) and the title of the work, journal citation and DOI.
Published under licence by IOP Publishing Ltd

recording RAW image [12]. Additionally, the camera could be launched from an external triggering device. It is extremely important to have the accurate camera channel's spectral sensitivity for pyrometric measurements, so an experimental set-up for its measurement was assembled (figure 1).

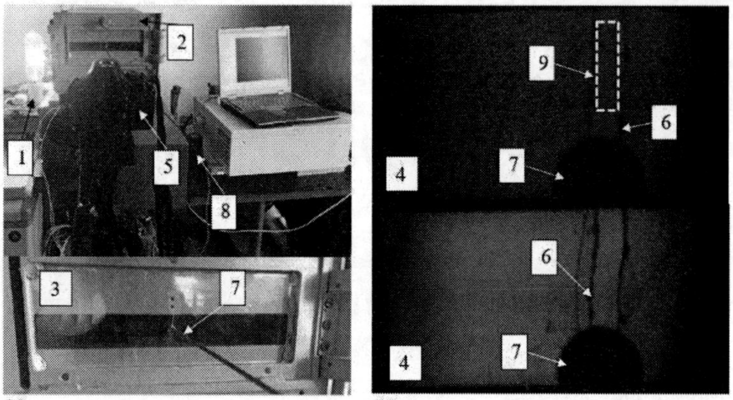

(a) (b)

Figure 1. Experimental set-up for measurement spectral sensitivity of digital camera: (a) apparatus; (b) images of the exit window at various positions of the diffraction grating.

Set-up included DSF-452 spectrograph, Solar S100 spectrometers, and SI 10-300 tungsten ribbon filament lamp as a broadband radiation source. The lamp (1) was located in front of the entrance slit of the DFS-452 spectrograph (2), and an emission spectrum (4) was observed in the output window of the spectrograph (3). A camera (5) was installed in front of the output of the spectrograph and shoot the spectrum (figure 1b) through a narrow split (6) in a Plexiglas plate installed on the DFS-452 output unit. Images were saved in the RAW format. Optical fiber (7) of the Solar S100 spectrometer (8) with flat-field correction was fixed under the split (9), and made it possible to control the current relative intensity and wavelength.

The emission spectrum in split (6) was changed by rotating the diffraction grating in the DFS-452 and was performed in 5 nm step by rotating the grating. A series of recorded images was decoded using the Dcraw [13], resulting in the formation of spatial arrays of 14-bit frame brightness values. Spectral frames for different camera channels (R, G, B) were analyzed in the Matlab R2020b. The resulting spectral sensitivity of the Canon 550D RGB-channels are shown in figure 2a. Additionally, it was found that the output signal from the pixels depends linearly on the exposure value (figure 2b).

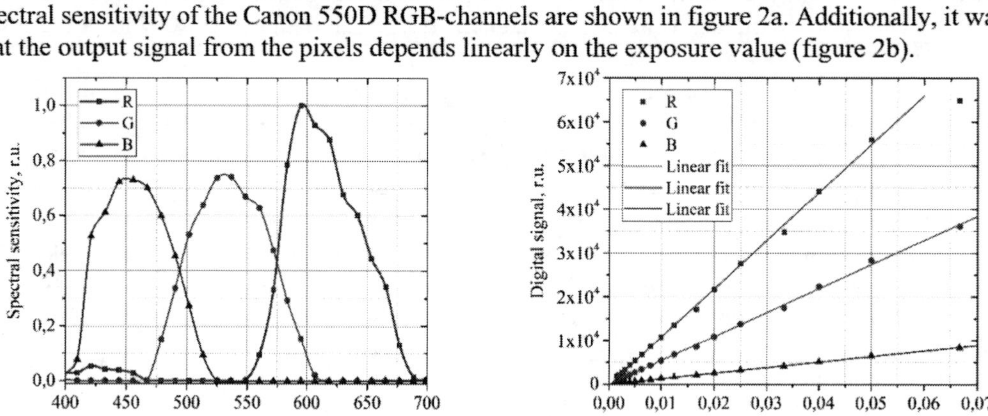

(a) (b)

Figure 2. Characteristics of a digital camera Canon 550D:
(a) spectral sensitivity of RGB channels; (b) digital output vs. exposition time

An Ocean Optics DH-3 CAL calibration source with known spectral and energy characteristics was used as a spectral brightness standard. As a result, the calibration coefficients of the camera RGB channels were found, which make it possible to convert a digital signal from pixels into absolute values of the spectral brightness.

2.2. Image post-processing

Recent high-speed optical emission study of plasma vortex rings has shown that the emission spectrum of the vortex ring is maintained in time, and condensed Al_2O_3 particles are a gray body emitter [7]. Figure 3 shows the characteristic emission spectrum of a vortex ring and the channels spectral sensitivities of Canon 550D digital camera.

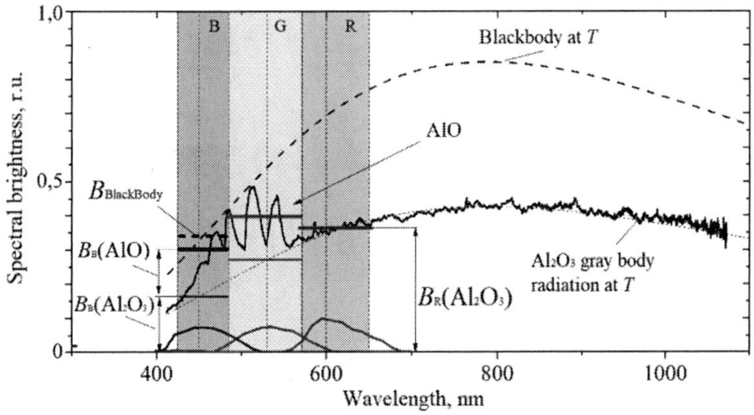

Figure 3. Typical emission spectrum of plasma vortex ring and RGB spectral sensitivity of Canon 550D.

The figure shows the spectral brightness corresponding to the various flow components such as Al_2O_3 and gas phase monoxide AlO. The total brightness recorded by the B-channel corresponds to the radiation of the condensed and gaseous phases:

$$B_{b\Sigma} = B_b(\mathrm{Al_2O_3}) + B_b(\mathrm{AlO}),$$

where $B_b(Al_2O_3)$ and $B_b(AlO)$ are average spectral brightness of Al_2O_3 and AlO in the channel B band (λ_B=450 nm, $\Delta\lambda_{0.5}$=85 nm).

The brightness of the long-wavelength R-channel (λ_R=600 nm; $\Delta\lambda_{0.5}$=75 nm) is due to the emission of only condensed gray body Al_2O_3

$$B_{R\Sigma} = B_R(\mathrm{Al_2O_3}).$$

In study it was shown that the molecular bands emission has a high optical density [8]. In this regard, the total spectral brightness of channel B is close to the brightness of the blackbody. This allows us to estimate the thermodynamic temperature by

$$T \approx \frac{C_2}{\lambda_B} \cdot \left(\ln\left[\frac{C_1}{\lambda_B^5 \cdot B_{B\Sigma}} + 1 \right] \right)^{-1}$$

where C_1 и C_2 are the first and second radiation constants.

Note that the brightness in different spectral ranges and temperature are functions of the coordinates (r, z) in the frame plane. The spectral brightness of Al_2O_3 and AlO can be determined from measurements of the long-wavelength channel R using the ratio for gray body:

$$B_B(\mathrm{Al_2O_3}) = B_R(\mathrm{Al_2O_3}) \cdot \left(\frac{\lambda_R}{\lambda_B} \right)^5 \frac{\exp\left[C_2/(\lambda_B T) \right] - 1}{\exp\left[C_2/(\lambda_R T) \right] - 1},$$

$$B_B(\mathrm{AlO}) = B_{B\Sigma} - B_B(\mathrm{Al_2O_3}).$$

2.3. Experimental details – vortex ring imaging

The experimental setup for producing plasma vortex rings and visualization its structures is shown in figure 4. A pulsed electrical explosion plasma generator was used to inject a plasma jet in atmospheric air. The design of the generator is described in detail in [14]. We used aluminum foil as plasma forming material. The generator was powered by a capacitor bank (U_0 = 3.9 kV, W_0 = 2.5 kJ).

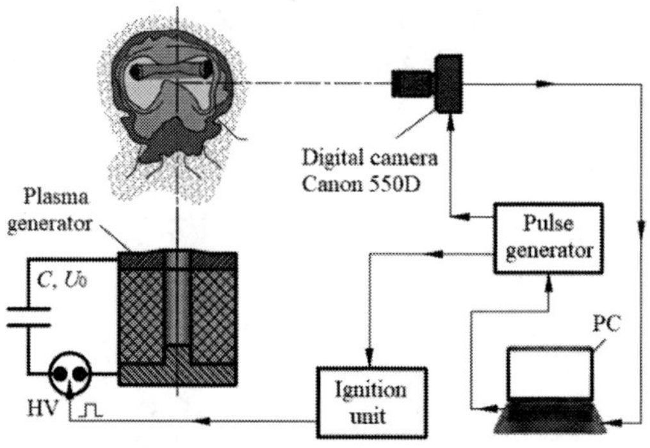

Figure 4. Scheme of the experimental stand.

The discharge was switched by a trigatron by applying a high-voltage pulse from the initiation unit. The delay time between the start signal and the initiation was no more than 1 μs. The discharge current is a damped sinusoid consisting of two half-waves (the quasi-half-period is ~ 50 μs). The maximum current is reached by ~ 28 μs and is equal to 60 kA.

Canon 550D has a trig delay τ_{del} (jitter), that varies from 70 to 200 ms. The lifetime of the vortex ring is several ms, so a synchronization system was used. Synchronization of processes was carried out using a pulse generator assembled on the basis of Arduino Uno.

Arduino generated a trigger signal on the Canon 550D, and after a jitter time τ_{del}, a frame exposure started (t_{exp}=250 μs). Programming the delay between camera launches and generator initiation made it possible to register various stages of the evolution of the plasma formation.

3. Results and discussion

Figure 5 shows the brightness temperature fields of the vortex formation in different spectral ranges for the moment of time corresponding to the maximum radiation power (~1.3 ms).

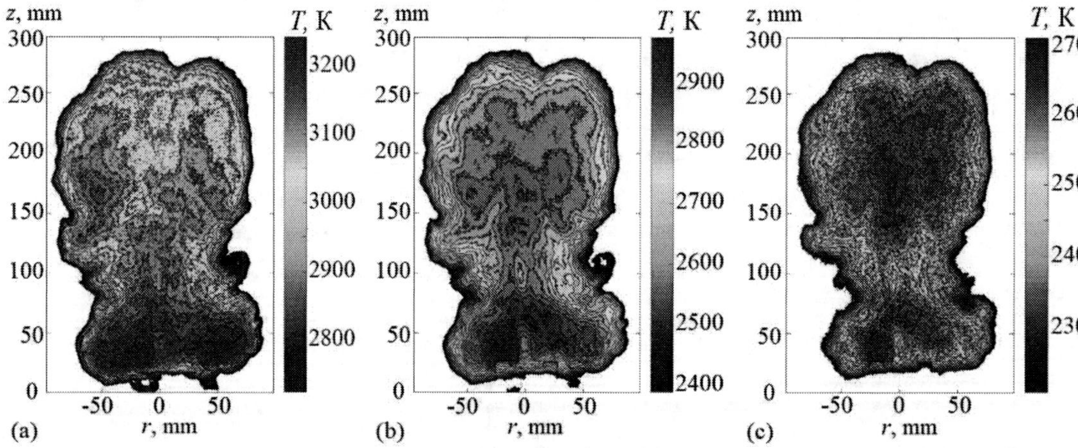

Figure 5. Distributions of plasma-vortex structure brightness temperatures in different spectral regions: (a) λ_B=450 nm, $\Delta\lambda_{0.5}$=85 nm; (b) λ_G=530 nm; $\Delta\lambda_{0.5}$=95 nm; (c) λ_R=600 nm; $\Delta\lambda_{0.5}$=75 nm.

The profiles of the brightness fields in different spectral ranges are significantly different, structurally inhomogeneous, with several brightness extrema, which is due to the complex gas-dynamic structure of the flow.

The maximum brightness temperatures are reached in the vortex tail and, depending on the spectral range, are 2700-3250 K with maximum values in the blue region of the spectrum. The brightness temperatures maps recorded by the spectral channels B and R, a vortex structure with maximum temperatures of ~3100–3250 K is clearly visualized. According to estimates made using the B channel image, vortex core occupies ~1/3 of the volume of the surrounding vortex atmosphere. It should be noted that in some measurements we have observed temperatures in the vortex core reaching values close to the temperatures of stoichiometric aluminum combustion ~ 3350–3400 K.

Figure 6 shows the brightness distributions of AlO and Al_2O_3 condensed phase for spectral channels B (λ_B=450 nm) and R (λ_R=600 nm). The non-linear dependence of the brightness on temperature and optical density does not allow one to directly correlate it with the concentrations of emitting particles. However, as analysis of the brightness distribution of various flow components shows, a significant part of AlO gas-phase monoxides is localized in the vortex core. The radial brightness distribution of AlO is similar to previously demonstrated in [8] and corresponds to the emission of a partially transparent vortex ring. The brightness distribution of Al_2O_3 is close to hyperbolic and is due to the radiative transfer of radiation in the spherical volume of the vortex atmosphere. Relatively weak brightness extrema observed against the background of a quasi-hyperbolic distribution (figures 6b) indicates the presence of a small amount of condensed phase in the vortex core.

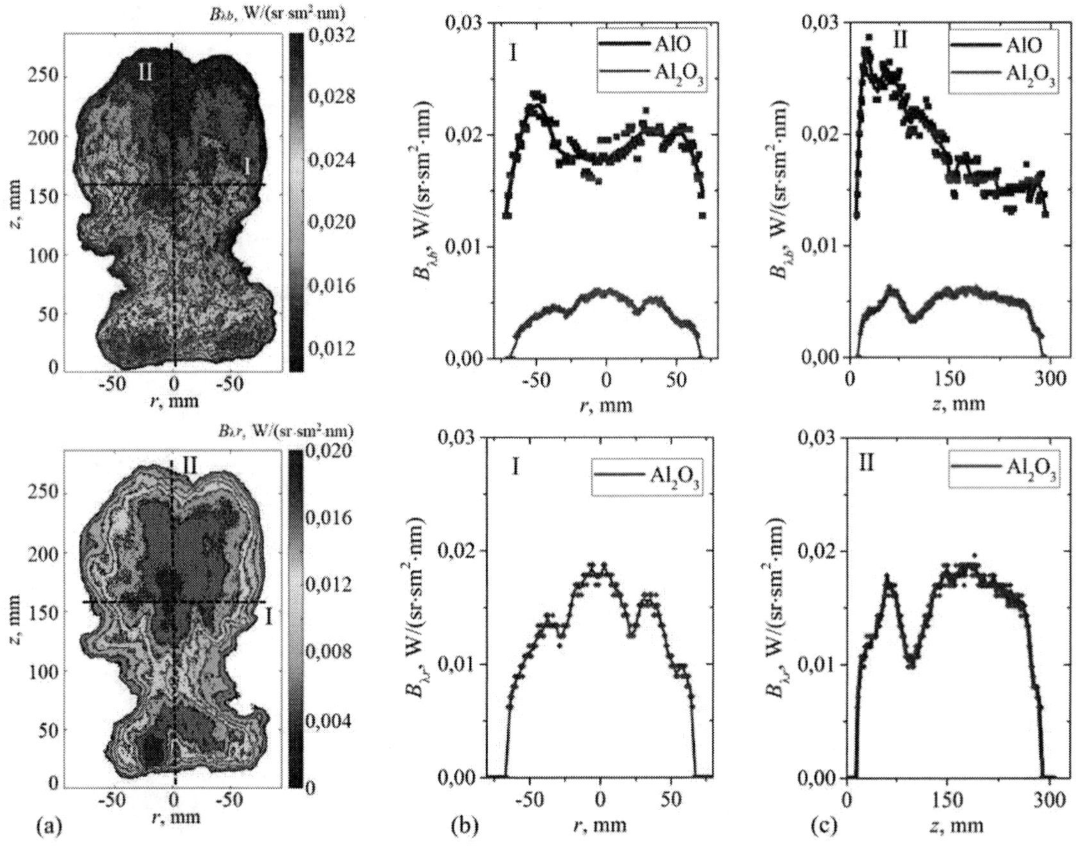

Figure 6. Spectral brightness distributions for λ_B=450 nm (top) and λ_R=600 nm (bottom): (a) two-dimensional field; (b) radial - along cross-section I; (c) axial - along cross-section II.

The reason for the structural inhomogeneities of radiation is probably the feature of turbulent vortex rings to hold the initial material of the plasma jet in vortex core. The observed effect was demonstrated for vortices in a chemically active environment [15-17].

The aluminum is held in the vortex core and reacts with atmospheric air, so high-temperature products of various phase compositions are produced. Emission AlO oxides, due to the anisotropy of diffusion across the vortex line (the effect described in [18]), are localized in the core. Condensed Al_2O_3 macroparticles leave the core under conditions of vortex motion, fill the vortex atmosphere, and subsequently being forced out into the vortex wake.

4. Conclusions

A digital camera and a three-color pyrometry technique have been applied to investigate the heterogeneous plasma vortex ring. Image post-processing algorithm has successfully been applied to visualize internal vortex structures. It has been established that the vortex core is characterized by a high concentration of optically dense gas-phase AlO oxides. The temperature level reached in the vortex core is 2700-3250 K. The vortex atmosphere consists of a low-temperature condensed Al_2O_3 phase, which leaves the vortex core due to vortex motion.

Acknowledgments

This work was supported by grant 18-29-21039 from the Russian Foundation for Basic Research.

References

[1] Zharnikov M H, Kamrukov A S, Kozhevnikov I V, Kozlov N P and Roslyakov I A 2008 *Appl.Phys. Letters* **34**(1) 74
[2] Bulanov S S et al 2008 *Tech. Phys. Lett.* **34** (1) 34
[3] Bulanov S S, Esiev R U, Kamrukov A S, Kozlov N P, Morozov M I and Roslyakov I A 2010 *Tech. Phys.* **55**(11) 1633
[4] Yusupaliev U, Shuteev S A, Vinke E E and Yusupaliev P U 2010 *Bull. of the Lebedev Phys. Inst.* **37**(8) 227–33
[5] Lodes A, Curry R D, Rudroff M E J, Schmidt M, Bauer A J R and Brown W 2011 *IEEE Pulsed Power Conf.* (Chicago: IEEE) 1077
[6] Lodes A and Curry R D 2012 *IEEE Int. Power Modulator and High Voltage Conf.* (San Diego: IEEE) 562-5
[7] Volodin L Y and Kamrukov A S 2019 *J. Phys.: Conf. Ser.* **1393** 012063
[8] Volodin L Y and Kamrukov A S 2020 *Proc. 7th Int. Congr. on Energy Fluxes and Radiat. Eff. (Tomsk)* (IEEE) pp 98-101
[9] Densmore J M, Biss M M, McNesby K L and Homan B E 2011 *Appl. Opt.* 2011. **50**(17) 2659–65
[10] Hossain M M, Lu G, Sun D and Yan Y 2013 *Meas. Sci. Technol.* **24**(074010) 1–10
[11] Jakob M, Lehnen P, Adomeit P and Pischinger S 2014 *Combust. Flame* **161**(11) 2825-41
[12] Sankaranarayanan A, Swami U, Chowdhury A and Kumbhakarna N 2021 *Rev. Sci. Instrum.* **92** 044905
[13] Coffin D 2008 DCRAW: Decoding raw digital photos in linux [Internet] Available from: https://www.dechifro.org/dcraw/
[14] Volodin L Y, Kotovrasov E Y and Fomin V E 2018 *J. Phys.: Conf. Ser.* **1058** 012022
[15] Maxworthy T 1974 *J of Fluid Mech.* **64**(2) 227-39
[16] Johari H 1995 *Phys. Fluids* **7**(10) 2420-27
[17] Ishizuka S, Yamashita T and Shimokuri D 2013 *Proc. of the Combust. Inst.* **34** 745-53
[18] Yusupaliev U, Shuteev S A, Vinke E E and Elenskii V G 2012 *Bull. of the Lebedev Phys. Inst* **39**(3) 89–94

Electric discharge generator of long-lived plasma formations in atmospheric pressure air

L Y Volodin[1], A S Kamrukov[1]

[1]Bauman Moscow State Technical University, Moscow, Russian Federation

volodinlu@yandex.ru

Abstract. The paper presents the results of a study of the developed pulsed plasma generator based on the cylindrical foil electric explosion. The generator is used to form large-scale toroidal plasma vortices in atmospheric air. The influence of the specific stored energy on the phase composition of the plasma flow and the lifetime of plasmoids using a high-speed camera was studied. It was found that large condensed particles in the flow do not participate in the vortex motion and cause the destruction of plasma vortices. The formation conditions of stable plasma vortices were determined.

1. Introduction

Long-lived formations are toroidal plasma vortices formed by pulsed injection of a supersonic plasma jets into atmospheric air. Its afterglow time is several orders of magnitude longer than the energy deposition time. Studies of this plasma structures are of significant scientific and practical interest for solving the problems of creating a powerful radiation source [1-3], a laboratory analogue of ball lightning [4], attenuation of shock waves during interaction with plasma [5]. The generation of plasma vortices is a complex and multifactorial process, that depends on the gas-dynamic characteristics of the plasma jet, its phase composition, etc. Formation conditions of vortex rings on plasma generators operating on gaseous plasma-forming materials are described in [6]. The use of metals as plasma-forming materials [1, 7] is of particular interest, which makes it possible to achieve a high efficiency of stored energy conversion into radiation. Under these conditions, the plasma flow contains a condensed phase, that affects the main characteristics of vortices, in particular, their stability, afterglow time, optical characteristics etc. [8, 9]. In the research [1] the plasma vortices were produced in less than half of the cases of discharge pulses. The causes leading to the instability of vortex generation process and the issues of their further stability were not considered in detail. However, it is important for solving the light source design problems. This work is aimed at studying operating modes of the experimental generator and the processes causing the stable transformation of the plasma flow into a vortex ring.

2. Experimental

We designed a pulsed plasma generator based on a localized electrical explosion of the cylindrical foil. The plasma generator (figure 1a) is formed by a flexible dielectric (polyvinyl chloride) channel (Ø10 x 35 mm) and two electrodes, one of which is made in the form of a nozzle. The use of a flexible dielectric channel wall supported by a metal shell makes it possible to achieve high strength of the generator. A metal cylindrical foil is installed in the channel before each experimental shot. The plasma generator is powered by a capacitive storage $C = 962$ μF. The operating voltage U_0 of the battery is 1–5 kV.

Content from this work may be used under the terms of the Creative Commons Attribution 3.0 licence. Any further distribution of this work must maintain attribution to the author(s) and the title of the work, journal citation and DOI.
Published under licence by IOP Publishing Ltd

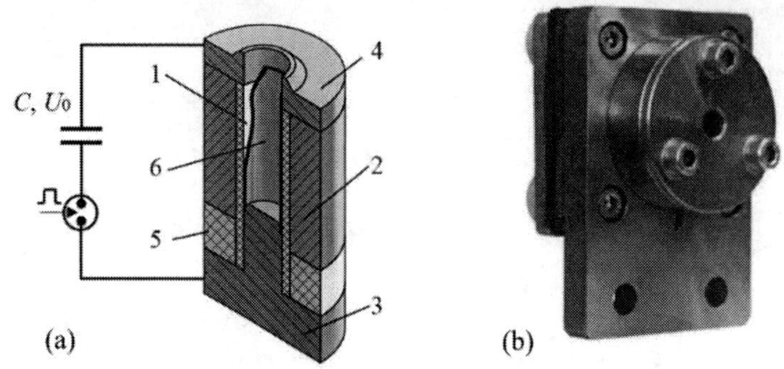

Figure 1. Electric discharge generator of heterogeneous plasma
a - pulse generator; b - scheme of the experimental stand
(a) schematic diagram: 1 – elastic dielectric (PVC); 2 - shell; 3 – rod cathode;
4 - anode-nozzle; 5 - dielectric; 6 - plasma-forming insert; (b) appearance.

A pulsed jet is formed when the installed in the channel foil is electrically exploded. The generator produces a heterogeneous plasma flow, and its phase composition is controlled by changing the mass of the plasma-forming material and the input energy. In this study we varied the value of the specific stored energy per unit mass of the plasma-forming substance $q = W_0/m$ (W_0 is the energy stored in the capacitor bank, m is the mass of the plasma-forming material). Aluminum was used as plasma forming material

The electrical characteristics were recorded using a Pearson 4418 current sensor and an AKTAKOM ASA 6039 high voltage divider. Registration of the vortex ring formation process, determination of the vortex ring lifetime (afterglow duration) and visual analysis of the flow composition were carried out using a Casio EX-F1 9 high-speed camera (600 fps).

3. Results

Figure 2 shows typical frames of the process of formation of plasma-vortex rings at various values of q. The obtained frames showed that for different values of q, three main modes of vortex formation are realized: unstable, partial stability and stable modes. For unstable mode (figure 2a) plasma jet transforms in an asymmetric fast-breaking structure and vortex ring is not produced. The partial stability mode (figure 2b) is characterized by the destruction of the vortex ring by condensed particles in the flow. A plasma toroidal vortex in the regime of stable formation is an azimuthally homogeneous structure with a lifetime of more than 13 ms (figure 2c).

The figure 3a shows the data for the lifetime of the plasma vortex, obtained from the processing of high-speed video recording. Filled dots correspond to cases of stable and partial stable vortex formation modes, unfilled dots correspond to unstable one. The most stable plasma vortices are produced at $q = 20 \ldots 30$ kJ/g and foil masses of 150 mg and 200 mg. The mode of unstable formation is typical for foil masses less than 100 mg, and partially stable - for masses of more than 200 mg.

An analysis of the electrical characteristics showed that in all the cases a subcritical damping regime with two current half-waves is realized (the damping parameter $\gamma = 0.73$, the duration of the first half-cycle is $T_{1/2} \sim 90$ μs), and the electric explosion occurs without a current quasi-pause. During an electric explosion, the input energy is spent on the phase changes of the conductor, its dispersion, the formation of liquid and vapor phases (W_{expl}). Further heating and expansion of the products of the electric explosion by a high-current discharge (W_{pl}). Variation of q in these studies led both to a change in the ratio between the explosion and plasma energies and to a change in the total efficiency of the energy input (figure 3b). In the graphs on the second abscissa scale the q values are normalized to the aluminum sublimation energy $Q_{subl} = 11.4$ kJ/g.

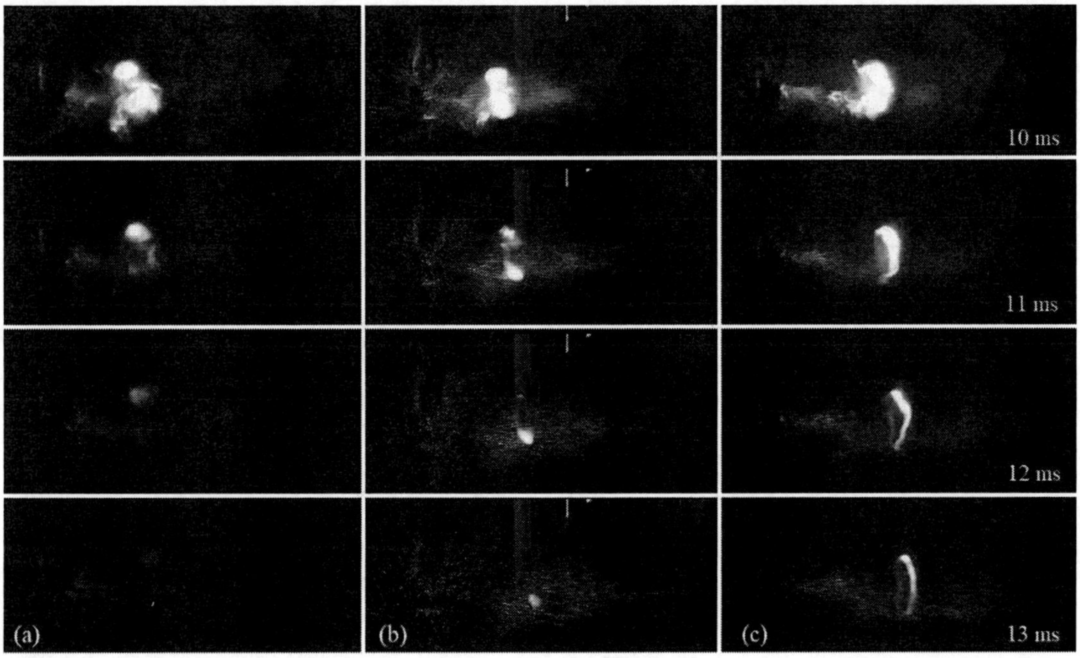

Figure 2. Typical frames of plasma vortex rings formation: (a) unstable formation (q=35-120 kJ/g); (b) partially stable (q=8-15 kJ/g); (c) stable (q=20-30 kJ/g).

With the mass of the plasma-forming foil m=250 mg and q=8…20 kJ/g (0.7…1.8 Q_{subl}), the overall efficiency of the energy input and the energy spent on heating and accelerating the plasma flow are reduced. In this mode the electric explosion occurs after the discharge current maximum, and a large number of large drops are observed in the flow, which are not involved in the vortex motion and disrupt the vortex formation process (figure 2b).

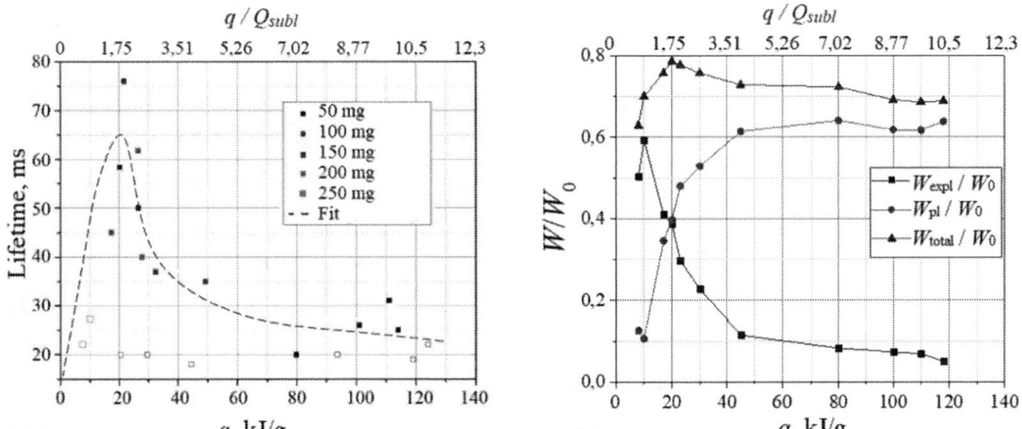

(a) (b)

Figure 3. The dependence of the lifetime (a) and the efficiency of energy transfer to the load (b) on the specific energy input q.

The specific energy spent on the explosion W_{expl}/W_0 decreases with the growth of q (decrease in m), thus the efficiency of metal destruction decreases. The specific energy spent on heating and accelerating the plasma W_{pl}/W_0 increases and the amount of the liquid phase in the flow decreases as well. The electric explosion time decreases and is similar to the discharge current maximum time at q=20-30 kJ/g (~1.8–2.6 Q_{subl}).

Figure 4 shows the electrical characteristics of one mode, in which the electric explosion occurs near the current maximum and the vortex formation is formed stable (q=30 kJ/g).

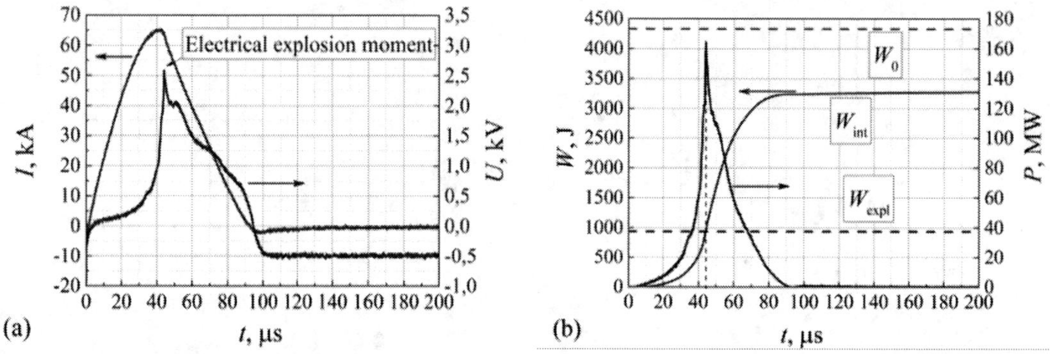

Figure 4. Electrical characteristics of the generator at U_0=3.0 kV; W_0=4.3 kJ; m=150 mg (q=30 kJ/g): (a) current and generator voltage; (b) electrical power and energy.

The maximum current is reached at ~ 40 μs and is 65 kA. The total energy release time is ~ 100 μs, the peak voltage on the generator exceeds 2.5 kV, the peak electric power is ~ 165 MW. W_{int} ~ 3.3 kJ is transferred to the load (W_{int} / W_0 ~ 76%). The energy consumption for the explosion is W_{expl} ~ 1 kJ, the efficiency of the energy input into the plasma is ~ 53% of the stored energy (W_{pl} ~ 2.3 kJ).

With a further increase in q more than 30 kJ/g (>2.6 Q_{subl}) the total efficiency of energy transfer to the load decreases as well as the density of the outflow and the formed plasma structure (figure 2a), despite a significant decrease in the droplet phase fraction in the flow and a high efficiency of energy transfer to plasma.

4. Conclusions

The modes of operation of a pulsed electric-discharge generator of heterogeneous aluminum plasma were studied in a wide range of values of the specific stored energy (from 0.7 to 10 sublimation energies of the plasma-forming material). The specific stored energy influence on the characteristics of electroexplosive processes, which cause the transformation of a plasma flow into radiating vortex structures, has been found. An empirical ratio for stable vortex generation has been obtained (q=20...30 kJ/g). If the electric explosion occurs near the current maximum, then the plasma jet contains the optimal number of drops involved in the vortex motion, and the jet transforms into a stable vortex. Deviation from the ratio leads to an increase in the proportion of a large droplet phase, which is not involved in the vortex motion and destroys the vortex ring. A critical decrease in the mass and density of the formed plasma jet leads to the rapid destruction of the vortex structure.

Acknowledgments

This work was supported by grant 18-29-21039 from the Russian Foundation for Basic Research.

References

[1] Zharnikov M H, Kamrukov A S, Kozhevnikov I V, Kozlov N P and Roslyakov I A 2008 *Appl. Phys. Letters* **34**(1) 74

[2] Bulanov S S et al 2008 *Tech. Phys. Lett.* **34** (1) 34

[3] Bulanov S S, Esiev R U, Kamrukov A S, Kozlov N P, Morozov M I and Roslyakov I A 2010 *Tech. Phys.* **55**(11) 1633

[4] Andrianov A M and Simitsyn V I 1976 *JETP Lett.* **24**(2) 56-9

[5] Lodes A, Curry R D, Rudroff M E J, Schmidt M, Bauer A J R and Brown W 2011 *IEEE Pulsed Power Conf.* (Chicago: IEEE) 1077
[6] Yusupaliev U, Yusupaliev P U and Shuteev S A 2007 *Tech. Phys.* **52**(7) 872-83
[7] Volodin L Y and Kamrukov A S 2020 *Proc. 7th Int. Congr. on Energy Fluxes and Radiat. Eff. (Tomsk)* (IEEE) pp 98-101
[8] Lodes A and Curry R D 2012 *IEEE Int. Power Modulator and High Voltage Conf.* (San Diego: IEEE) 562-5
[9] Volodin L Y and Kamrukov A S 2019 *J. Phys.: Conf. Ser.* **1393** 012063

Journal of Physics: Conference Series **2270** (2022) 012060 doi:10.1088/1742-6596/2270/1/012060

OPTIMIZATION OF THE CONDITIONS FOR THE SYNTHESIS OF ZINC OXIDE NANOPARTICLES BY PLASMA CHEMICAL METHOD

S M Shavaleeva[1], Z A Uktamov[1], E A Erlingayte[1], K S Nuriddinov[1]

[1]Kazan National Research Technical University named after A. N. Tupolev - KAI, Kazan, Russia

shsm@mail.ru

Abstract. Zinc oxide nanoparticles (ZnO) have a lot of useful properties, thanks to which they are used in various fields of science, technology and medicine. The properties of such nanoparticles largely depend on the method of their synthesis. This paper presents the results of optimization the conditions for the synthesis of zinc oxide nanoparticles by the plasma chemical method, as well as, clarifying the structure of the obtained samples. It is shown that the lack of air and the duration of the discharge affect the size and structure of the resulting particles. Analysis of the data obtained shows that synthesis in a vacuum chamber on graphite electrodes located inside a galvanived pipe at a pressure of 66, 7 k PA, a voltage of 20 V, DC power of 75 A is optimal for obtaining zinc oxide nanorods by the plasma chemical method.

1. Introduction

The development of technological bases, simple and powerful methods for obtaining nanoscale ordered structures of zinc oxide with reproducible properties is of great interest in order to create effective materials with specified properties as a functional material [1-3].

Hollow objects in which the size of one of the measurements falls into the nanometer range are of particular interest [4]. A high degree of anisotropy of properties, the effects of quantum localization, a large specific surface area and selective adsorption on the inner walls allows them to be attributed to promising materials. There is also an increased interest in quasi-one- dimensional filamentous crystal, thin and thick zinc oxide films, which are seen as promising in the field of creating miniature optoelectronic systems, physical-chemical sensors.

Zinc oxide (ZnO) nanoparticles, which have useful semiconductor, optical, piezoelectric and catalytic properties, are used in various areas of science, technology and medicine: in displays, gas sensors, piezoelectric devices, as well as catalysts [5].

The photocatalytic properties of zinc oxide are particular interesting in the area of reducing the technogenic negative impact on the environment. When cleaning industrial effluents from organic toxicants, the method of heterogeneous photocatalysis can be applied [6]. A photocatalyst based on an oxide semiconductor material, including zinc oxide, increases the cleaning efficiency under the influence of electromagnetic radiation [7]. Non- toxicity, chemical resistance (including at high temperatures), high reactivity of zinc oxide in the photocatalytic oxidation of organic pollutants, the possibility of modifications of structural, morphological surface and optical properties allows us to consider it as a promising catalyst that retains its photocatalytic activity after several cycles of reuse [8, 9].

Content from this work may be used under the terms of the Creative Commons Attribution 3.0 licence. Any further distribution of this work must maintain attribution to the author(s) and the title of the work, journal citation and DOI.
Published under licence by IOP Publishing Ltd

2. Literature Review

The useful properties of such nanoparticles largely depends on the method of their synthesis, which differ from each other, in relative ease of implementation, the ability to control the size, morphology, composition and structure of the product, environmental safety and economic feasibility. Currently, solid-phase, liquid- phase, gas- phase methods of synthesis of nano- and microparticles of zinc oxide have been made regarding the relative arrangement of atoms and their electronic structure.

Liquid- phase methods (method of precipitation from aqueous solutions of Zn salts hydrotermal method, microemulsion method) [10] make it possible to obtain homogeneous rod- shaped zinc oxide particles with a narrow size distribution.

For example, during the synthesis by the hydrothermal method at a reaction temperature 120 C and a duration of 24 hours, ZnO nanoparticles with a rod diameter of 450+_ 50 nm and a length of over 10 microns were obtained [11].

Gas- phase methods make it possible to obtain a wide nomenclature of ZnO structures: nanotubes, nanorods, nanocircles, nanofilms, nanolegs etc. Depending on the goal setting, it is possible to vary the geometry of the synthesis chambers, preionization methods, etc. [12].

Pure, structurally homogeneous, monocrystalline nanofilms of Zn, Sn, In, Cd, and Ga oxides with a rectangular cross section and constant dimensions were synthesized by the method of thermal sublimation [13]. The authors describe the structure of ZnO, as a series of alternating planes consisting of tetrahedral ions O and Zn arranged alternately along one axis. They form polar surfaces, leading to spontaneous polarization along the axis and the appearance of a dipole moment and a divergence in surface energy. The authors note that the process of thermal evaporation is very sensitive to the concentration of oxygen in the system, which affects not only the volatility of the starting material, but also the stoichiometric composition of the gas phase, as well as the formation of the product.

In the paper [14], the authors demonstrated the formation of the heterostructure of coaxial Zn-ZnO nanorods by the pyrolysis of zinc acetylacetone and suggested that the single- crystal structure of ZnO is a shell of a core of pure zinc, which can be considered as the basis for the subsequent spontaneous formation of single-crystal nanotubes of ZnO.

At the same time, a number of authors claim that although low- valent zinc compounds Zn are extremely rare, compounds containing the Zn-Zn bond can be obtained and are quite stable under the conditions studied [15, 16].

In the previous work [17], the authors of this article reviewed the results of plasma chemical synthesis of zinc oxide nanoparticles in an argon arc on a graphite electrode under conditions of oxygen deficiency. The obtained samples were examined by electron microscopy. The analysis of elements showed that the samples contain 70 wt.% Zn, 23 wt.% O, 5,6 wt.% C. It was assumed that under these conditions carbon nanotubes coated with zinc oxide in the form of a "pipe in a pipe"or" Russian matryoshka" were obtained.

3. Experimental part

The aim of these research was to optimize the conditions for the synthesis of zinc oxide nanoparticles by the plasma chemical method, as well as to refine the structure of the obtained samples.

The experiment was carried out in a vacuum chamber for plasma chemical synthesis of nanostructures, the scheme of which is shown in Fig.1.

As a source of Zn, a galvanized iron tube with a diameter of 19mm was used inside which an arc discharge took place. Graphite, molybdenum and zinc rods were used as electrodes, a copper plate or micron mesh made of stainless steel were used as a substrate, a power source was a DC generator.

An observation in the wall of the water - cooled vacuum chamber allowed to observe the experiment, a special device made it possible to adjust the distance between the electrodes.

During the experiments, the pressure inside the chamber, the gas surrounding, the current and voltage varied. Experimental data on the optimization of the conditions for the synthesis of zinc oxide nanoparticles by plasma chemical method are given in Table 1. The obtained samples were examined by transmission electron microscopy method (TEM).

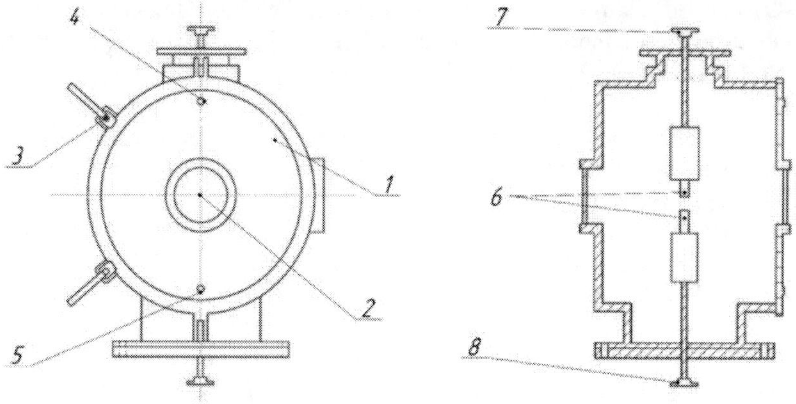

Figure 1. Vacuum chamber for plasmachemical synthesis of nanostructures. 1- front door; 2 - viewing window for observation in the front door; 3 - door lock; 4 - water supply pipe; 5- water outlet pipe; 6 - electrodes; 7 - upper regulator of the distance between the electrodes.

Table 1 shows that experiments N 5, 6, and 7 turned out to be effective.

Table 1. Conditions for the synthesis of ZnO nanoparticles by plasmachemical method.

Experiment conditions		Experiment number						
		1	2	3	4	5	6	7
Electrode material		Graphite		Graphite and Zn-galvanized pipe		Graphite and molybden	Graphite	Graphite
Air surrounding		Air				Argon+ Air	Argon+ Air	Air
Source of zinc		ZnO (s)	Zn-galvanized pipe					
Pressure, kPa		38	100	100	100	66,7	66,7	66,7
Current strength, A		15-30	0,1	0,1	0,1	60	75	75
Voltage, B		30	5000	5000	5000	20	20	20
Discharge duration, sec		50	30	30	30	90	40	60
Result		Negative				Plates	Nanorods	
Quantity						A few	Many	Many
Colour						Transparent	White	White
Linear dimensions	*d*= diametr					-	120-150 нм	70-300 нм
	l= length/ width					10-30 mkm/ 5-10 mkm	10^6 nm	$1\text{-}2*10^4$ nm
	h= thickness					0,75 mkm	-	-

Figures 2 and 3 show electronic images of the nanostructures obtained in experiment N 7.

Figure 2. Zinc oxide nanostructures, magnification by 1500 times

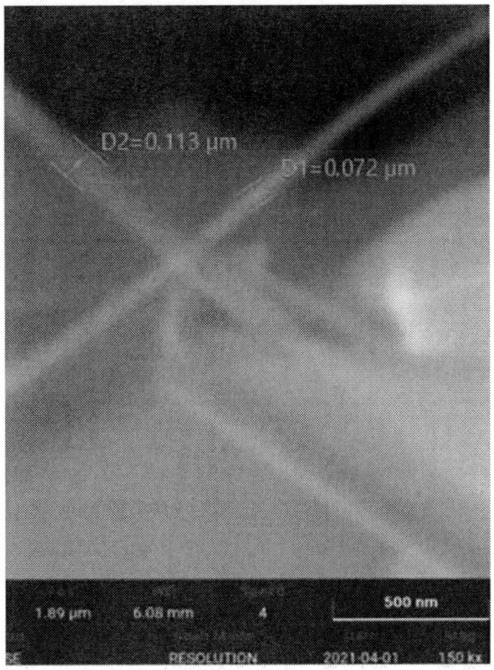

Figure 3. Zinc oxide nanostructures, magnification by 150000 times

4. Conclusions

It can be seen that the lack of air and the duration of the discharged affect the size and structure of the resulting particles: in an argon- air environment depending on the duration of the discharge, the formation of both lammellar particles and nanorods with a very high length/meter ratio is possible. Nanorods with a wide range of particle size dispersion and a lower length/meter ratio are obtained in the air.

The data of the elemental analysis obtained during the experiment of №7 nanorods were also analyzed. Elemental composition 70 weight % Zn, 23 weight % O, 5,6 weight % assumes the general chemical formula of the obtained particles - Zn_2O_3C. The presence of carbon may indicate the formation of a carbon nanotube coated with zinc oxide. However, the developed specific surface area and high adsorption capacity of the obtained samples, as well as the low percentage of carbon atoms characteristics of all nanostructures, suggest that it may be carbon dioxide CO_2 adsorbed from the air. Deducting CO_2 from the general formula, we get the formula of the resulting substance Zn_2O. This indicates two possible variants of the structure of the resulting substance: either it is the structure of a ZnO coaxial nanorode, in which a core of pure zinc is covered with a thin regular layer of zinc oxide, or it consists of Zn(I) oxide. Clarification and verification of the assumption made will be the subject of the further research.

Thus, the analysis of the data obtained showed that the synthesis in a vacuum chamber on graphite electrodes located inside a galvanized pipe at pressure of 66.7 kPa, a voltage of 20 V, a DC power of 75 A with a charge duration of 60 seconds is optimal for obtaining zinc oxide nanorods by the plasma chemical method.

References

[1] Burakov, V. S. Morphology and optical properties of zinc oxide nanostructures synthesized by thermal and electric discharge sputtering. [Text]/V.S.Burakov, N.V.Tarasenko, Ye.A.Nevar, M.I.Nedelko // Journal of technical physics. - 2011. - N 2.- P. 89- 97.
[2] Plasma-Chemical Decomposition of Hydrocarbons on the Basis of the Micro-Arc Discharge with Disc Electrodes Rotating in the Bulk of Raw Materials/ Saifutdinova, A.A., Sofronitskiy, A.O., Timerkaev, B.A., Saifutdinov, A.I. Russian Physics Journal, 2020, 62(11), стр. 2132–2136 / DOI 10.1007/s11182-020-01957-0
[3] Growing of Carbon Nanotubes from Hydrocarbons in an Arc Plasma/ Timerkaev, B.A., Ganieva, G.R., Kaleeva, A.A., Israfilov, Z.K., Sofronitskii, A.O. Journal of Engineering Physics and Thermophysics, 2019, 92(5), стр. 1248–1252/ DOI 10.1007/s10891-019-02040-3
[4] Gorbik P.P. Production and research of zinc oxide nanotubes]/P.P.Gorbik, I.V.Dubrovin, G.N.Kashin // Metallophysical latest technologies. - 2016. - V.38.N 3. - P.341- 351.
[5] Avdeeva A.V. Obtaining rod- shaped zinc oxide nanoparticles by deposition [Text]/A.V.Avdeeva, S.Tsan, A.G.Muradova, Ye.V.Yurtov // Chemical Technology. – 2014. - V.15. Issue 12. - P. 723- 728.
[6] Lathasree S. Heterogeneous photocatalytic mineralization of phenols in aqueous solutions / S.Lathasree, B.Nagesware Rao, V.Sivashankar, K.Sadasivam // Journal of Molecular Catalysis A: Chem. – V. 223, – 2004. – pp. 101–105.
[7] Oppenlander T. Photochemical Purification of Water and Air // Weinheim: Wiley-VCH, – 2003. – P. 368.
[8] Sosnin I.M. Modified zinc oxide nanoobjects designed for photocatalytic post- treatment of wastewater under the influence of visible light / I.M.Sosnin, D.A.Ulesova, A.A.Vikarchuk, A.Ye.Romanov // VIII Psysical materials science international school with elements of a scientific school for young people Actual problems of strength LIX, Tolyatti ,- 2017. - P.52 - 53.
[9] Ulesova D.A., Heterogeneous photocatalysis in wastewater treatment process from organic pollutants. / D.A.Ulesova, I.M.Sosnin, A.Ye.Romanov, A.A.Vikarchuk // International forum of the sixth international environmental congress ELPIT- 2017, Samara, - 2017. - P.199- 203.
[10] Vokhmintsev K.V. Regulates of formation of nanosystems based on ZnO and Bi2O3 and their physicochemical properties Dissertation for the degree of candidate of Chemical Science. - Moscow, IMMS named after A.A.Baikova RAS, 2015. - 182c.
[11] Hsu K.C., Liao J.D, Fu Y.S / Hydrothermal Synthesis of ZnO Nanorods using the HMT Surfactant // Integrated Ferroelectrics. — 2013. —V.143. — P. 97-106.
[12] X. TY Asadullin, IG Galeyev, BA Timerkayev Journal of Physics: Conference Series T.669 №1, C.012001
[13] Zhong Lin Wang / Zinc oxide nanostructures: growth, properties and applications //J. Phys.: Condens. Matter 16 (2004) —P.829–858.
[14] Jih-Jen Wua and Sai-Chang Liu, Chien-Ting Wu and Kuei-Hsien Chen, Li-Chyong Chen / Heterostructures of ZnO–Zn coaxial nanocables and ZnO nanotubes // applied physics letters volume 81, number 7 12 august 2002. P. 1312-1314.
[15] J. F. Van der Maelen, E. Gutiérrez-Puebla, Á. Monge, S. García-Granda, I. Resa, E. Carmona, M. T. Fernández-Díaz, G. J. McIntyre, P. Pattison and H.-P. Weber / Experimental and theoretical characterization of the Zn-Zn bond in [Zn2(5-C5Me5)2] /Acta Cryst. (2007). B63, 862-868
[16] Stephan Schulz, Daniella Schuchmann Ulrich Westphal and Michael Bolte. Dizincocene as a Building Block for Novel Zn–Zn-Bonded Compounds? Organometallics 2009, 28, 5, 1590–1592
[17] Timerkaev B.A., Felsinger V.S., Kaleeva A.A., Erlingayte E.A., Uktamov J.A., Nuriddinov H.S. // H.S."Plasma-chemical synthesis of zinc oxide nanotubes", Journal of Physics: Conference Series, 1870 (2021), 012003, IOP Publishing, doi:10.1088/1742-6596/1870/1/012003

About graphite evaporation dynamics caused by radiation in near zone of magnetoplasma compressor discharge

A V Pavlov[1], T S Shchepanyuk[1], E O Chebykin[1], A S Skriabin[1], V D Telekh[1]

[1]Bauman Moscow State Technical University, Moscow, Russian Federation

telekh@bmstu.ru

Abstract. The results of processes above the surface of graphite samples which exposed with powerful vacuum ultraviolet (VUV) radiation experimental studies are presented. The characteristic radiation exposure time was 3...100 μs, the radiation energy in the vacuum region of the spectrum was 1...2 kJ. The main characteristics of the processes were recorded on direct, shadow photographs and interferograms. The temporal dynamics of the expansion of the plasma layer formed during evaporation of the flat graphite targets surface by VUV radiation, as well as the dynamics of buffer gas compression and shock wave motion are analyzed. It is shown that the temperature of the carbon plasma above the samples surface reached 1 eV. Research is relevant in the fields of space, plasma technologies, the development of the element base of microelectronics and fundamental studies of the powerful radiation interaction with matter.

1. Introduction

The phenomenon of light erosion is realized in a wide range of devices and technical applications. For example, light erosion (ablation) can occur when the surface of materials is treated with plasma [1], [2] and radiation [3], [4] flows, during the formation and modification of nano-structured relief on the surfaces of various construction materials [5], in ablative pulsed plasma thrusters and electric propulsion systems [6], in plasma dynamic sources of UV/VUV/EUV radiation [7], [8], etc. Of great interest is the impact of powerful radiation on the matter when a photons energy higher than the ionization potential of the samples atoms (i.e. radiation of the vacuum ultraviolet spectrum range). For many such devices, the plasma-forming substance is polymers, dielectrics, special heat-protective coatings, etc. These materials are ablated under the influence of radiation from the discharge, and the efficiency of such devices is largely determined by the ability to control the dynamics of the mass flow of the ablating material. At the same time, the determination of the characteristics of the light-erosion plasma and its spatial and temporal dynamics is important for studying the gas-dynamic "response" from the surface. This explains the interest in these studies.

2. Experimental setup

The study of the interaction of high-power fluxes of short-wave radiation of the UV and VUV spectral range with graphite samples was carried out on an experimental setup described in [9]. This setup was created on the basis of a plasmodynamic discharge – the discharge of an erosion type magnetoplasma compressor *1*, *2*, *3* (Figure 1) [7, 9, 10], which was installed in a vacuum chamber *4*. The source of radiation in such a discharge is a plasma focus, which is formed on the axis of symmetry near the central

Content from this work may be used under the terms of the Creative Commons Attribution 3.0 licence. Any further distribution of this work must maintain attribution to the author(s) and the title of the work, journal citation and DOI.
Published under licence by IOP Publishing Ltd

electrode [11]. Such discharges are characterized by high spectral-brightness characteristics of radiation in the UV and VUV spectral regions and have an extended glow region [8, 11].

Plasmodynamic, radiative and electrical properties of this MPC have been presented in a number of articles and reviews [10 – 12]. These properties describe both the features of the discharge itself with a good efficiency of ≈ 70…90%, and radiation with high-performing spectral-brightness characteristics in the UV and VUV spectral regions with a light emission efficiency of ≈ 40…60% of the energy deposited into the plasma.

The experimental methods were similar to our previous works [9, 13]. The vacuum chamber was pumped out to a pressure of $p \leq 1$ Pa, then filled with background gas (in this series of experiments we used Ne). This procedure was repeated after each discharge, since erosion processes change the radiation spectrum of discharge. Capacitive storage *5* (a battery of low-inductance capacitors with a capacity of 18 μF, 25 kV) was switched with the MPC using a thyratron *6* (Pulsed Technology Ltd., Ryazan, Russia). Current waveforms were recorded with Pearson current monitor 110 (*7*) and Tektronix TDS 2024b oscilloscope.

Graphite bars (with dimensions of 30 mm×50 mm and a thickness of 10 mm) *8* were used as samples for the VUV exposure. They were installed with their long side along the discharge axis as shown in Figure 1. Thus, the nearest sample end to the MPC barrel received 2…2.5 times more energy than the far end. This made it possible, among other things, to register different modes of expansion of a vapor-plasma flow when moving from one end of the target to another. The pressure of background gas (neon) in the chamber was p_0=5.32·10^4 Pa (a density was ρ_0=0.9 kg/m^3). The charging voltage of the capacitor was U_0 = 20 kV that corresponded to a stored electrical energy of 3.6 kJ.

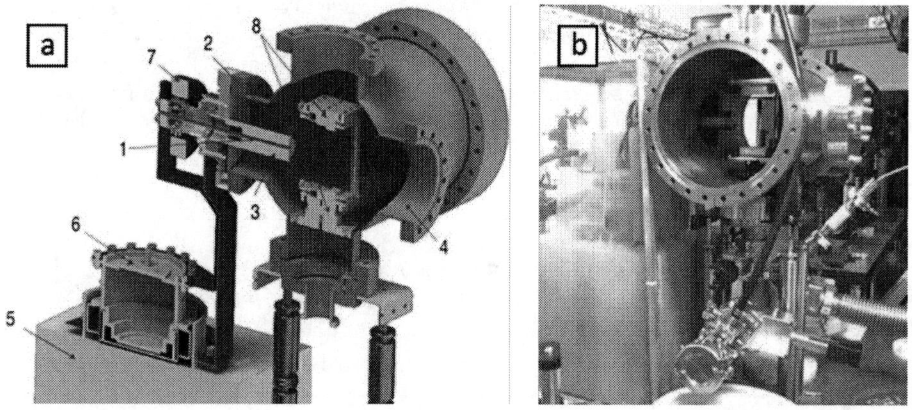

Figure 1. 3D model (a) and photo (b) of experimental set up: 1 – central electrode, 2 – external electrode, 3 – ablative dielectric, 4 – vacuum chamber, 5 – capacitor, 6 – thyratron, 7 – current monitor, 8 – samples holders.

Optical laser diagnostics was used for visualization and quantitative study of a gas-dynamics response from the graphite samples under the VUV radiation. It included double-exposure holographic interferometry and the Toepler's schlieren scheme in the bright field mode [9, 14 – 16].

The method of plasma flows laser diagnostics used in our research was described in [9]. The peculiarity of the technique used is that we get a hologram of a focused image [15] with visualization of large optical fields (in our case 150 mm) with a constant resolution (no more than 50 microns).

Note that the emission spectrum of the MPC discharge respect to time. At first, the main radiation energy is concentrated in the visible spectrum range. We assume that under the influence of this radiation, the surface of the samples is heated, and their diffuse evaporation also takes place. At about 5 μs after the discharge start, a plasma focus is formed, emitting in the VUV range (this moment should correspond to the maximum discharge current). Under the influence of VUV radiation, carbon vapor ionization occurs (ionization potential I_C = 11.25 eV, the buffer gas transmission boundary is determined by the ionization potential I_{Ne} = 21.6 eV). Photoionization of carbon vapor leads to the formation of a plasma layer above the surface of the samples. This plasma, expanding, creates a "plasma piston", above which there was a layer of buffer gas compressed by the impact and a shock wave.

3. Results discussion

Figure 2 shows interferograms of near-surface layers above a graphite sample obtained with different delays (6 and 8 μs) from the beginning of the discharge. The moment of the photo was determined by oscillograms (Figure 3), in which channel 1 is a photodiode signal that captures a pulse of probing laser radiation (the moment of receiving the interferogram), and channel 2 is a discharge current oscillogram (attenuated by 10).

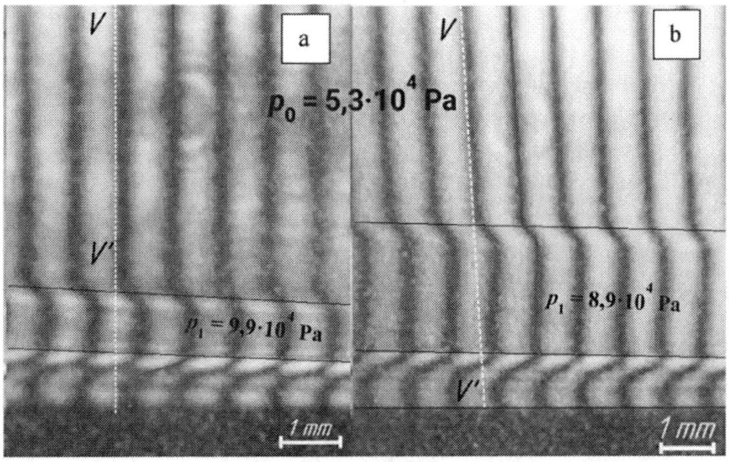

Figure 2. Graphite evaporation interferograms (a – at 6 μs, b – at 8 μs)

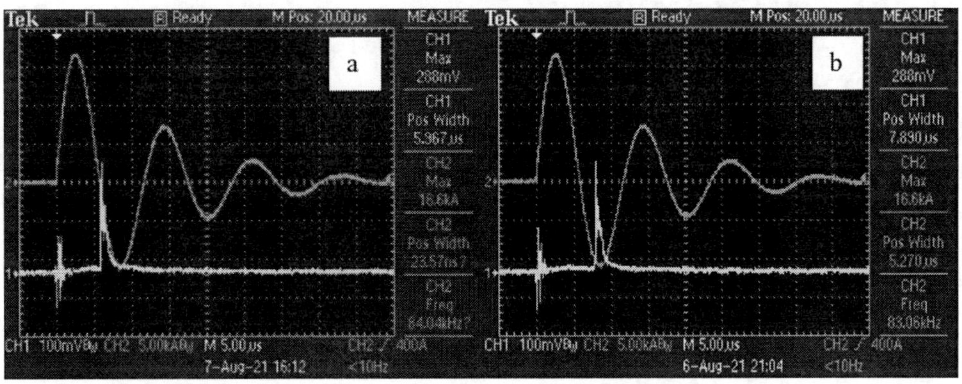

Figure 3. Current oscillogramms (a – at 6 μs, b – at 8 μs)

The analysis of the obtained interferograms (Figure 2) makes it possible to determine the parameters of both the shock wave and the shock-compressed layer and the plasma layer formed from the vapors of the sample material as a result of effective photoionization. The general principles of obtaining quantitative information from the phase inhomogeneities interferograms are described in the literature [15, 16]. The feature of the experimental scheme is that the length of the probing laser radiation optical path in the "phase inhomogeneity" is constant, which greatly simplifies the processing of interferograms. The interpretation of interferograms in our experimental scheme is described in detail in [13].

The features of the shift of interference fringes over a graphite sample (Figure 2) are associated with plasma formation above the surface and expansion of the carbon plasma in the background gas (in neon). A detailed description of the characteristic areas of interferograms is given in [13]. The upper part of the interferograms is the undisturbed region corresponding to the background gas at the initial pressure ($p_0 \approx 5.3 \cdot 10^4$ Pa). The next shift of the fringes to the right (jump) this is a sharp change in the density of the gas in the shock wave. Below is a layer of shock-compressed gas. Since the interference fringes in the shock-compressed gas are smooth, this indicates a slight change in its pressure in height. After that, the fringes turn to the left. The area from this turn to the sample surface is the plasma layer. The bending of the fringes in this region shows a change in the refractive index (n-1) in the plasma layer and reflects the complex dependence of the thermodynamic parameters of the plasma on the height above the graphite sample.

At the time of the interferogram registration, the background gas occupies the volume from the contact boundary to the shock wave. At the initial moment, the background gas occupied the entire volume from the sample surface to the shock wave. Assuming that the contact boundary moves at speeds less than the local speed of sound, we assume the process isobaric.

From Figure 2a we determine that at 6 μs the compression ratio of Ne is approximately 1.5 times, and at 8 μs (Figure 2b) the compression ratio is approximately 1.4 times. The presence of interferograms makes it possible to refine the compression degree using the relationship of the fringes shift with the change in gas density (due to changes in the refractive index).

So, at the 6 μs, the background gas is compressed by about 1.44 times, at 8 μs, the compression ratio is approximately 1.35. Using the shock adiabat, knowing the ratio ρ_0/ρ_1, the pressure ratio p_1/p_0 can be determined as [16]:

$$\frac{p_1}{p_0} = \left(\frac{\gamma+1}{\gamma-1} - \frac{\rho_0}{\rho_1}\right)\left(\frac{\gamma+1}{\gamma-1}\cdot\frac{\rho_0}{\rho_1} - 1\right)^{-1}$$

Under the implemented conditions for a monatomic gas (with an adiabatic index of $\gamma = 1.67$) for $p_1/p_0 \approx 1.44$, the absolute value of the pressure behind the shock wave front was $p_1 \approx 9.9 \cdot 10^4$ Pa.

Knowing the position of the shock wave front for two consecutive time moments, we determine its velocity $V_{SW} \sim 700$ m/s. At the same time, the speed of sound in neon is $V_s \sim 455$ m/s. Thus, the Mach number is approximately $M \sim 1.5$.

Under the assumption of local thermodynamic equilibrium, the change in the refractive index in the plasma (the shift of interference fringes in the plasma) and the plasma parameters have an unambiguous relationship [17]:

$$\Delta n = \frac{L}{\lambda}\left[2\pi \sum \alpha_x N_x - 4{,}49 \cdot 10^{-14}\lambda^2 \alpha_e\right] n_0$$

here N_x is the partial concentration of atoms or ions, α_e is the ionization degree, n_0 - plasma concentration, α_x is the dipolar polarizability of the atom or ion [18]. Thus, the maximum plasma temperature is about 10 kK (1 eV) inside the plasma layer.

4. Conclusion

Processing of the interferograms made it possible to determine the parameters of graphite plasma during evaporation of a samples by VUV radiation. A method of diagnostics and quantitative description of the parameters of plasmodynamic structures arising from the interaction of powerful broadband radiation on the surface of structural materials is proposed. Information about the processes is obtained as a result of a limited series of experiments (practically - a single one). As a result it is possible to determine both parameters of generated flows (pressure, density in shock-compressed layer, etc.) and thermodynamic parameters inside plasma. So, plasma temperature above graphite target surface (in approximation of local thermodynamic equilibrium) was about 10 kK.

Acknowledgments
This work was performed using research facilities cluster "Beam-M" of Bauman Moscow State Technical University following the government task by the Russian Ministry of Education and Science (0705-2020-0046).

References

[1] Astashynski V M et al. 2020 *High Temperature Material Processes* **24 (2)** 99-107
[2] Shanenkov I et al. *Ceramics International* 2021 **47(5)** 6884-95.
[3] Lippert T 2005 *Plasma Processes Polym.* **2** 525-46.
[4] Skurat V 2003 *Nuclear Instruments and Methods in Physics Research, Section B: Beam Interactions with Materials and Atoms* **208(1-4)** 27-34
[5] 2006 *EUV Sources for Lithography* / ed. V Bakshi. Wash. Bellingham: SPIE Press, 2006.
[6] Popov G A and Antropov N N 2006 *Acta Astronautica* **59** 175
[7] Kamrukov A S et al. 1989 *High Temperature* **27 (1)** 141–55.
[8] Protasov Yu S et al. 2014 *Enciclopedia of Low Temperature Plasma. Vol IX-4. Plasma aerodynamics*. Moscow: Yanus-K 383-436
[9] Pavlov A V, Protasov Yu Yu, Telekh V D, Shchepanuk T S 2019 *Scientific Visualization* **11 (3)** 111-25
[10] Kozlov N P, Protasov Y S 1978 *Phys. Lett. A* **67 (3)** 191
[11] Kamrukov A S, et al. 1984 *Sov. J. Quantum Electron* **14 (8)** 1094 http://dx.doi.org/10.1070/QE1984v014n08ABEH005371
[12] Protasov Yu S, Chuvashev S N, Shchepanyuk T S, Kutyrev M V 1992 *High Temperature* **30 (2)** 236
[13] Pavlov A V, Shchepanuk T S, Chebykin E O, Protasov Yu Yu, Telekh V D 2021 *Journal of Physics: Conference Series* **2064** 012078
[14] Settles G S 2001 *Schlieren and Shadowgraph Techniques*. Springer-Verlag, Berlin Heidelberg, Germany 143-63
[15] 2005 *Handbook of Holographic Interferometry: Optical and Digital Methods* / ed. T Kreis. WILEY-VCH Verlag GmbH & Co. KGaA
[16] Ostrovskaya G V 2008 *Technical Physics* **53 (9)** 1103
[17] Raizer Yu P 2011 Introduction to Fluid Dynamics and Shock Wave Theory for Physicists (Dolgoprudny: Publishing House Intellect) [in Russia]
[18] Peter Schwerdtfeger, Jeffrey K 2018 Nagle Table of static dipole polarizabilities of the neutral elements in the periodic table. *Molecular Physics* DOI: 10.1080/00268976.2018.1535143.

Interaction of powerful plasma jets with Mo/Si multilayers

A S Skriabin[1], V D Telekh[1], A V Pavlov[1], A G Turyanskiy[2], V M Senkov[2], P A Tsygankov[3], F F Parada-Becera[3]

[1] Bauman Moscow State Technical University, 2-ya Baumanskaya 5, 105005, Moscow, Russia
[2] Lebedev Physical Institute of the Russian Academy of Sciences, Leninskiy Avenue 53, 119991, Moscow, Russia
[3] Industrial University of Santander, Carrera 27 #Calle 9, Bucaramanga 680002, Colombia

Abstract. The paper presents some results of experimental studies of Mo/Si multilayer degradation under UV exposure from compressed plasma jets in air. Such high current plasma flows generate broadband powerful UV radiation fluxes. A single exposure caused a slight decrease of reflectivity from 0.54 to 0.47 due to the coating contamination by the debris from the plasma light emitter.

1. Introduction

The high current emitters based on compressed plasma jets are sources of powerful broadband radiation (with fluxes of $\approx 10^{20}$–10^{22} photon/s per a single pulse) [1]. Ponderomotive forces lead to formation of a high temperature plasma focus with a brightness temperature of ≈30–100 kK [2]. The maximal spectral brightness lies within the range of ≈30–100 nm. Operation of the minimal wavelength is possible by a choosing of background gas in the discharge chamber [3]. The presented emitters can be a basis of novel systems for EUV lithography.

Mo/Si multilayers are used as mirrors for transport of EUV/VUV fluxes from the emitters to the irradiated surface [4]. We note that such bilayers can degrade due to coating contamination, partial evaporation, etc. The previously reported study presented some results of temperature damage of Mo/Si bilayer with formation of molybdenum silicides [5]. The high pressure plasma light sources emit such high broadband radiation fluxes which are characterized by a global temperature and stress fields in the irradiated substrates. A complex nature of interaction (mechanical impact, VUV exposure) provides a lot of transformation processes including morphology changing, chemical reaction, surface degradation etc. This can decrease reflectivity and reduce the mirror quality. So, the aim of the present study is experimental evaluation of degradation efficiency.

2. Experimental procedures

Mo/Si multilayers were prepared by magnetron sputtering on fused silica bars (15×8×3 mm). The corresponding experimental vacuum setup is presented in figure 1. A total thickness was 3420.0 nm. Bilayer thickness was 11.4 nm at the bilayer number of 300.

Irradiation of the Mo/Si multilayers was carried out with a coaxial plasma accelerator – magnetoplasma dynamic compressor with a capacitive source. This is a high current (I>100 kA) setup which generates strong shock waves, high brilliance UV/VUV radiation fluxes (with a brilliance

Content from this work may be used under the terms of the Creative Commons Attribution 3.0 licence. Any further distribution of this work must maintain attribution to the author(s) and the title of the work, journal citation and DOI.
Published under licence by IOP Publishing Ltd

temperature of ≈30–100 kK) etc. The samples were installed in the chamber which was fulfilled by a background gas (neon and air). Pressure in the chamber was ≈300–400 torr. A single radiation exposure was studied. A stored energy in the capacitor was ≈3.6 kJ. The original and irradiated coatings were studied with X-ray reflectometry (XRR, CDP systems, Moscow) and electron microscopy (TESCAN VEGA 3 XMU, Czech Republic) with a system of energy dispersive spectroscopy EDX (AZTEC Energy Analysis System). Interpretation of XRR spectra was made with the technique [6].

Figure 1. Experimental setup for Mo/Si multilayer preparing

3. Results and discussion

UV/VUV exposure in neon with the maximum quanta energy of 21.56 eV (discharge in neon) led to a significant surface degradation [1]. Surface erosion was detected. The coating was damaged. But no changes were detected for irradiation by soft quanta for discharge in air (≈6 eV). Figure 2 presents XRR spectra of original and Mo/Si multilayers irradiated in air. We found that even a single exposure in air led to a slight decrease of reflectivity from 0.54 to 0.47.

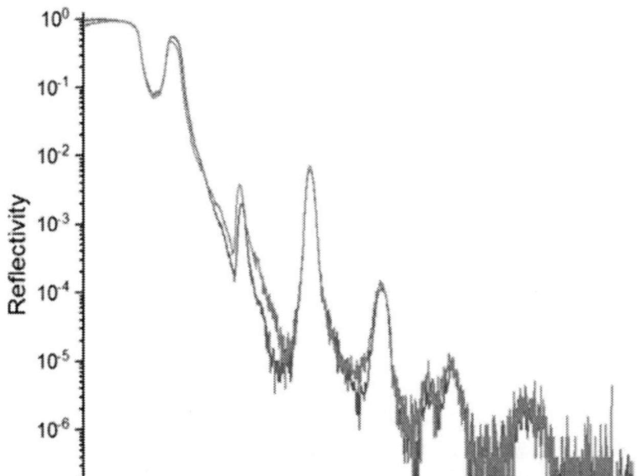

Figure 2. XRR spectra of original (red) and irradiated in air (violet)

Such degradation was caused by a local damage of the surface by molten debris from electrodes. But sufficient changes of Mo/Si bilayer thickness were not found. The typical EDX spectrum of irradiated in air Mo/Si multilayers is presented in figure 3. This contains the spectral lines of Fe, Cr and Ni. Their appearance was caused by contamination of condensed vapors which were generated by electrode evaporation.

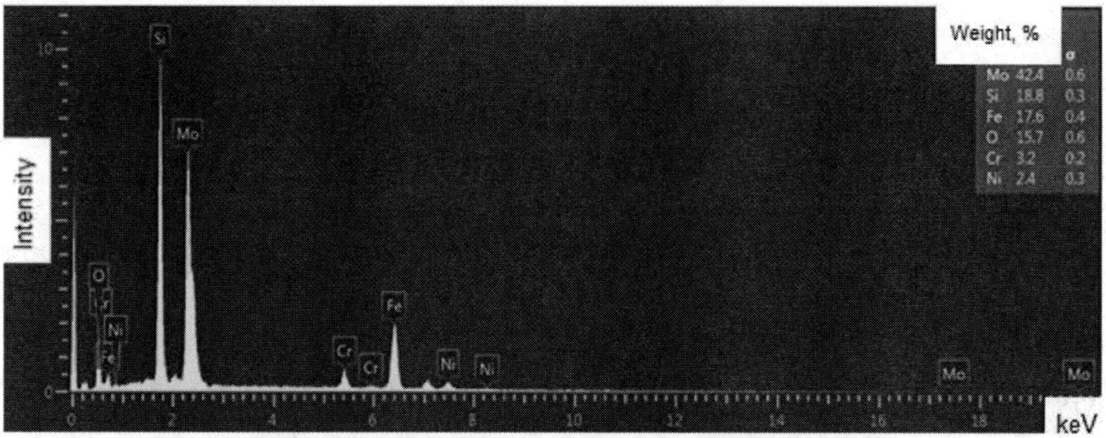

Figure 3. EDX spectra of Mo/Si multylayer irradiated in air

4. Conclusion

Coaxial plasma accelerators (so called magneto plasma compressors) are sources of strong shocks and broadband radiation (including hard UV and VUV quanta). Irradiation of materials (including thin multilayers) leads to a complex of phase and chemical transformations and modification. Exposure of powerful UV/VUV radiation in air (with a maximal energy of ≈6 eV) caused minimal changes in the studied Mo/Si multilayers. As found, a single exposure decreased a reflectivity from 0.54 to 0.47 due to surface damage by debris emission and vapor condensation.

Acknowledgments

The reported study was funded by RFBR and ROSATOM (project № 20-21-00087). This work was performed using research facilities cluster "Beam-M" of Bauman Moscow State Technical University.

References

[1] Telekh V D et al. 2022 *Coating* **12(2)** 290
[2] Anevsky S et al. *Nucl. Instrum. Methods Phys. Res. Sect. A Accel. Spectrometers Detect. Assoc. Equip.* 1989 **282** 714–715.
[3] Couprie M. *J. Electron Spectrosc. Relat. Phenom.* 2014 196 3–13
[4] Trost M. et al. *Opt. Express* 2013 **21** 27852–27864.
[5] Fedorenko A.I. et al. Thermally induced structural and phase transformations of Mo-Si and MoSi 2 -Si X-ray multilayer mirrors. In Proceedings of the X-ray Optics and Surface Science, Moscow, Russia, 10 November 1994; pp. 15–24.
[6] Popov N.L. et al. *Semiconductors* 2003 **37** 675–680.

Journal of Physics: Conference Series **2270** (2022) 012063 doi:10.1088/1742-6596/2270/1/012063

Finite-difference scheme and algorithm for solving a two-dimensional system of Maxwell's equations for inductively coupled plasma

A V Gerasimov, A P Kirpichnikov, R N Gaynullin, Y D Lapteva, T M Pham

Kazan National Research Technological University, Kazan, Russia

aleks_gerasimov@mail.ru

Abstract. The paper presents a numerical scheme for solving a two-dimensional system of Maxwell equations for a cylindrical coordinate system.

1. Introduction

Although plasma production using inductively coupled radio frequency (high frequency (HF)) the discharge has been used for many decades [1-3], inductively coupled plasma (ICP) has recently aroused significantly increased interest in its new applications. For many years, high-pressure ICP has been widely studied to use it for spectral chemical analysis, for chemical synthesis using plasma, for the synthesis of nanostructures as a gas-discharge light source, as well as for crystal growth and plasma deposition. Conditions of absorption of electromagnetic energy in the conducting zone determine the profile of the magnetic component of the inductor's electromagnetic flux. For other conditions, the electrical conductivity of the gas will determine the crossing rate of the magnetic field strength. At atmospheric pressures and electron concentrations, the plasma of the discharge channel is thermally equilibrium. The conductivity is a single-valued function of pressure and temperature. Considering such an exact dependence, it is possible to carry out calculations within Maxwell's system of equations, which describes the characteristics of the electromagnetic field for the plasma torch inductor and obtain information about the thermal and electromagnetic characteristics of the ICP.

To describe the electromagnetic fields of a high-frequency induction (RF) discharge, we use a system of two-dimensional Maxwell equations written in a cylindrical coordinate system:

$$
\begin{aligned}
&\frac{1}{r}\frac{\partial(rE_\varphi)}{\partial r}=\frac{\omega}{c}H_z\sin(\varphi_{Hz}-\varphi_E)\\
&\frac{\partial\varphi_E}{\partial r}=-\frac{\omega}{c}\frac{H_z}{E_\varphi}\cos(\varphi_{Hz}-\varphi_E)\\
&\frac{\partial H_z}{\partial r}=\frac{\partial H_r}{\partial z}\cos(\varphi_{Hz}-\varphi_{Hr})+H_r\frac{\partial\varphi_{Hr}}{\partial z}\sin(\varphi_{Hz}-\varphi_{Hr})-\frac{4\pi}{c}\sigma E_\varphi\cos(\varphi_{Hz}-\varphi_E)\\
&\frac{\partial\varphi_{Hz}}{\partial r}=-\frac{1}{H_z}\frac{\partial H_r}{\partial z}\sin(\varphi_{Hz}-\varphi_{Hr})+\frac{H_r}{H_z}\frac{\partial\varphi_{Hr}}{\partial z}\cos(\varphi_{Hz}-\varphi_{Hr})+\frac{4\pi}{c}\sigma\frac{E_\varphi}{H_z}\sin(\varphi_{Hz}-\varphi_E)\\
&\frac{1}{r}\frac{\partial(rH_r)}{\partial r}=-\frac{\partial H_z}{\partial z}\cos(\varphi_{Hz}-\varphi_{Hr})+H_z\frac{\partial\varphi_{Hz}}{\partial z}\sin(\varphi_{Hz}-\varphi_{Hr})\\
&\frac{\partial\varphi_{Hr}}{\partial r}=-\frac{1}{H_r}\frac{\partial H_z}{\partial z}\sin(\varphi_{Hz}-\varphi_{Hr})-\frac{H_z}{H_r}\frac{\partial\varphi_{Hz}}{\partial z}\cos(\varphi_{Hz}-\varphi_{Hr})
\end{aligned}
\tag{1}
$$

Content from this work may be used under the terms of the Creative Commons Attribution 3.0 licence. Any further distribution of this work must maintain attribution to the author(s) and the title of the work, journal citation and DOI.
Published under licence by IOP Publishing Ltd

The values H_z from the system are obtained experimentally. The system above comprises seven unknowns $E_\varphi, \varphi_E, \varphi_{Hz}, H_r, \varphi_{Hr}, \sigma, H_z$ and six equations. Then the system is closed and electromagnetic values inside the discharge chamber of the plasma torch can be calculated.
We use a system of equations like system (1), written with respect to derivatives in the z coordinate, for numerical calculations of the electromagnetic and thermophysical parameters of an RF plasma in a finite-size inductor. As the result the system of equations has the following form:

$$\begin{cases} \frac{\partial E_\varphi}{\partial z} = -\frac{\omega}{c} H_r \sin(\varphi_{Hr} - \varphi_E) \\ \frac{\partial \varphi_E}{\partial z} = \frac{\omega}{c} \frac{H_r}{E_\varphi} \cos(\varphi_{Hr} - \varphi_E) \\ \frac{\partial H_r}{\partial z} = \frac{\partial H_z}{\partial r} \cos(\varphi_{Hz} - \varphi_{Hr}) - H_z \frac{\partial \varphi_{Hz}}{\partial r} \sin(\varphi_{Hz} - \varphi_{Hr}) + \frac{4\pi}{c} \sigma E_\varphi \cos(\varphi_{Hr} - \varphi_E) \\ \frac{\partial \varphi_{Hr}}{\partial z} = \frac{1}{H_r} \frac{\partial H_z}{\partial r} \sin(\varphi_{Hz} - \varphi_{Hr}) + \frac{H_z}{H_r} \frac{\partial \varphi_{Hz}}{\partial r} \cos(\varphi_{Hz} - \varphi_{Hr}) - \frac{4\pi}{c} \sigma \frac{E_\varphi}{H_r} \sin(\varphi_{Hr} - \varphi_E) \\ \frac{\partial H_z}{\partial z} = \frac{1}{r} \frac{\partial (r H_r)}{\partial r} \cos(\varphi_{Hz} - \varphi_{Hr}) + H_r \frac{\partial \varphi_{Hr}}{\partial r} \sin(\varphi_{Hz} - \varphi_{Hr}) \\ \frac{\partial \varphi_{Hz}}{\partial z} = \frac{1}{H_z} \frac{1}{r} \frac{\partial (r H_r)}{\partial r} \sin(\varphi_{Hz} - \varphi_{Hr}) - \frac{H_r}{H_z} \frac{\partial \varphi_{Hr}}{\partial r} \cos(\varphi_{Hz} - \varphi_{Hr}) \end{cases} \tag{2}$$

2. Method of solution

Consider a finite-difference scheme of approximation for the first, second and sixth equations presented in system (2). For the first equation we have (2):

$$\frac{E_{\varphi(i,j)} - E_{\varphi(i,j-1)}}{\Delta z} = -\frac{\omega}{c} H_{r(i,j-1)} \sin\left(\varphi_{Hr(i,j-1)} - \varphi_{E(i,j-1)}\right).$$

Where from:

$$E_{\varphi(i,j)} = E_{\varphi(i,j-1)} - \Delta z \frac{\omega}{c} H_{r(i,j-1)} \sin\left(\varphi_{Hr(i,j-1)} - \varphi_{E(i,j-1)}\right). \tag{3}$$

For the second equation of the system (2), we can write:

$$\frac{\varphi_{E(i,j)} - \varphi_{E(i,j-1)}}{\Delta z} = \frac{\omega}{c} \frac{H_{r(i,j-1)}}{E_{\varphi(i,j-1)}} \cos\left(\varphi_{Hr(i,j-1)} - \varphi_{E(i,j-1)}\right).$$

Then:

$$\varphi_{E(i,j)} = \varphi_{E(i,j-1)} + \Delta z \frac{\omega}{c} \frac{H_{r(i,j-1)}}{E_{\varphi(i,j-1)}} \cos\left(\varphi_{Hr(i,j-1)} - \varphi_{E(i,j-1)}\right).$$

We will rewrite the sixth equation of system (2) in a different form:

$$\frac{\partial \varphi_{Hz}}{\partial z} = \frac{1}{H_z} \frac{1}{r} \left(\frac{\partial H_r}{\partial r} + \frac{H_r}{r} \right) \sin(\varphi_{Hz} - \varphi_{Hr}) - \frac{H_r}{H_z} \frac{\partial \varphi_{Hr}}{\partial r} \cos(\varphi_{Hz} - \varphi_{Hr}).$$

To approximate the derivative $\frac{\partial H_r}{\partial r}$ we use the central difference. Then the finite-difference analog of this equation is written as:

$$\frac{\varphi_{Hz(i,j)} - \varphi_{Hz(i,j-1)}}{\Delta z} = \frac{1}{H_{z(i,j-1)}} \left(\frac{H_{r(i+1,j-1)} - H_{r(i-1,j-1)}}{2\Delta r} + \frac{H_{r(i,j-1)}}{r_{(i)}} \right) \times \\ \times \sin\left(\varphi_{Hz(i,j-1)} - \varphi_{Hr(i,j-1)}\right) - \frac{H_{r(i,j-1)}}{H_{z(i,j-1)}} \frac{\varphi_{Hr(i,j-1)} - \varphi_{Hr(i-1,j-1)}}{\Delta r} \cos\left(\varphi_{Hz(i,j-1)} - \varphi_{Hr(i,j-1)}\right). \tag{4}$$

Denoted $r(i) = \Delta r(i-1)$, finally, you can write:

$$\varphi_{Hz(i,j)} = \varphi_{Hz(i,j-1)} + \frac{\Delta z}{\Delta r}\frac{1}{H_{z(i,j-1)}}\left[\left(\frac{H_{r(i+1,j-1)} - H_{r(i-1,j-1)}}{2} + \frac{H_{r(i,j-1)}}{i-1}\right)\times\right.$$
$$\left.\times\sin\left(\varphi_{Hz(i,j-1)} - \varphi_{Hr(i,j-1)}\right) - H_{r(i,j-1)}\left(\varphi_{Hr(i,j-1)} - \varphi_{Hr(i-1,j-1)}\right)\cos\left(\varphi_{Hz(i,j-1)} - \varphi_{Hr(i,j-1)}\right)\right] \tag{5}$$

To find the transverse component of the magnetic field H_r we use the fifth equation of the system (1). It is equivalent to the corresponding equation of the system (2). Let's write it a little differently:

$$\frac{H_r}{r} + \frac{\partial H_r}{\partial r} = -\frac{\partial H_z}{\partial z}\cos(\varphi_{Hz} - \varphi_{Hr}) + H_z\frac{\partial \varphi_{Hz}}{\partial z}\sin(\varphi_{Hz} - \varphi_{Hr}).$$

We write a finite-difference analog of this equation using a half-weight scheme:

$$\left[\frac{H_{z(i,j)} + H_{z(i-1,j)}}{2}\cdot\frac{\varphi_{Hz(i,j)} - \varphi_{Hz(i-1,j)}}{\Delta r} + \frac{H_{r(i,j)} - H_{r(i,j-1)} + H_{r(i-1,j)} - H_{r(i-1,j-1)}}{2\Delta z}\times\right.$$
$$\times\sin\left(\frac{\varphi_{Hz(i,j)} + \varphi_{Hz(i-1,j)} - \varphi_{Hr(i,j)} - \varphi_{Hr(i-1,j)}}{2}\right) - \frac{H_{r(i,j)} + H_{r(i-1,j)}}{2}\times$$
$$\times\frac{\varphi_{Hr(i,j)} - \varphi_{Hr(i,j-1)} + \varphi_{Hr(i-1,j)} - \varphi_{Hr(i-1,j-1)}}{2\Delta z}\times$$
$$\times\cos\left(\frac{\varphi_{Hz(i,j)} + \varphi_{Hz(i-1,j)} - \varphi_{Hr(i,j)} - \varphi_{Hr(i-1,j)}}{2}\right) + \frac{H_{z(i,j)} + H_{z(i-1,j)}}{2}\times$$
$$\frac{H_{r(i,j)} + H_{r(i-1,j)}}{2\left(r_{(i)} + \frac{\Delta r}{2}\right)} + \frac{H_{r(i,j)} - H_{r(i-1,j)}}{\Delta r} = -\left(\frac{H_{z(i,j)} - H_{z(i,j-1)} + H_{z(i-1,j)} - H_{z(i-1,j-1)}}{2\Delta z}\right)\times$$
$$\times\left(\frac{\varphi_{Hz(i,j)} - \varphi_{Hz(i,j-1)} + \varphi_{Hz(i-1,j)} - \varphi_{Hz(i-1,j-1)}}{2\Delta z}\right)\sin\left(\frac{\varphi_{Hz(i,j)} + \varphi_{Hz(i-1,j)} - \varphi_{Hr(i,j)} - \varphi_{Hr(i-1,j)}}{2}\right)$$

After simple transformations, you can write an expression for

$$H_{r(i,j)}:\ H_{r(i,j)} = \left(1 - \frac{1}{i-1}\right)H_{r(i-1,j)} + \left(1 - \frac{1}{2(i-1)}\right)\Delta r\left[\frac{H_{z(i,j)} + H_{z(i-1,j)}}{2}\times\right.$$
$$\left(\varphi_{Hz(i,j)} - \varphi_{Hz(i,j-1)} + \varphi_{Hz(i-1,j)} - \varphi_{Hz(i-1,j-1)}\right)\sin\left(\frac{\varphi_{Hz(i,j)} + \varphi_{Hz(i-1,j)} - \varphi_{Hr(i,j)} + \varphi_{Hr(i-1,j)}}{2}\right) -$$
$$\left.-\left(\frac{H_{z(i,j)} - H_{z(i,j-1)} + H_{z(i-1,j)} - H_{z(i-1,j-1)}}{2\Delta z}\right)\cos\left(\frac{\varphi_{Hz(i,j)} + \varphi_{Hz(i-1,j)} - \varphi_{Hr(i,j)} - \varphi_{Hr(i-1,j)}}{2}\right)\right]. \tag{6}$$

As you can see, the unknown value on the right side of this expression is $\varphi_{Hr}(i,j)$. To find it, we use the fourth equation of the system (1), substituting the value σ obtained from the third equation of the same system. Then this equation is rewritten as:

$$\frac{\partial \varphi_{Hz}}{\partial r} = -\frac{1}{H_z}\frac{\partial H_r}{\partial z}\sin(\varphi_{Hz} - \varphi_{Hr}) + \frac{H_r}{H_z}\frac{\partial \varphi_{Hr}}{\partial z}\cos(\varphi_{Hz} - \varphi_{Hr}) +$$
$$+\frac{\frac{\partial H_r}{\partial z}\cos(\varphi_{Hz} - \varphi_{Hr}) + H_r\frac{\partial \varphi_{Hr}}{\partial z}\sin(\varphi_{Hz} - \varphi_{Hr}) - \frac{\partial H_z}{\partial r}}{H_z\cos(\varphi_{Hz} - \varphi_{Hr})}\sin(\varphi_{Hz} - \varphi_E).$$

Or:

$$\left(\frac{\partial \varphi_{Hz}}{\partial r}H_z + \frac{\partial H_r}{\partial z}\sin(\varphi_{Hz} - \varphi_{Hr}) - H_r\frac{\partial \varphi_{Hr}}{\partial z}\cos(\varphi_{Hz} - \varphi_{Hr})\right)\cos(\varphi_{Hz} - \varphi_E) +$$
$$\left(\frac{\partial H_z}{\partial r} - \frac{\partial H_r}{\partial z}\cos(\varphi_{Hz} - \varphi_{Hr}) - H_r\frac{\partial \varphi_{Hr}}{\partial z}\sin(\varphi_{Hz} - \varphi_{Hr})\right)\sin(\varphi_{Hz} - \varphi_E) = 0.$$

We write the finite-difference analog of the last equation using the half-weight scheme:

$$\left[\frac{H_{z(i,j)}+H_{z(i-1,j)}}{2}\cdot\frac{\varphi_{Hz(i,j)}-\varphi_{Hz(i-1,j)}}{\Delta r}+\frac{H_{r(i,j)}-H_{r(i,j-1)}+H_{r(i-1,j)}-H_{r(i-1,j-1)}}{2\Delta z}\times\right.$$
$$\times\sin\left(\frac{\varphi_{Hz(i,j)}+\varphi_{Hz(i-1,j)}-\varphi_{Hr(i,j)}-\varphi_{Hr(i-1,j)}}{2}\right)-\frac{H_{r(i,j)}+H_{r(i-1,j)}}{2}\times$$
$$\times\frac{\varphi_{Hr(i,j)}-\varphi_{Hr(i,j-1)}+\varphi_{Hr(i-1,j)}-\varphi_{Hr(i-1,j-1)}}{2\Delta z}\times$$
$$\left.\times\cos\left(\frac{\varphi_{Hz(i,j)}+\varphi_{Hz(i-1,j)}-\varphi_{Hr(i,j)}-\varphi_{Hr(i-1,j)}}{2}\right)\right]\cos\left(\frac{\varphi_{Hz(i,j)}+\varphi_{Hz(i-1,j)}-\varphi_{E(i,j)}-\varphi_{E(i-1,j)}}{2}\right)+$$
$$+\left[\frac{H_{z(i,j)}-H_{z(i-1,j)}}{\Delta r}-\frac{H_{r(i,j)}-H_{r(i,j-1)}+H_{r(i-1,j)}-H_{r(i-1,j-1)}}{2\Delta z}\times\right.$$
$$\times\cos\left(\frac{\varphi_{Hz(i,j)}+\varphi_{Hz(i-1,j)}-\varphi_{Hr(i,j)}-\varphi_{Hr(i-1,j)}}{2}\right)-\frac{H_{r(i,j)}+H_{r(i-1,j)}}{2}\times$$
$$\left.\times\frac{\varphi_{Hr(i,j)}-\varphi_{Hr(i,j-1)}+\varphi_{Hr(i-1,j)}-\varphi_{Hr(i-1,j-1)}}{2\Delta z}\sin\left(\frac{\varphi_{Hz(i,j)}+\varphi_{Hz(i-1,j)}-\varphi_{Hr(i,j)}-\varphi_{Hr(i-1,j)}}{2}\right)\right]\times$$
$$\times\sin\left(\frac{\varphi_{Hz(i,j)}+\varphi_{Hz(i-1,j)}-\varphi_{E(i,j)}-\varphi_{E(i-1,j)}}{2}\right)=0. \quad (7)$$

In this equation, the unknown quantity is also $\varphi_{Hr(i,j)}$.

In general, equation (7) can be written as:

$$F\left(H_{z(i,j)},H_{z(i-1,j)},\varphi_{Hz(i,j)},\varphi_{Hz(i-1,j)},\varphi_{E(i,j)},\varphi_{E(i-1,j)},H_{r(i,j)},H_{r(i-1,j)},H_{r(i,j-1)},H_{r(i-1,j-1)},\right.$$
$$\left.\varphi_{Hr(i,j-1)},\varphi_{Hr(i-1,j)},\varphi_{Hr(i-1,j-1)},\varphi_{Hr(i,j)}\right)=0.$$

3. Results and conclusion

The algorithm for solving a system of discrete equations is as follows. First, a step is taken along the z coordinate and the radial distributions are found from equations (3-5) $E_{\varphi(i,j)},\varphi_{E(i,j)},\varphi_{Hz(i,j)}$. Then equations (6) and (7) are solved together. Moreover, equation (7) is solved by dividing the segment in half. From the solution of these equations, the radial distributions of quantities are found $H_{r(i,j)},\varphi_{Hr(i,j)}$. Then the z-coordinate step is taken again, and the same procedure is repeated. This is how the radial profiles of values are calculated $E_\varphi,\varphi_{\dot{O}},\varphi_{D\acute{y}},H_r,\varphi_{Hr},\sigma$ for each section of the inductor. The boundaries of the calculated area lay on the axis of the inductor in the range of r=0 cm to r=3.2 cm close to the wall of the inductor at the r coordinate. At the z coordinate from z=0 cm in the Central section of the inductor and up to z=7 cm from the output of the inductor. The step along the r and z coordinates was assumed to be uniform and was equal to 0.1 cm. The results of solving the system of equations near the axis (up to r=1.1 cm) were replaced for each section by the results of a calculation, which are based on analytical dependencies $E_\varphi(r,z),\varphi_E(r,z),H_z(r,z),\varphi_{Hz}(r,z),H_r(r,z),\varphi_{Hr}(r,z)$ [4-5].

Radial distributions of values are obtained $H_z,E_\varphi,\ \varphi_E,\varphi_{Hz},\varphi_{Hr},\Delta\varphi_1=\varphi_{Hz}-\varphi_E,\Delta\varphi_2=\varphi_{Hz}-\varphi_{Hr}$ for different sections of the inductor. Then the value of the specific electrical conductivity σ was calculated and using the dependence σ(T), the radial temperature distributions for each section of the inductor were found.

The proposed mathematical model allows us to calculate the main parameters of the plasma in the discharge and restore its structure, using minimal experimental information about the profile of the longitudinal magnetic field in the discharge.

References

[1] Rykalin N.N., Sorokin L.M. Metallurgical RF plasmatrons. Electro- and gas dynamics. M.: Nauka, 1987. 164 p.

[2] Dresvin S.V. Fundamentals of the theory and calculation of high-frequency plasmatrons. L.:Energoatomizdat, 1991. 312 p.

[3] Spectral analysis of pure substances/ Ed. Silberstein H. I. St. Petersburg: Chemistry, 1994. 335 p.

[4] Kirpichnikov A. P., The structure of quasistationary electromagnetic field of a high-frequency induction discarge near the plsmoid axis // Teplofizika Vysokikh Temperatur (High Temperature) 1995. V. 33 No 1 (1995) P. 139-140.

[5] A. P. Kirpichnikov, A. V. Gerasimov, The structure of a hight frequency induction discarge near the plasmod axis in the case on an inductor on finit dimension //PIazmotekhnologtya-95 (Plasma Technology-95) Zaporozh'e, (1995) P. 28-29.

Journal of Physics: Conference Series **2270** (2022) 012064 doi:10.1088/1742-6596/2270/1/012064

COMPARISON OF THE SMALLEST SQUARES AND SMALLEST MODULES METHODS IN MODELING PROCESSING OF MATERIALS BY PLASMA OF A GAS-VAPOR DISCHARGE WITH LIQUID ELECTRODES

R R Takseitov[1], R K Galimova[1], Z Y Yakupov[1]

[1]Kazan National Research Technical University, A.N. Tupolev - KAI, Kazan, Russia

zymat@bk.ru

Abstract. This paper describes the application of the least squares and least modulus methods to obtain the values of the power parameters of the central Lennard-Jones potential when modeling processes in the technology of processing materials with vapor-gas discharge plasma with a liquid electrode.

1. Introduction.

The mathematical apparatus is a universal tool for describing technical problems [1-5]. Therefore, the progress of engineering and technology (technical sciences, engineering practice) is closely connected with the development of mathematics. The emergence and improvement of a large number of numerical methods of mathematics is associated with applied needs. The paper compares the possibilities of using two mathematical methods - the method of least squares (MLS) [6] and the method of least modules (MLM) [7] - for processing the results of a technical experiment to study the characteristics of one of the modifications of the electric discharge. As in other areas of engineering and technology [8], the experimental study of processes in low-temperature plasma may be accompanied by the need to solve problems based on the extremal principle associated with one of the areas of mathematics - the theory of optimal control. The possibilities of obtaining low-temperature plasma by igniting an electric discharge between solid metallic and liquid non-metallic (electrolytes with the addition of inorganic and organic impurities) electrodes are interesting from the point of view of developing technologies for improving the performance properties of materials with different physical and mechanical properties. A large number of statistical data that appear during the experiments described in this paper require generalization [9–10]. Most often, MLS is used to process experimental statistical information. MLM is used less often, but its statistical properties are more often better [1, 6–7]. When using the least squares, the conditions are determined for which the values of the square of the known function are the smallest. In this general form, the method is widely used for the mathematical description of scientific and technical problems (in physics, mechanics, mathematical statistics, in the theory of errors in the processing of experimental data, in solving problems of numerical analysis). The method makes it possible to use others, simpler and more convenient in research, for an approximate representation of the given functions. The described

Content from this work may be used under the terms of the Creative Commons Attribution 3.0 licence. Any further distribution of this work must maintain attribution to the author(s) and the title of the work, journal citation and DOI.
Published under licence by IOP Publishing Ltd

advantages of the method determined its choice for processing the results of a technical experiment to study the characteristics of the investigated modifications of the electric discharge [11-14].

Among a large number of factors affecting the results of the technological process of processing surfaces of products by a vapor-gas discharge between solid metal and liquid non-metal electrodes, there are the electric current in the discharge and the voltage drop in the interelectrode gap [6-7, 9-14]. MLS and MLM, just optimal for consideration in the processing of experimental information with a combination of several factors of influence. As in all types and varieties of chemical technology and electrical technology, the use of various modifications of the vapor-gas discharge between solid metal (or liquid non-metal) and liquid electrodes is associated with a large number of complex, time-consuming experiments. It is for this reason that the involvement of the mathematical apparatus in describing the results of experiments is a useful tool in the research work in this area, and the mathematical methods we have chosen for comparison allow us to build probabilistic-statistical models that allow us to make predictions in complex studies [1-14].

2. The relevance of research.

The work is devoted to the study of two mathematical methods (MLS and MLM) for describing the process of processing substances using an electric discharge in a vapor-air environment and the possibility of generalizing the results of the study for the technological description of various modifications of a vapor-gas discharge between solid metal (or liquid non-metallic) and liquid electrodes. At the same time, the degree of correspondence between calculated and experimental data depends on the selection of constants that determine the intensity of physical and chemical processes.

3. Research methods.

The scheme of the technological experiment is shown in fig. 1, where it is indicated: 1. Electrochemical cell; 2. Electrolyte (liquid electrode); 3. Solid state metal electrode; 4. Metal plate. The material processing technology involves the simultaneous action of both an electric discharge and electrochemical processes on it [6-7, 9-14].

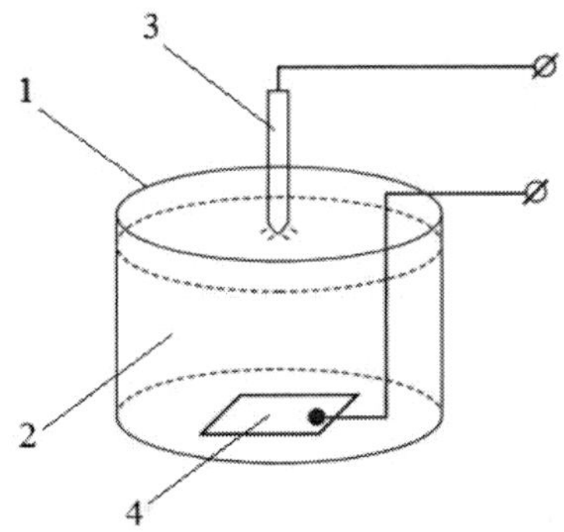

Figure 1. Scheme of technological experiment.

When a part is exposed to low-temperature plasma in the interelectrode gap in the pre-breakdown mode and during breakdown, hydrogen and oxygen are released on the metal electrode. This process is accompanied by the release of heat, the transition of the gas-liquid state into a vapor-gas state with the formation of a layer of gas and liquid. The need to take into account a large number of processes greatly complicates the mathematical modeling of the considered physical system [6-7, 9-14].

The coefficients of diffusion and viscosity of gas mixtures were obtained on the basis of a strict kinetic theory of gases using the central potential of Lennard-Jones (8-6) [6-7, 15-17]. MLS and MLM made it possible to obtain the force parameters of some real gases. For comparison, we used the values of binary collision integrals obtained in experiments on the scattering of molecular beams in the temperature range 20000–50000 K [17].

4. Results. The obtained values of power parameters for some real gases are presented in Table 1:

Table 1. Values of power parameters

Substance	MLS		MLM	
	$\sigma, \mathring{A}$	$\varepsilon / \kappa, K$	$\sigma, \mathring{A}$	$\varepsilon / \kappa, K$
H_2	2.786	34	2.812	36
O	2.695	117	2.578	112
N_2	3.586	96	3.593	86
O_2	3.619	117	3.683	119
NO	3.324	124	3.154	129
CO	3.435	101	3.412	105

Below, in fig. 2 shows potential curves for molecular hydrogen. Here, the number 1 denotes the curve corresponding to the Lennard–Jones potential (8–6) with the parameters obtained using the MLM (Table 1); the number 2 is the classical Lennard-Jones potential (12–6) with the parameters taken from the handbook [17]; 3 – potential of Lennard-Jones type (8–6) with parameters obtained using MLS (Table 1); 4 – experimental data on molecular beam scattering [17]. It can be seen that the potential of the Lennard-Jones type (8–6) is in better agreement with the potential curve obtained from the data on molecular beam scattering. At the same time, the data obtained using the MLS are in better agreement with the experiment at high temperatures, and those obtained using the MLM are at low temperatures.

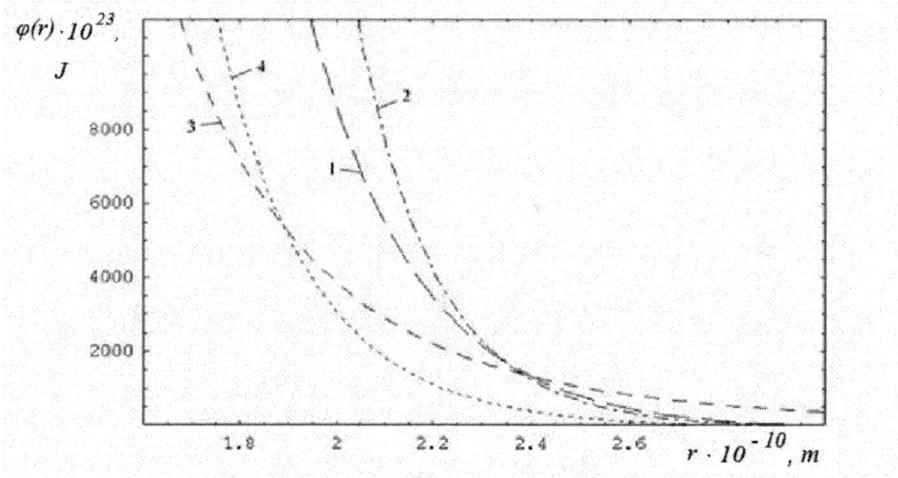

Figure 2. Potential curves for molecular hydrogen,
φ(r) is the Lennard-Jones potential; r - intermolecular distance

5. Conclusions.

The choice of algorithm for solving the problem of interpolation, curve approximation or smoothing depends on the nature of the data being processed and on the desired output result. When choosing an algorithm to use on a computer, the limitations of that device must be taken into account. Often there is a contradiction between the accuracy and time of computing on a computer, on the one hand, and the required amount of memory, on the other. MLS and MLM require the solution of a system of linear algebraic equations. As the order of least squares approximation increases, the number of equations in the system increases. In the MLS problem, the number of equations increases with the increase in the considered points. This increase in the size of the system leads to an extension of the execution time and a significant increase in the amount of memory for arrays required when obtaining a solution [1, 6-7]. Taking into account these features of the two mathematical methods and comparing the data on the power parameters of the proposed central potential of the Lennard-Jones type (8–6) (Fig. 2) allows us to choose the most appropriate model for an adequate description of the technological process associated with the use of a vapor-gas discharge.

References

[1] Yakupov Z. Ya., Galimova R. K. Basic concepts of research of Lyapunov transformation groups. *Bulletin of the Kazan national technical University named after A. N. Tupolev,* PP. 171-176. V. 73, No. 4. 2017

[2] Yakupov Z. Ya., Galimova R. K. Methods of least squares and least modules in scientific and technical calculations. 2-nd ed., corrected and additional. St. Petersburg: *SUPER Publishing House*, 2021. 160 p.

[3] Mardanov R. R., Kayumov R.R., Akhatov M.F., Gazizova A.I., Gaisin A.F. Modification of the surface of polyethylene by low-temperature plasma and liquid electrodes// *Journal of Physics: Conference Series, LTP Coatings* 2019, 1588 (2020) 012028

[4] Gaisin A.F., Son E.E., Efimov A.V., Gil'mutdinov A.K., Kashapov N.F. Spectral diagnostics of plasma discharge between a metal cathode and liquid anode// *High Temperature*, 2017, 55(3), pp. 457–460

[5] Gaisin A.F., Son E.E., Petryakov S.Y. Radio-frequency capacitive discharge with flowing liquid electrodes at reduced gas pressures // *Plasma Physics Reports*, 2017, 43(7), pp. 741–7484

[6] Gaisin A.F., Kashapov N.F. Investigation of Physical Processes in the Gas Discharge Region between Liquid Electrodes // *Journal of Applied Mechanics and Technical Physics*, 2018, 59(4), pp. 591–593

[7] Takseitov R. R., Galimova R. K., Yakupov Z. Ya. On the issue of modeling the processing of materials by plasma of a vapor-gas discharge with liquid electrodes// *Journal of physics: series of conferences 1923(1), 012012,* 2021

[8] Gaisin A.F., Kashapov N.F., Kuputdinova A.I., Mukhametov R.A. Discharge between Liquid Jet and Metallic Electrodes// *Technical Physics*, 2018, 63(5), pp. 695–699

[9] Akhatov M. F., Kayumov R. R., Valeeva R. R., Mardanov R. R. Wastewater disinfection methods// *Journal of Physics: Conference Series, LTP Coatings*, 1588 (2020) 012005

[10] Akhatov M. F., Kayumov R. R., Mardanov R. R. Saifutdinova I. I. Voltage drop in the jet electrolytic cathode// *Journal of Physics: Conference Series, LTP Coatings*, 1588 (2020) 012004

Journal of Physics: Conference Series **2270** (2022) 012064 doi:10.1088/1742-6596/2270/1/012064

[11] Galimzyanov I.I., Gaisin A.F., Fakhrutdinova I.T., Shakirova E.F., Akhatov M.F., Kayumov R.R. Characteristics of the Development of Electric Discharge between Jet Anode and Liquid Cathode// *High Temperature*, 2018, 56(2), pp. 296–298

[12] Fedotov A.I., Basyrov R.S., Abdullazyanov R.É., Vagapov G.V. Practical Implementation of Monitoring and Ice Melting on 110 – 120 kV Overhead Transmission Lines Based on the Dip Angle of the Wire// *Power Technology and Engineering,* 2019, 53(4), pp. 508–515

[13] Gaisin A.F., Basyrov R.S., Son E.E. Model of glow discharge between an electrolytic anode and a metal cathode// High Temperature, 2015, 53(2), pp. 188–192

[14] Mardanov R. R., Kayumov R. R., Akhatov M. F., Gaisin A. F. A review of use of barrier discharge for modification of surface of polymers. *Journal of Physics: series of conferences 1588 (2020) 012029, LTP Coatings 2019*

[15] *Modern mathematics for engineers* / Ed. E. F. Beckenbach; per. from English under the general ed. I.N. Vekua. M.: Publishing house of foreign countries. Lit. 1959. 500 p.

[16] Ignat'ev V.N., Takseitov R. R. High-temperature coefficients of viscosity and diffusion of some gases and gas mixtures calculated on the basis of central potential. *Izvestiya Vysshikh Uchebnykh Zavedenij. Aviatsionnaya Tekhnika*, 2004 (4). 20-22.

[17] Alemasov V.E., Dregalin A.F., Tishin A.P., Khudyakov V. A. Thermodinamic and thermophisical properties of combustion products. Handbook in five volumes. Vol. 1. Moscow: VINITI, 1972. 267 p.

AUTHOR INDEX

Abdulkadieva, M. 208
Abdullin, A A 140
Akhatov, M F 13, 42, 79, 137
Akhverdiyev, R F 238
Akishev, Yu. 208
Alekseeva, T. 208
Amirov, I I 183
Andreev, S N 178
Antonova, M V 37
Ara, J 188
Asadullin, T Y 149
Avetisyan, L. 208
Baranova, L 17, 204
Barsuk, V. 208
Baysova, B 17, 204
Belov, Y D 183
Bernatskiy, A V 178
Bigeyeva, K F 238
Chapaeva, L V 264
Chebykin, E O 287
Chernukha, M. 208
Danilaev, M P 82
Demin, S A 173, 196
Doronina, Maria A. 158
Dushkin, G A 178
Egamov, M K 45
Elenev, A A 196
Erlingayte, E A 282
Ermolaeva, S. 208
Fadeev, S A 26
Fairushin, I I ....
Fairushin, I. I. 1, 162
Farkhutdinov, A R 101
Fayzulloev, I K 45
Feltsinger, V S 229
Filinov, A.V. 188
Fukina, O V 264
Galeev, I G 149
Galimova, R K 13, 300
Ganieva, G R 234
Gaynullin, R N 295
Gerasimov, A V 295
Gevorgyan, R K 32, 140
Gilmutdinov, A K 115
Gluchshenko, N 53
Gorbatov, S 53
Grebenshchikova, M M 267
Hakimov, D S 22
Ibatullina, A R 37
Ivanov, I 53
Juraev, H. S. 67
Juraev, H. Sh. 106
Juraeva, G.H. 67
Kamrukov, A S 277
Kayumov, R R 42, 79, 137
Khabibullin, R A 169, 201
Khairullina, R R 145
Kharlamov, V 53
Khayytov, M. 264
Khazeev, K I 137
Khusnutdinov, R M 145, 245
Kirillov, D V 166, 260
Kirpichnikov, A P 295
Kozaev, A S 250
Krasina, I V 37
Kuklin, V A 82
Lapteva, Y D 295
Loginov, N A 13
Loginova, I M 42
Makhsudov, B I 45
Makhsudov, B.I. 106
Maksimov, R A 26
Mamatkulova, N.O. 106
Manegin, D S 166
Mansurov, R N 115
Mardanov, R R 13, 42, 79
Masyanov, A A 241
Matlabi, J 125
Medzhidov, I 53
Mironov, M M 267
Mironov, V V 250
Mirzoeva, Q 125
Morozov, M V 115
Nagulin, K Y 115
Naumov, V V 183
Negodyaev, A S 250
Nekrasov, I K 63
Nizameev, A A 13, 79, 137
Nizamova, D K 238
Nuriddinov, K S 282
Oshenko, I I 129
Pankova, E A 264
Parada-Becera, F F 292
Parsanov, A S 37
Pavlov, A V 287, 292
Petrova, O A 218, 224

Petrukhina, D 53
Petryakov, A 208
Pham, T M 295
Pikuz, S A 178
Plotnikova, O P 166
Pogosyan, V M 4
Rajabov, A R 22
Rajabova, S S 9
Rakhmatullina, G R 238, 264
Riazanov, V A 166
Safarov, M M 9, 22, 91, 121, 125
Saidzoda, Q B 91
Samsonova, E S 26, 63
Savinov, S Y 178
Sayfutdinov, A I 218, 224
Semenov, A A 91
Senkov, V M 292
Shaikhattarov, R R 32, 140
Shakirov, B R 234
Shakirov, R B 32
Shamsutdinov, R S 218, 224, 229, 234
Shavaleeva, S M 282
Shchepanyuk, T S 287
Shemakhin, A Yu 63
Shemakhin, E Yu 26, 63
Shemakhin, Yu A 26
Shishko, V 53
Sidorov, I N 82
Skriabin, A S 287, 292
Smirnov, S A 129
Sokolenko, O N 4
Strunin, V 17, 204
Suleimanov, R M 140
Sysolyatina, E. 208
Takseitov, R R 300
Tazmeev, A K 97
Tazmeev, B K 4, 57, 87, 133
Tazmeev, G K 57, 87
Tazmeev, K K 87, 97, 133
Tazmeeva, R N 4, 57, 133
Telekh, V D 287, 292
Terentev, T N 26, 63
Tikhonov, A 53
Tikhonov, V 53
Tikhonova, V P 238
Tilikin, I N 178
Tilloeva, T R 9, 121
Timerkaev, B A .... , 32, 140, 218, 224, 229, 234
Titov, N. 208
Tkachenko, I.M. 188
Tkhorik, O 53
Tskhai, S N 178
Tsokalo, I E 63
Tsybulevsky, V V 57
Tsygankov, A A 153
Tsygankov, P A 292
Tukmakov, D A 73
Tumanov, N A 166, 260
Turutin, V Y 229
Turyanskiy, A G 292
Tyurin, V A 26, 63
Uktamov, Z A 282
Vakhitov, R R 229
Valeeva, D R 149
Vasilieva, E. 208
Volodin, L Y 271, 277
Vorob'ev, E V 260
Voznesenskiy, E F 26
Yakupov, Z Y 300
Yarullin, Dinar T 158, 256
Yunusov, M B 245
Yunusov, V A 173, 196
Zalyalieva, A A 32, 229, 234
Zaripova, M A 9, 121
Zhelonkin, Y O 26
Zheltukhin, V S 63
Zimin, S P 183